MEDICAL INTELLIGENCE UNIT 25

Nutritional Support in Cancer and Transplant Patients

Rifat Latifi, M.D.
Department of Surgery
Virginia Commonwealth University
Medical College of Virginia Hospitals
Richmond, Virginia, U.S.A.

Ronald C. Merrell, M.D., F.A.C.S.
Department of Surgery
Virginia Commonwealth University
Medical College of Virginia Hospitals
Richmond, Virginia, U.S.A.

CRC Press
Taylor & Francis Group
Boca Raton London New York

CRC Press is an imprint of the
Taylor & Francis Group, an informa business

Nutritional Support in Cancer and Transplant Patients

Medical Intelligence Unit

CRC Press
Taylor & Francis Group
6000 Broken Sound Parkway NW, Suite 300
Boca Raton, FL 33487-2742

First issued in hardback 2019

ISBN-13: 978-1-58706-049-6 (hbk)
ISBN-13: 978-1-58706-079-3 (pbk)

Visit the Taylor & Francis Web site at
http://www.taylorandfrancis.com

and the CRC Press Web site at
http://www.crcpress.com

Library of Congress Cataloging-in-Publication Data

Nutritional support of cancer and transplant patients / [edited by] Rifat Latifi, Ronald C. Merrell.

p. ; cm. -- (Medical intelligence unit)

Includes bibliographical references and index.

ISBN 1-58706-049-3 (hardcover)

1. Cancer -- Patients -- Nutrition. 2. Transplantation of organs, tissues, etc. -- Patients -- Nutrition. 3. Immunosuppression -- Diet therapy. I Latifi, Rifat. II. Merrell, Ronald C. III. Series.

[DNLM: 1. Neoplasms -- diet therapy. 2. Parenteral Nutrition. 3. Enteral Nutrition. 4. Transplantation. QZ 266 N9765 2001]

RC268.45.N89 2001

616.99′40654 -- dc21 00-037182

Dedication

To Stanley J. Dudrick, M.D., the surgical scientist who made parenteral nutrition possible. We are indebted to his genius, his energy and his vision.

CONTENTS

EDITORS

Rifat Latifi

Department of Surgery
Virginia Commonwealth University
Medical College of Virginia Hospitals
Richmond, Virginia, U.S.A.
Chapters 6, 12

Ronald C. Merrell

Department of Surgery
Virginia Commonwealth University
Medical College of Virginia Hospitals
Richmond, Virginia, U.S.A.
Chapter 6

CONTRIBUTORS

Steve F. Abcouwer
Division of Surgical Oncology
Massachusetts General Hospital
Boston, Massachusetts, U.S.A.
Chapter 3

Giacomo Basadonna
The Department of Surgery
Division of Transplant
Yale University
New Haven, Connecticut, U.S.A.
Chapter 12

S. V. Beath
The Birmingham Children's Hospital
and University of Birmingham
Birmingham, U.K.
Chapter 11

Barrie P. Bode
Department of Surgery
St. Louis University
St. Louis, Missouri, U.S.A.
Chapter 3

Carlo Cangiano
Department of Internal Medicine
Laboratory of Clinical Nutrition
Viale dell 'Universita
Rome, Italy
Chapter 8

Antonia Cascino
Department of Internal Medicine
Laboratory of Clinical Nutrition
Viale dell 'Universita
Rome, Italy
Chapter 8

Antonio Cayco
Section of Nephrology
Yale University School of Medicine
New Haven, Connecticut, U.S.A.
Chapter 13

Matthew E. Cohen
Section of Digestive Diseases
Yale University School of Medicine
New Haven, Connecticut, U.S.A.
Chapters 4, 5

Colette Coudray-Lucas
Laboratoire de Biochemie A
Hotel-Dieu de Paris, AP-HP
Laboratoire de Biologie de la Nutrition
Pharmacy School
Paris V University
Paris, France
Chapter 10

Susan T. Crowley
Section of Nephrology
Yale University School of Medicine
New Haven, Conneticut, U.S.A.
Chapter 13

Luc Cynober
Laboratoire de Biochimie A
Hotel-Dieu de Paris, AP-HP
Laboratoire de Biologie de la Nutrition
Pharmacy School
Paris V University
Paris, France
Chapter 10

John Damreis
Department of Surgery
Oregon Health Sciences University
Portland, Oregon
Chapter 6

Rosemarie L. Fisher
Section of Digestive Diseases
Yale University School of Medicine
New Haven, Connecticut, U.S.A.
Chapter 4

Richard Formica
Section of Nephrology
Yale University School of Medicine
New Haven, Conneticut, U.S.A.
Chapter 13

Narihide Goseki
The First Department of Surgery
Tokyo Medical and Dental University
School of Medicine
Tokyo, Japan
Chapter 9

Wendy Holmes
Clinical Nurse Specialist
University of Massachusetts
Memorial Hospital
Worchester, Massachusetts, U.S.A.
Chapter 14

S. Janes
The Birmingham Children's Hospital
and University of Birmingham
Birmingham, U.K.
Chapter 11

Gretchen R. Kilmartin
Food and Nutrition Services
Yale-New Haven Hospital
New Haven, Connecticut, U.S.A.
Chapter 14

Cheng-Mao Lin
Division of Surgical Oncology
Massachusetts General Hospital
Boston, Massachusetts, U.S.A.
Chapter 3

Colette Coudray-Lucas
Laboratoire de Biologie de la Nutrition
Pharmacy School
Paris V University
Paris, France
Chapter 10

Amadeo Marcos
The Department of Surgery
University of Rochester
Rochester, New York, U.S.A.
Chapter 12

Michael M. Meguid
Surgical Metabolism and Nutrition
Laboratory
Neuroscience Program
Department of Surgery
SUNY Health Science Center
Syracuse, New York, U.S.A.
Chapter 8

Maurizio Muscaritoli
Department of Internal Medicine
Laboratory of Clinical Nutrition
Viale dell 'Universita
Rome, Italy
Chapter 8

Hassan A. Naama
Department of Surgery
Hospital of the University of Pennsylvania
University of Pennsylvania School of Medicine
Philadelphia, Pennsylvania, U.S.A.
Chapter 1

Takeshi Nagahama
The First Department of Surgery
Tokyo Medical and Dental University
School of Medicine
Tokyo, Japan
Chapter 9

Joel M. Rappeport
Yale School of Medicine
Yale-New Haven Hospital
New Haven, Connecticut, U.S.A.
Chapter 14

Wiley W. Souba
Department of Surgery
Milton S. Hershey Medical Center
Hershey, Pennsylvania, U.S.A.
Chapter 3

Filippo Rossi-Fanelli
Maurizio Muscaritoli
Department of Internal Medicine
Laboratory of Clinical Nutrition
Viale dell 'Universita
Rome, Italy
Chapter 8

Ezra Steiger
Department of Surgery
Cleveland Clinic Foundation
Cleveland, Ohio, U.S.A.
Chapter 6

Michael H. Torosian
Department of Surgery
Hospital of the University of Pennsylvania
University of Pennsylvania School of Medicine
Philadelphia, Pennsylvania, U.S.A.
Chapters 1, 2, 7

PREFACE

Much has been learned, great developments have occurred, and so much has been written about cancer and transplantation in the last 2-3 decades. Yet, to our knowledge, no mono graph or book has addressed nutrition support of cancer and transplant patients together. Both cancer and transplant patients may suffer from a great deal of malnutrition and associated complications. Both have significant immune-related issues that need to be addressed, and both may have significant nutritional and metabolic deficiencies, which could be modulated with different nutrient substrates. The development of modern and sophisticated surgical techniques, powerful chemotherapy and radiation protocols will prove to be nothing but abstract achievements if they are applied to a severely malnourished, cachectic and dying cancer or transplant patient.

Nutrition support of critically ill patients has entered a new era of nutri-pharmaceutics. This has occurred because of our recent knowledge regarding the role of the gut in immunity and the role of nutrient substrates in modifying the immune response after severe injury. The question to be answered is not should critically ill patients be fed, rather when to feed and what to feed them. Both cancer and transplant patients with their physiologic insult should be treated as critically ill, since patients from both groups often end up in the intensive care unit after major operations—malnourished, cachectic and severely ill.

One of the areas of major concern over the past decade has been the adjunctive use of nutrition support in the treatment of patients with cancer. A vast amount of data and experience has been acquired in understanding the interactions of neoplastic disease and nutrient regimens while treating and studying cancer patients. However, the specific role of nutrient substrates in cancer only recently has received needed attention. Because of the debilitating nature of most oncologic processes on the body cell mass, the proportion of patients with malignant disease experiencing a significant degree of malnutrition is larger than the number of hospitalized patients without malignant disease. Cancer-related malnutrition has a poor prognosis but may be treated or prevented to some extent with TPN or enteral nutrition. For this reason, it is especially important that the treatment of cancer patients be accompanied by adjunctive nutritional support in order to achieve the best possible therapeutic results with the lowest morbidity and mortality and to prevent death from starvation. While it is clearly known that most patients with significant malignant disease are malnourished, the role of nutrition support in patients with cancer is controversial. Moreover, the method of correcting or preventing the malnutrition and associated complications is still controversial. Enteral nutrition support is more physiologic and has additional benefits; however, in patients who suffer from gastrointestinal cancer it is not always practical. For this reason, one depends on TPN to prevent, correct malnutrition or restore nutrition status in this group of patients. Determining the best method for nourishing the cancer and transplant patient depends on the patient's nutritional status, the level and the degree of residual gastrointestinal function and the type and magnitude of oncologic surgical therapy that the patient had undergone. When the use of the enteral route is contraindicated as in the presence of severe gastrointestinal dysfunction such as intestinal obstruction, prolonged ileus, upper gastrointestinal bleeding, and/or intractable vomiting or diarrhea, then parenteral nutritional should be instituted as a means of nutritional rehabilitation. Currently, there is strong evidence to suggest that nutrition support is of benefit only in severely malnourished cancer patients or in those whose treatment toxicity will preclude oral or enteral intake for longer than one week. Regardless of postulated or demonstrated tumor-induced abnormalities in the intermediary metabolism of the host, the predominant factor in the development of cancer cachexia is an imbalance between nutrient intake and host nutrient requirements, which can be treated beneficially in severely malnourished patients by enteral or parenteral nutrition, or both. While it has become common knowledge that enteral nutrition is better than parenteral nutrition, the comparison of these two techniques is not justifiable. Most enteral nutrition formulations today used for critically ill patients contain increased amount of peptides, arginine, glutamine, nucleoside and nucleotide, branched-chain-amino acids, taurine, and omega 3-fatty acids, while TPN contain sugar, protein and lipids as omega 6-fatty acids. Comparing the two techniques of nutritional support, especially their effect on the immune system, is like comparing apples and oranges.

The ultimate goal in the nutritional management of patients with cancer and transplant is the same as with all other patients. Provision of optimal nutrition to all patients under all conditions at all times, as long as there is a reasonable chance for curing or improving the quality of life of that patient is the goal. TPN, and for that matter nutrition support, however, should not be used in cancer patients who are completely unresponsive to therapy and in whom extraordinary measures to provide nutrients can serve only to prolong unrelieved suffering and inevitable death. Nutritional support of patients with cancer, nonetheless remains an important aspect of therapy of these patients.

We asked experts from around the world to address the nutrition support in cancer and transplant patients in this unique monograph. The book is divided into two parts: Part I deals with nutrition support in cancer patients, including the specific role of nutrition on immunity, cancer cachexia, and the role of different substrates. Part II addresses nutrition in transplant patients.

The first two Chapters deal with the immunologic role of nutrition and cancer cachexia. Chapter 3 elegantly and extensively reviews the nutritional implications; its biochemistry and the role of one of the most studied amino acids in clinical practice—glutamine. This is followed by two Chapters of nutrition support of patients with head and neck cancer and nutrition support of patients with gastrointestinal cancer. Chapters 6 and 7 review the role of total parenteral nutrition on perioperative nutritional support and cell cycle kinetics. While the plasma amino acids profile in cancer patients and the role of L-methionine is addressed in great details in Chapters 8 and 9, the role of ornithine alpha-ketoglutarate administration on surgical, trauma and cancer-bearing patients is reviewed in Chapter 10.

Part Two of this book starts with the review of nutritional support in small bowel transplantation. This Section elegantly describes the process of recovery of small bowel from the ischemia and preservation, weaning from parenteral nutrition support and establishment of normal diets. In addition, monitoring techniques and methods, as well the as complications of this dramatic and desperate surgical intervention are described. Chapter 12 on liver failure and liver transplant patients addresses hepatic encephalopathy and role of certain amino acids, nutrition assessment techniques and metabolic changes following liver transplantation. It offers some practical advice on how to establish nutrition support routes in these very ill patients.

Nutrition support in renal transplantation, including metabolic abnormalities in renal failure, is described on Chapter 13. This monograph ends with a Chapter on total parenteral nutrition in bone marrow transplant patients.

It is our hope that everyone involved in this critical portion of care of these patients will find this to be an important and useful reference. Generations that will follow us will find current techniques of nutrition support and dietary regimens for cancer and transplant patient at best primitive. Molecular biology of nutrient substrates and their role in the disease process and the biology of the cancer and immune system will most likely render this information outdated. Nutri-pharmaceutics will become applicable in cancer and transplant patients, just as it has become in the critically ill and septic trauma patients with multiple system organ failure. To feed or not to feed these patients will clearly not be a question. The main question to be answered will be what and how to feed them? What we are going to feed these patients will be a matter of future developments.

This book has been made possible because of the timely, relevant and significant contributions of the authors, to whom we are sincerely grateful.

Rifat Latifi, M.D.
Ronald C. Merrell, M.D., F.A.C.S.

CHAPTER 1

Immunologic Role of Nutrition

Hassan A. Naama and Michael H. Torosian

Modulation of the immune system by specific nutrients is a well-established phenomenon that has tremendous clinical potential. Diets deficient in specific substrates (e.g., arginine, glutamine, nucleotides, etc.) have been shown to suppress certain immune functions and restoring these substrates can reconstitute immune functions. Animal models remain the major source of data concerning the immunomodulatory role of nutrients. The general lack of large, well-controlled, randomized, human clinical trials is an obvious deficit in nutritional immunology. Firstly, it is difficult to extrapolate the findings of animal research data to the human situation. Secondly, one must differentiate between physiological replacement of specific nutrients and pharmacologic administration of nutrients to manipulate the immune system.[1] In this chapter we will discuss those nutrients that have been shown to be of potential clinical relevance to the surgical patient. Basic biochemistry, pertinent animal studies and clinical trials of specific nutrients will be reviewed as well.

Arginine

Arginine is an amino acid which is not considered essential for adult mammals as it can be synthesized in vivo via urea cycle intermediates. Arginine plays an important role in protein synthesis, urea cycle metabolism and the synthesis of the high-energy compounds, creatine and creatine phosphate. Two pathways of arginine metabolism have been identified as potential sources of critical mediators of arginine immunomodulatory actions in vivo. The so-called "arginase" pathway, in which arginine is converted to urea and ornithine, generates polyamines by the action of ornithine decarboxylase on the latter compound. This route of polyamine synthesis was proposed as the mechanism whereby arginine augments lymphocyte mitogenesis.[2] Induction of arginase was also proposed as the effector pathway in arginine-dependent macrophage-mediated tumor cell cytotoxicity.[3] A second more recently described pathway of arginine metabolism leads to the generation of reactive nitrogen intermediates—principally nitric oxide and its peroxynitrite derivatives. These reactive nitrogen intermediates are derived from arginine through the action of nitric oxide synthase.[4] It is postulated that macrophage-mediated tumor cytotoxicity and fungistasis are mediated through this pathway.[5] Over the last few years, extensive research has been conducted on the biology and pathology of nitric oxide and its role in cellular metabolism. Nitric oxide serves as a messenger in the central and peripheral nervous system and as a mediator of cerebral hypoxic injury under ischemic conditions.[6,7] In the vascular endothelium nitric oxide is thought to be the active moiety of the endothelium-derived relaxation factor which mediates vascular vasodilatation.[8] Nitric oxide is

Nutritional Support in Cancer and Transplant Patients, edited by Rifat Latifi and Ronald C. Merrell.

also believed to regulate the secretagogic effects of arginine on the pituitary, adrenal and pancreatic islets.[9-12]

Studies by Barbul et al demonstrated that supplemental arginine in injured rats resulted in accelerated wound healing and abrogated thymic involution that occurs with injury. Arginine supplementation in uninjured animals was associated with an increase in thymic weight. This response was accompanied by an increase in total lymphocyte counts and enhanced blastogenic response.[13-15] In a burn model utilizing guinea pigs with 30% body surface burned, the group receiving a 2% arginine-supplemented diet showed improved delayed-type hypersensitivity reaction and decreased pustule size following subdermal Staphylococcal inoculation.[16] Arginine supplementation in the tumor-bearing host may inhibit tumor growth and reduce metastatic spread in immunogenic tumors.[17] Reynolds et al showed that arginine supplementation in mice bearing either immunogenic or non-immunogenic tumors resulted in enhanced splenocyte mitogenesis and IL-2 production in comparison to control mice on regular diets.[18] However, tumor growth was inhibited only in the immunogenic tumor model, suggesting augmentation of antitumor immunity with arginine supplementation. In contrast, Yeatman et al reported beneficial effects of dietary arginine depletion on liver metastases in a murine model.[19] This effect was attributed to the dependence of this particular tumor model on arginine for optimal growth in vitro and in vivo.

In clinical trials, supplemental arginine in normal human volunteers led to significant increases in peripheral blood lymphocyte proliferation to mitogens and in the T-helper:T-suppressor cell ratio.[20,21] To determine the clinical effect of supplemental arginine, Daly et al in a controlled randomized trial, studied arginine-supplemented versus glycine-supplemented total parenteral nutrition (TPN) in cancer patients undergoing major surgery. Mean lymphocyte proliferation and T-helper subsets were significantly elevated in the arginine-supplemented group.[22] A subsequent study showed that this effect was dependent on the administration of arginine with other amino acids, as administration of arginine as the sole amino acid source did not augment T-cell function. This finding implied a complex pattern for the immunomodulatory role of arginine, including interaction with other dietary elements to mediate the observed immunologic effects.[23] These clinical effects were confirmed in a larger randomized prospective study involving 85 patients, who underwent surgery for gastrointestinal malignancies. Patients were randomized to receive an enteral diet supplemented with a mixture of arginine, ω6/ω3 fatty acids and nucleotides, or a standard isocaloric enteral diet serving as a control. Enteral feeding was started by jejunostomy tube in the immediate postoperative period and continued for at least seven days. A significant improvement in lymphocyte mitogenic response in the supplemented group was observed. In addition, a 70% reduction in infectious and wound complications and a 22% reduction in hospital stay occurred in diets supplemented with oral arginine. This demonstrated a significant enhancement of lymphocyte mitogenic response in these patients. However, no change in T-helper:T-suppressor cell ratio was detected. A small number of patients were reported in this study, although the trend is encouraging. Nevertheless, more studies are needed to assess the value of arginine supplementation in this group of patients. Clearly, arginine has great potential as an immunomodulator and may prove useful in catabolic conditions such as severe sepsis and postoperative stress.

Glutamine

Glutamine is quantitatively the most abundant amino acid in the circulation and in the intracellular free amino acid pool. Rose in 1938 determined that glutamine can be synthesized de novo in the body and hence was a nonessential amino acid.[25] Eagle had earlier reported that glutamine was required in high concentration in tissue culture media for optimal growth of cultured mammalian cells, a practice which continues today.[26] Immune cells such as lymphocytes and macrophages have been demonstrated to utilize glutamine at a rate of 100 nmol/mg protein/hour. This rate of glutamine utilization is increased upon mitogenic stimulation of lymphocytes.[27,28] In fact, lymphocytes exhibit an increased mitogenic response in a dose-dependent fashion with increasing glutamine concentration.[29] Macrophage phagocytosis, protein and mRNA synthesis and IL-1 secretion have been shown to be increased in a dose-dependent manner with increasing glutamine concentration in vitro.[30,31] Research in our laboratory has shown that macrophage super oxide, nitric oxide production, Candida killing and tumor cytotoxicity are enhanced with increasing glutamine concentration in vitro.

In rats, glutamine-supplemented TPN has been found to improve peritoneal and alveolar macrophage function.[32] Glutamine metabolism is known to be altered under stress conditions such as sepsis, injury and following major surgery.[33,34] For instance, major surgery can significantly decrease circulating glutamine levels.[35] The classification of glutamine as a nonessential amino acid has recently been challenged in these stress states and has led to its reclassification as a "conditionally" essential amino acid.[36,37] A conditionally essential nutrient is not required in healthy mammals, as de novo synthesis is able to meet tissue demand. However, under conditions of sepsis, trauma or other catabolic insults, tissue demands exceed de novo synthetic capacity, resulting in a relative glutamine deficiency. Glutamine then becomes "conditionally" essential and is required from exogenous sources to meet the excess demand.

The intestinal epithelium exhibits a high rate of glutaminase activity and accounts for 40% of glutamine metabolized by the body.[38] It is not surprising, therefore, that the gut depends on glutamine metabolism as a major source of energy. The fall in plasma glutamine concentration following major surgery, sepsis or glutamine-free TPN administration leads to mucosal atrophy and damage with increased bacterial translocation and endotoxin transmigration across the intestinal epithelium. Glutamine supplementation under these conditions can restore gut morphology and perhaps restore its barrier function.[39] In a model of orthotopic small bowel transplantation, rats receiving glutamine-supplemented TPN had improved gut structure, glucose absorption and decrease in bacterial translocation compared with control animals.[40] Following radiotherapy or chemotherapy for tumors, glutamine supplementation has been shown experimentally to improve the response to therapy and reduce treatment-related intestinal mucosal damage.[41,42]

In the tumor-bearing host, it has been postulated that glutamine may represent an anabolic substrate that enhances tumor growth.[43] Although in vitro tumor cell growth can be stimulated with glutamine supplementation, in vivo animal studies have failed to confirm this phenomenon. Studies utilizing sarcoma and mammary adenocarcinoma models in rats have demonstrated that glutamine supplementation can improve host nutritional status without enhancing tumor growth.[44-46]

Glutamine-supplemented TPN in postoperative patients can preserve the intracellular glutamine pool and improve nitrogen balance in the supplemented group compared to controls.[47] The most comprehensive trial of glutamine efficacy in a clinical setting is that of Zeigler et al in a randomized double-blinded prospective study, bone marrow transplant patients receiving glutamine-supplemented TPN were compared to patients receiving standard TPN. The glutamine-supplemented group demonstrated a significant improvement in nitrogen balance, significantly shorter hospital stay and a significant reduction in infectious complication.[48] Although most enteral feeding formulas contain a significant amount of glutamine, parenteral glutamine supplementation remains limited. Glutamine is not included in commercially available TPN solutions, primarily due to the relative instability of glutamine in such solutions. Glutamine analogues or innovative parenteral formulations may overcome this problem in the future to enable glutamine to be included as an integral component of TPN regimens.

Lipids

During illness, sepsis or following injury, the major source of energy is the body's peripheral fat stores. The early response to injury is manifested by lipolysis accompanied by an increase in plasma-free fatty acids and increased rate of fatty acid oxidation. The long-chain polyunsaturated fatty acids (PUFA) have attracted a great deal of attention because of their immunomodulatory properties and their role as a critical fuel substrate. PUFA act as substrates for plasma membrane constituents, regulate cholesterol transport and metabolism and function as second messengers and inflammatory immunologic mediators (eicosanoids, prostaglandins and leukotrienes). Unsaturated lipids are classified into three broad categories and these lipids can be subsequently metabolized to additional bioactive compounds. These categories include the ω9, ω6 and ω3 PUFA. The ω9 oleic acid is the main constituent of olive and canola oils, ω6 linoleic acid is the main fatty acid in corn, cotton and safflower oils and ω3 linolenic acid is the active ingredient in fish and menhaden oils. Most of the scientific investigation has focused on the ω6/ω3 PUFA, as reduced ratios of these lipids were thought to be responsible for the lower incidence of western-type diseases in Eskimos. Identification of such dietary relationship and the adoption of chemopreventive diets may help reduce the incidence of coronary artery disease, arthritis and specific cancers (e.g., breast, colorectal and ovarian). Fernandez et al demonstrated that total fat intake could influence the progression of autoimmune disease in NZB mice. Animals in the high fat diet group had earlier onset of autoimmune disease and shorter life span compared to animals in the low fat group.[49] PUFA supplementation of the diet is known to decrease lymphocyte mitogenic response and ability of lymphocytes to lyse tumor cells in vitro.[50] Lymphocyte production of IL-2 was also reduced by diets containing high ω3 lipids.[51] ω3 and ω6 fatty acids exhibit different effects on macrophage production of TNF.[52] Resident peritoneal macrophages from mice fed diets high in ω3 lipids produced more TNF and less PGE2 than mice fed ω6 lipid. ω6 and ω3 PUFA metabolism occurs by either the cyclo-oxygenase or the lipo-oxygenase enzyme systems. Arachidonic acid is formed from ω6 precursors by the action of phospholipase A2 (PLA2) and gives rise to the prostaglandins 2 (PGE2) series and leukotriene 4 (LT4) series. The exact role that LT and PG play in modulating the immune system is not completely known, but isolated effects of specific moieties have been identified. PGE1 and PGE2 generally inhibit T-cell mitogenesis and macrophage functions.[53] Other studies have demonstrated that PGE2 is important in augmenting the initial immune response. In addition, antibodies against PGE1 early in the immune response suppress cell-mediated immunity in a rat model of cellular immunity.[54] It has been suggested that LT and PG might act sequentially in given situations to determine the outcome of an immune response. It is possible that the LT are important in generating an immune response and the PG functions subsequently to downregulate the mounted response.[55] PGs and their precursor AA are known to induce many of the signs of inflammation and potentiate the inflammatory effects of IL-1, histamine and bradykinin. PGI2, E1, E2 and D2 cause vasodilatation, edema and hyperalgesia.[56,57] In addition, PG12, E1 and E2 increase cAMP and lead to downregulation of neutrophil functions (lysosomal enzyme release, chemotaxis, margination and adherence to vascular endothelium).[58]

Eicasopentaenoic acid (EPA) is derived from ω3 PUFA and has structural similarities to arachidonic acid but very different functional properties. EPA is metabolized by the same enzyme systems (i.e., cyclo-oxygenase and lipo-oxygenase) as AA but gives rise to PGE3 and LT5 series. In general, products of the 3 and 5 series are characterized as being anti-inflammatory while those of the PG2 and LT4 are predominantly proinflammatory.[59] This mechanism explains, in part, the beneficial effects of fish oil or ω3 lipids for treating inflammatory conditions such as arthritis and autoimmune diseases. It is known that diets rich in fish oil derived fatty acids inhibit LTB4 production in monocytes and neutrophils with resultant impairment in chemotaxis and adherence to endothelial cells.[60] In studies by Tate et al it was shown that urate crystal induced inflammation can be significantly attenuated by fish oil in a rat model.[61,62] This was accompanied by a decrease in the number of inflammatory cells at the site, decrease in PGE2 levels and a decrease in neutrophil phagocytosis. In guinea pigs, administration of diets rich in ω3 fatty acids attenuated the septic response and improved survival following challenge with endotoxin.[63,64] Fish oil supplementation has also been shown to attenuate the effect of myocardial ischemia-reperfusion injury and preserve cardiac function in rats.[65,66]

Supplementation of humans with fish oil leads to suppression of blood monocyte generation of IL-1 and TNF in response to endotoxin.[67] Western-world type of cancers including colorectal, breast and ovarian are relatively rare in Far Eastern and African countries. This epidemiologic difference has been attributed, in part, to dietary factors, especially excess fats in the western diet. It has been shown in rodent studies that diets mimicking human fat consumption lead to the promotion of mammary tumors in mice and rats.[68] However, studies examining different types of fats on the genesis of tumors found that, depending on the lipid type, different tumors can be inhibited. Mehanden (fish) oil supplemented diet was found to reduce the incidence and the size of mammary tumors in rodents.[69,70] In nude mice bearing human colon carcinoma, it was found that ω3 supplementation resulted in the greatest decrease in tumor volumes compared to ω6 supplemented mice.[71] This was confirmed in a recent study, which compared tumor volume and metastases from human breast tumor xenografts in nude mice.[72] Mice receiving ω3 fatty acid supplementation had reduced tumor growth and metastases compared to ω6 supplemented mice. This reinforces the idea that ω6 fatty acids are detrimental and tumorigenic, while the ω3 fatty acids may be beneficial in reducing tumor incidence.[73] An interesting experimental approach was utilized in a rabbit model of hepatic tumors.[74] Hepatic artery injection of linoleic or alpha or gamma linolenic acids with a carrier led to a significant tumor response and prolongation of survival in all the fatty acids groups compared to the groups injected with the carrier alone. This study demonstrates the potential clinical utility of fatty acids as anticancer drugs.

A recent development in lipid biochemistry is the design of structured lipids (SL) by the Babayan group. These are lipid emulsions, consisting of various combinations of medium-chain triglycerides (MCT) and ω3 and ω6 long chain triglycerides (LCT) constructed on a single glycerol molecule. These structured lipids, in theory, incorporate the advantages of medium chain and long-chain triglycerides in a single molecule which is easily absorbed, easily metabolized and perhaps less toxic than standard lipid emulsions.[75] Preliminary studies in animal models have shown that enteral or parenteral administration of various SL preparations (75% MCT and 25% LCT or MCT and ω3 PUFA) can improve nitrogen balance, increase protein synthesis, reduce energy expenditure and improve metabolic and cardiovascular parameters in response to stress.[76,77] These same attributes may render this form of lipid supplementation an attractive alternative to existing formulations and clearly requires further clinical evaluation.

Nucleotides

Nucleotides are the building blocks of DNA, RNA and various high-energy compounds (ATP, GTP, etc.). Nucleotide metabolites (cAMP, cGMP) also play a central role in cell signaling and messenger functions. Under most circumstances cellular generation of these compounds from precursors occurs in the de novo synthetic pathway. Alternatively, there is an active salvage pathway that resynthesizes nucleotides from their breakdown products. Cells with very high metabolic rates and proliferative activities exhibit varying degrees of dependence on exogenous sources of nucleotides for their optimal functioning. Exogenous nucleotides may be required for optimal growth of intestinal epithelium and lymphocotes.[78,79] In animal studies utilizing diets deficient or supplemented with nucleotides for 3 weeks prior to an infectious challenge with either *C. albicans* of *S. aureus*, survival was prolonged in the nucleotide-supplemented groups. An LD60 in normal diet group was 100% lethal in the nucleotide-free group indicating an increased susceptibility to infection. Nucleotide-supplemented animals demonstrated improved splenic macrophage bactericidal capacity.[80,81] Animals maintained on nucleotide-free diets and subjected to allogenic heart transplantation had depressed cellular immunity shown by enhanced graft survival.[82] In further experiments, nucleotide-free diets resulted in defective delayed hypersensitivity reaction in vivo, postulated to result from delayed maturation of lymphocytes.[83] Human mononuclear cells supplemented in vitro with nucleotide/RNA mixture demonstrated enhanced production IgG and IgM when stimulated with mitogen.[84]

The above studies led to the trial of nucleotides as nutrient supplements in clinical settings. Two clinical studies have been reported using nucleotides as one of several clinical supplements tested. Cerra et al studied the effect of Impact (arginine, ω3 and nucleotide mixture) compared to Osmolyte (isocaloric enteral formula) supplementation in a critical care setting on immune function. Lymphocyte mitogenic response to PHA and Con A was significantly improved and there was a tendency towards shorter hospital stay in the Impact group (not statistically significant).[85] Daly et al in a larger randomized prospective trial comparing Impact and Osmolyte diets in patients post surgery for malignant gastrointestinal disease found a 70% reduction in infectious and wound complications and a 22% reduction in hospital stay in the Impact group.[24] However, this diet contained other putative dietary immunomodulators including arginine and ω3 PUFA, so the effect of specific nutrients can not be specifically determined. Nevertheless, this trial does indicate the potential for dietary modulation of the host immune system resulting in improved clinical outcome.

Vitamins

Vitamins are essential for normal metabolism and synthesis of lipids, carbohydrates and proteins. Vitamins or their precursors need to be supplied exogenously to prevent the development of deficiency states. Since vitamins are involved in an extensive array of biological reactions, it is not surprising that vitamin deficiency states are associated with immune functional defects. The case, however, for immune stimulatory effects of mega doses of vitamins is not clear.

Vitamin A deficiency is known to impair antibody response, lymphocyte mitogenesis, decrease natural killer cell activity and interferon-gamma production.[86-88] Animals on diets containing high levels of vitamin A showed increased rate of skin graft rejection, increased macrophage tumoricidal activity and phagocytosis and increased delayed-type hypersensitivity reaction.[89-91] In lung cancer patients receiving high doses of vitamin A, enhanced T-cell mitogenesis and reversal of postoperative immunosuppression was documented.[92] Similarly, deficiency in the B group of vitamins resulted in decreased antibody response, thymic atrophy, decreased T-cell mitogenesis and decreased cytotoxic T-cell function.[93,94] Vitamin C levels in mononuclear-phagocytes are known to decrease following surgical insults and burns.[95] This decrease is postulated to be, in part, responsible for impaired macrophage and neutrophil functions under these conditions. Normal volunteers receiving mega doses of vitamin C showed enhanced neutrophil motility and chemotaxis. This finding was reproduced in patients with chronic granulomatous disease and Chediak-Higashi syndrome receiving vitamin C.[96,97] Activated T and B cells express receptors for vitamin D on their surface.[98] The consequence of vitamin D interaction with its receptor is the inhibition of IL-2 production and mitogenesis in stimulated peripheral blood mononuclear cells, inhibition of cytotoxic T-cell generation and inhibition of immunoglobin production.[99,100] Supplemental vitamin E is thought to enhance phagocyte function, cellular and humoral immunity.[101] Thus, there is ample scientific evidence to implicate a role for several vitamins in clinical immunology.

Minerals

Like vitamins, minerals and trace elements participate in diverse biological reactions as catalysts or part of enzyme complexes. Certain mineral deficiency states lead to immunosuppression and altered immune functions. However, the case for immunostimulatory effects of mega doses remains unclear.

Copper deficiency in animal models leads to suppression of T- and B-cell mitogenesis, decreased macrophage activity and impaired cellular immunity.[102-104] In humans, copper deficiency leads to increased incidence of bacterial infections and bronchopneumonia.[105] Children with Menkes syndrome-congenital disease resulting in copper deficiency—die prematurely, often from pneumonia.[106] Iron deficiency states lead to decreased lymphocyte numbers, suppressed T-cell mitogenesis, impaired cellular and humoral immunity and impaired neutrophil-mediated bacterial killing.[107-109] Selenium is an important component of the enzyme glutathione peroxidase (GP), which acts as an antioxidant.[110] It is thought that the observed immune-enhancing effects of selenium are the result of the ability of GP to protect cells of the immune system against oxidative damage.[111] Finally, zinc deficiency is known to cause thymic involution and a decrease in splenic weight, decreased lymphocyte count and defective cellular immunity.[112] Both T- and B-cell mitogenesis are suppressed by zinc deficiency.[113] In acrodermatitis enteropathica—congenital defect in zinc metabolisms—patients exhibit defective cellular immunity remedied by zinc supplementation.[114] Similar defects were seen in humans receiving TPN-deficient in zinc with reversal by zinc supplementation.[115] In malnourished children, zinc supplementation alone led to enhanced IgA production, restored skin test reactivity and reduced incidence of infection than placebo group.[116] This suggests that zinc deficiency may play a major role in inducing immunodeficiency accompanying malnutrition. Further clinical trials of minerals and trace elements are warranted to determine the clinical utility of these substrates in immunonutrition.

Conclusion

The ability to manipulate the immune system by dietary means has great potential in our effort to improve outcome in clinical settings. Arginine, glutamine, fatty acids and the nucleotides have already shown efficacy in various clinical trials. The optimal combination of these nutrients has yet to be determined to achieve optimal clinical response. Many other potential immunomodulatory nutrients remain to be investigated. The vitamins, minerals, trace elements and other putative immunomodulators have shown effects in in vitro systems or animal studies. Their role in surgical nutrition remains to be determined in clinical trials. Nutritional immunology is an evolving field with tremendous potential for improving the prospect of critically-ill patients. The extensive spectrum of basic and clinical research that is currently underway will certainly transform the field of nutritional immunology into future clinical practice.

References

1. Alexander JW, Peck MD. Future prospects for adjunctive therapy: Pharmacologic and nutritional approaches to the immune system modulation. Crit Care Med 1990; 18(suppl 2):S159.
2. Klein D, Morris D. Increased arginase activity during lymphocyte mitogenesis. Bichem Biophys Res Communication 1978; 81(1):199.
3. Currie GA. Activated macrophages kill tumor cells by releasing arginase. Nature 1978; 273:758.
4. Beckman, JS, Beckman TW, Chen J et al. Apparent hydroxyl radical production by peroxynitrite: Implication for endothelial injury from nitric oxide and superoxide. Proc Natl Acad Sci 1990; 87:1620.
5. Hibbs JBJ, Traintor RR, Vavrin Z. Macrophage cytotoxicity: Role for L-arginine deaminase and imino nitrogen oxidation to nitrite. Science 1987; 235:473.

6. Schuman EM, Madison DV. A requirement for the intercellular messenger nitric oxide in long-term potentiation. Science 1991; 254:1503.
7. Nowicki JP, Duval D, Poignet H et al. Nitric oxide mediates neuronal death after focal cerebral ischemia in the mouse. Eur J Pharmacol 1991; 204:339.
8. Rees DD, Palmer R, Moncada S. Role of endothelium-derived nitric oxide in the regulation of blood pressure. Proc Natl Acad Sci USA 1989; 86:3375.
9. Schmidt HHHW, Warner TD, Ishii K et al. Insulin secretion from pancreatic (cells caused by L-arginine-derived nitrogen oxides. Science 1992; 255:721.
10. Rakoff JS, Siver TM, Sinha YN et al. Prolactin and growth hormone release in response to sequential stimulation by arginine and TRF. J Clin Endocrinol 1973; 37:64.
11. Imms FJ, London DR, Neame RLB. The secretion of catecholamines from the adrenal gland following arginine infusion in the rat. J Physiol (Lond) 1969; 200:55.
12. Utsumi M, Makimura H, Ishihara K et al. Determination of immunoreactive somatostatin in rat plasma and responses to arginine, glucose and glucagon infusion. Diabetologica 1979; 17:319.
13. Barbul A, Rettura J, Levenson, SM et al. Arginine: Thymotropic and wound healing promoting agent. Surg Forum 1977; 28:101.
14. Barbul A, Wasserkrug HL, Seifter E et al. Immunostimulatory effects of arginine in normal and injured rats. J Surg Res 1980; 29:228.
15. Barbul A, Wasserkrug HL, Sisto DA. Thymic and stimulatory action of arginine. JPEN 1980; 4:446.
16. Saito H, Trocki O, Wang S. Metabolic and immune effects of dietary arginine supplementation after burn. Arch Surg 1987; 122:784.
17. Tachibana K, Mukai K, Hiraoka I et al. Evaluation of the effect of arginine-enriched amino acid solution on tumor growth. JPEN 1985; 9:428.
18. Renolds JV, Zhang SM, Thom AK et al. Arginine as an immunomodulator. Surg Forum 1987; 38:415.
19. Yeatman TJ, Risley G, Brunson M. Depletion of dietary arginine inhibits growth of metastatic tumor. Arch Surg 1991; 126:1376.
20. Barbul A, Lazaron SA, Efron DT et al. Arginine enhances wound healing and lymphocyte immune response in humans. Surgery 1990; 108:331.
21. Barbul A. The role of arginine as an immune modulator. In: Cunningham-Rundles S, eds. Nutrient modulation of the immune response. New York: Marcel Dekker, Inc., 1993:47.
22. Daly JM, Reynolds J, Thom A et al. Immune and metabolic effect of arginine in the surgical patient. Ann Surg 1988; 208:122.
23. Sigal R, Shou J, Daly J. Parenteral arginine infusion in humans: Nutrient substrate or Pharmacologic agent. JPEN 1992; 16:423.
24. Daly JM, Liberman MD, Goldfine J et al. Enteral nutrition with supplemental arginine, RNA and omega-3 fatty acids in postoperative patients: Immunologic metabolic and clinical outcome. Surgery 1992; 112:56.
25. Rose WC. The nutritive significance of the amino acids. Physiol Rev 1938; 18:109.
26. Eagle H. Nutritional needs of mammalian cells in tissue cultures. Science 1935; 122:501.
27. Ardawi MSM, Newsholme EA. Glutamine metabolism in lymphocyte of the rat. Biochem J 1983; 212:835.
28. Newsholme P, Gordon S, Newsholme EA. Rates of utilization and fates of glucose, glutamine, Pyruvate, fatty acids and ketone bodies by mouse macrophages. Biochem J 1987; 1989:121.
29. Szondy Z, Newsholme EA. The effect of glutamine concentration on the activity of carbamoyl phosphate synthetase II and on the incorporation of 3h-thymidine into DNA in rat mesenteric lymphocytes stimulated by phytohemagglutinin. Biochem J 1989; 261:979.
30. Parry-Billings M, Evans J, Calder PC et al. Does glutamine contribute to immunosuppression after major burns? Lancet 1990; 336:523.
31. Wallace C, Keast D. Glutamine and macrophage function. Metabolism 1992; 41(9):1016.
32. Shou J, Lappin J, Minnard E et al. Total parenteral nutrition, bacterial translocation and host immune function. Am J Surg 1994; 167:145.
33. Kapadia CR, Muhlbacher F, Smith RJ. Alteration in glutamine metabolism in response to operative stress and food deprivation. Surg Forum 1982; 33:19.
34. Ardawi M, Majzoub M. Glutamine metabolism in skeletal muscle of septic rats. Metabolism 1991; 40:155.

35. Parry-Billings M, Baigrie R, Lamont P et al. Effects of major and minor surgery on plasma glutamine and cytokine levels. Arch Surg 1992; 127:1237.
36. Lacey J, Wilmore DW. Is glutamine a conditionally essential amino acid? Nutr Rev 1990; 48:297.
37. Smith RJ. Glutamine metabolism and its physiologic importance. JPEN 1990; 14(4 supplement):40s-44s.
38. Pinkus LM, Windmueller HG. Phosphate-dependent glutaminase of small intestine and localization and role in intestinal glutamine metabolism. Arch Biochem Biophys 1977; 182:506.
39. Burke D, Alverdy JC, Aoys E et al. Glutamine supplemented TPN improves gut immune function. Arch Surg 1989; 124:1396.
40. Zhang W, Frankel WL, Singh A et al. Improvement of structure and function in orthotopic small bowel transplantation in the rat by glutamine. Transplantation 1993; 56(3):512.
41. Fox AD, Kripke SA, DePaula J et al. Effect of a glutamine supplemented enteral diet on methotrexate-induced enterocolitis. JPEN 1988; 12:325.
42. Klimberg VS, Souba WW, Dolson DJ et al. Prophylactic glutamine protects the intestinal mucose from radiation injury. Cancer 1990; 66:62.
43. Kovacevic Z, Morris HP. The role of glutamine in the oxidative metabolism of malignant T-cells. Cancer Res 1972; 32:326.
44. Austgen TR, Dudrick PS, Sitren H et al. The effects of glutamine-enriched total parenteral nutrition on tumor growth and host tissues. Ann Surg 1992; 215(2):107.
45. Klimberg SV, Souba WW, Salloum RM et al. Glutamine-enriched diets support muscle glutamine metabolism without stimulating tumor growth. J Surg Res 1990; 48:319.
46. Bartlett D, Torosian M. Effect of oral glutamine on tumor and host growth. Ann Surg Oncol 1993;in press.
47. Hammarqvist F, Wernerman J, Ali R et al. Addition of glutamine to total parenteral nutrition after elective abdominal surgery spares free glutamine in muscle, counteracts the fall in muscle protein synthesis and improves nitrogen balance. Ann Surg 1989; 209:455.
48. Ziegler TR, Young LS, Benfell K et al. Clinical and metabolic efficacy of glutamine-supplemented parenteral nutrition after bone marrow transplantation. Ann Intern Med 1992; 116:821.
49. Fernandes G, Yunis EJ, Smith J et al. Dietary influence on breeding behavior, haemolytic anemia and longevity in NZB mice. Proc Soc Exp Biol Med 1972; 139:1189.
50. Erickson KL, Adams DA, McNeill CJ. Dietary lipid modulation of immune responsiveness. Lipids 1983; 18:468.
51. Maydani SN, Endres S, Wood MM et al. Oral (ω-3) fatty acid supplementation suppresses cytokine production and lymphocyte proliferation: comparison between young and older women. J Nutr 1991; 121:547.
52. Hardardottir I, Kinsella J. Tumor necrosis factor production by murine resident peritoneal macrophages is enhanced by dietary ω-3 polyunsaturated fatty acids. Biochem Biophys Acta 1991; 1093:187.
53. Schultz RM, Pavlidis NA, Stylos WA et al. Regulation of macrophage tumoricidal function: A regulatory role for prostaglandins of the E series. Science 1978; 202:320.
54. Mertin J, Stackpoole A. Anti-PGE antibodies inhibit in vivo development of cell-mediated immunity. Nature 1981; 62:293.
55. Ninnemann JL. Prostaglandins and immunity. Immunol Today 1984; 5:170.
56. Higgs GA, Palmer RMJ, Eakins KE et al. Arachidonic acid metabolism as a source of inflammatory mediators and its inhibition as mechanism of action for anti-inflammatory drugs. Mol Aspects Med 1984; 4:275.
57. Willis AL, Cornelsen M. Repeated injections of prostaglandin E2 in rat paws induces chronic swelling and marked decrease in pain threshold. Prostaglandins 1973; 3:353.
58. Korchak HM, Abramson SB. The role of arachidonic acid metabolites in the regulation of neutrophil functions. In: Goodwin JS, eds. Prostaglandins and immunity. Boston: Martinus Nijhoff, 1985:179.
59. Homsy J, Morrow WJW, Levy JA. Nutrition and autoimmunity: A review. Clin Exp Immunol 1986; 65:473.
60. Lee TH, Hoover RL, Williams DJ et al. Effect of dietary enrichment with eicosapentaenoic and docosahexaenoic acids on in vitro neutrophil and monocyte leukotriene generation and neutrophil function. N Engl J Med 1985; 312:1217.
61. Tate GA, Mandell BF, Schumacher HR Jr. et al. Suppression of acute inflammation by 15-methyl prostaglandin E1. Lab Invest 1988; 59(2):192.

62. Tate G, Mandell BF, Laposata M et al. Suppression of acute and chronic inflammation by dietary gamma linolenic acid. J Rheumatol 1989; 16(6):729.
63. Pomposelli JJ, Mascioli EA, Bistrian BR et al. Attenuation of the febrile response in guinea pigs by fish oil enriched diets. JPEN 1989; 13:136.
64. Mascioli EA, Iwasa Y, Trimbo S et al. Endotoxin challenge after menhaden oil diet: Effects on survival of guinea pigs. Am J Clin Nutr 1989; 49:277.
65. Yang B, Saldeen TG, Nichols WW et al. Dietary fish oil supplementation attenuates myocardial dysfunction and injury caused by global ischemia and reperfusion in isolated rat hearts. J Nutr 1993; 123(12):2067.
66. Yang BC, Saldeen TG, Bryant JL et al. Long-term dietary fish oil supplementation protects against ischemia-reperfusion-induced myocardial dysfunction in isolated rats hearts. Am Heart J 1993; 126(6):1287.
67. Endres S, Ghorbani R, Nelley VE et al. Dietary ω-3 polyunsaturated fatty acids suppress synthesis of IL-1 and tumor necrosis factor. N Engl J Med 1987; 317:397.
68. Tinsley IJ, Schmitz JA, Pierce DA. Influence of dietary fatty acids on the incidence of mammary tumors in the C3H mouse. Cancer Res 1981; 41:1460.
69. Jurkowski JJ, Cave Wt. Dietary effects of menhaden oil on the growth and membrane lipid composition of rat mammary tumors. JNCI 1985; 74:1145.
70. Gabor H, Abraham S. Effect of dietary menhaden oil on tumor cell loss and the accumulation of mass of a transplantable mammary adenocarcinoma in BALB/c mice. JNCI 1986; 76:1223.
71. Sakaguchi M, Imray C, Davis A et al. Effects of dietary ω-3 and saturated fats on growth rates of the human colonic cancer cell lines SW-620 and LS 174T in vivo in relation to tissue and plasma lipids. Anticancer Res 1990; 10(6):1763.
72. Rose DP, Connolly JM. Effects of dietary omega-3 fatty acids on human breast-cancer growth and metastases in nude-mice. JNCI 1993; 85:1743.
73. Kromhout D. The importance of ω-6 and ω-3 fatty acids in carcinogenesis. Med Oncol Tumor Pharmacother 1990; 7(2-3):173.
74. Hayashi Y, Fukushima S, Kishimoto S et al. Anticancer effects of free polyunsaturated fatty acids in an oily lymphographic agent following intrahepatic arterial administration to a rabbit bearing VX-2 tumor. Cancer Res 1992; 52(2):400.
75. Babayan VK. Medium chain triglycerides and structured lipids. Lipids 1987;22:417.
76. Mok KT, Maiz A, Yamazaki K et al. Structured medium-chain and long-chain triglyceride emulsion are superior to physical mixtures in sparing body protein in the burned rat. Metabolism 1984; 33:910-915.
77. DeMichele SJ, Karlstad MD, Bistrian BR et al. Enteral nutrition with structured lipid: Effect on protein metabolism in thermal injury. Am J Clin Nutr 1989; 50:1295.
78. LeLeiko NS, Martin BA, Walsh M et al. Regulation of types I, II, III and IV procallagen mRNA synthesis in clucocorticoid-mediated intestinal development. Gastroeterology 1987; 93:1014.
79. Strauss PR, Henderson JF, Goodman MG. Nuclosides and lymphocytes: An overview. Proc Soc Exp Biol Med 1985; 179:413.
80. Fanslow WC, Kulkarni AD, Van Buren CT. Effect of nucleotide restriction and supplementation on resistance to experimental candidiasis. JPEN 1988; 12:49.
81. Kulkarni AD, Fanslow WC, Van Buren CT. Influence of dietary nucleotide restriction on bacterial sepsis and phagocytic cell function in mice. Arch Surg 1986; 121:169.
82. Van Buren CT, Kulkarni A, Schandle VB et al. The influence of dietary nucleotides on cell-mediated immunity. Transplantation 1983; 36:350.
83. Rudolph FB, Fanslow WC, Kulkarni AD et al. Effect of dietary nucleotides on lymphocyte maturation. Adv Exp Med Biol 1986; 1986:497.
84. Jyonouchi H, Zhang L, Tomita Y. Studies of immunomodulating actions of RNA/nucleotides. RNA/nucleotides enhance in vitro immunoglobulin production by human peripheral blood mononuclear cells in response to T-dependent stimuli. Pediatr Res 1993; 33(5):458.
85. Cerra FB, Lehman S, Konstantinides N et al. Effect of enteral nutrient on in vitro tests of immune function in ICU patients: A preliminary report. Nutrition 1990; 6(1):84-7.
86. Pasatiempo AMG, Taylor CE, Ross AC. Vitamin A status and the immune response to pneumococcal polysaccharide: Effects of age and early stage retinol deficiency in reats. J Nutr 1990; 121:556.
87. Micksche M, Cerni C, Kokron O et al. Stimulation of immune response in lung cancer patients by vitamin A therapy. Oncology 1977; 34:234-238.

88. Bowman TA, Goonewardene MI, Pasatiempo AG et al. Vitamin A deficiency decreases natural killer cell activity and interferon production in rats. J Nutr 1990; 120:1264.
89. Floersheim G, Bollog W. Accelerated rejection of homografts by Vitamin A acid. Transplantation 1974; 14:564.
90. Tachibana K, Sone S, Tsubura E et al. Stimulatory effect of vitamin A tumoricidal activity of rat alveolar macrophages. Br J Cancer 1984; 49:343.
91 .Miller K, Maisey J, Malkovsky M. Enhancement of contact sensitization in mice fed a diet rich in vitamin A acetate. Int Arch Allergy Appl Immunol 1984; 1984:120.
92. Mickshe M, Cerni C, Kokron O et al. Stimulation of immune response in lung cancer patients by vitamin A therapy. Oncology 1977; 34:234.
93. Bendichd A, Cohen M. B vitamins: Effects on specific and nonspecific immune responses. In: Chandra RK, eds. Nutrition and Immunology. New York: Alan R. Liss, 1988.
94. Hollingsworth JW, Carr J. 3H-Uridine incorporation as a T-lymphocyte indicator in rats. Cell Immunol 1973; 8:270-279.
95. Oberritter H, Glatthaar B, Moser U et al. Effect of functional stimulation on ascorbate content in phagocytes under physiological and pathological conditions. Int Arch Allergy Appl Immunol 1984; 81:46.
96. Anderson R, Oosthuizen R, Martiz R et al. The effects of increasing weekly doses ascorbate on certain cellular and humoral immune functions in normal volunteers. Am J Clin Nutr 1980; 33:71.
97. Boxer LA, Watanabe AM, Rister M et al. Correction of leukocyte function in Chediak-Higashi syndrome by ascorbate. N Engl J Med 1976; 295:1041.
98. Provvedini D, Tsoukas C, Deftos L et al. 1,25-dihydroxyvitamin-D3 receptors in human leukocytes. Science 1983; 221:1181.
99. Lemire J, Adams JS, Sakai R et al. 1,25-dihydroxyvitamin-D3 suppresses proliferation and immunoglobulin production by normal human peripheral blood mononuclear cells. J Clin Invest 1984; 74:657.
100. Rigby W, Stacy T, Fangar W. Inhibition of T-lymphocyte mitogenesis by 1,25-dihydroxyvitamin-D3 (calcitriol). J Clin Invest 1984; 74:1451.
101. Tengerdy RP. The role of vitamin E in immune response and disease resistance. Ann N Y Acad Sci 1990; 587:24.
102. Lukasewycz OA, Prohaska JR. Lymphocytes from copper-deficient mice exhibit decreased mitogen reactivity. Nutr Res 1983; 3:335.
103. Lukasewycz OA, Prohaska JR. Immunization against transplantable leukemia impaired in copper-deficient mice. J Nat Cancer Inst 1982; 69:489.
104. Lukasewycz OA, Prohaska JR. Immune response in copper deficiency. Ann N Y Acad Sci 1990; 587:147.
105. Al-rashid RA, Spangler J. Neonatal copper deficiency. N Engl J Med 1971; 285:841.
106. Pedroni E, Bianchi E, Ugazio AG et al. Immunodeficiency and steely hair. Lancet 1975; 1:1303.
107. Chandra RK, Au B, Woodford G et al. Iron status, immune response and susceptibility to infection. In: Kies H, Eds. Iron Metabolism. Amesterdam: Elsevier/Excerpta Medica, 1977.
108. Chandra RK. Iron and immunocompetence. Nutr Rev 1976; 34:129.
109. Kochanowski BA, Sherman AR. Decreased antibody formation in iron-deficient rat pups-effect of iron repletion. Am J Clin Nutr 1985; 41:278.
110. Forstrom JW, Zakowski JJ, Tappel AL. Identification of the catalytic site of rat liver glutathione peroxidase as selenocysteine. Biochemistry 1978; 17:2639.
111. Levander OA. Clinical consequences of low selenium intake and its relationship to vitamin E. Ann NY Acad Sci 1982; 393:70.
112. Fernandes G, Nair M, Onoe K et al. Impaired cell-mediated immunity functions by dietary zinc deficiency in mice. Proc Natl Acad Sci USA 1979;76:457.
113. Zanzonica P, Gernandes G, Good RA. The differential sensitivity of T-cell and B-cell mitogenesis to in vitro zinc deficiency. Cell immunol 1981;60:203.
114. Olseke JM, Westphal ML, Shore S et al. Correction with zinc therapy of depressed cellular immunity in acrodermatitis enteropathica. Am J Dis Child 1979;133:915.
115. Allen JI, Kay NE, McClain CJ. Severe zinc deficiency in humans: Association with a reversible T-lymphocyte dysfunction. Ann Int Med 1981;95:154.
116. Castillo-Duran C, Heresi G, Fiseberg M et al. Controlled trial of zinc supplement during recovery from malnutrition: Effects on growth and immune function. Am J Clin Nutr 1987;45:602.

CHAPTER 2

Cancer Cachexia: Etiology, Treatment and Future Research

Michael H. Torosian

Cancer patients have the highest prevalence of malnutrition of any group of hospitalized patients. Catabolism secondary to malignancy and antineoplastic therapy contribute to the development of malnutrition in the cancer patient. Widespread metabolic and nutritional abnormalities have been documented in the cancer patients including changes in protein, carbohydrate, lipid and energy metabolism.[1-3] Nutritional status is an important independent factor for prognosis in cancer patients with malnutrition being associated with increased morbidity and mortality.[4,5]

Despite retrospective studies which suggest reduced morbidity in malnourished cancer patients receiving nutrition support, the use of nutrition support in the cancer patient remains controversial. Prospective, randomized trials have, in general, failed to document a significant reduction in complications in cancer patients undergoing therapy and receiving nutrition support.[1,2,5,6] Only in severely malnourished patient populations undergoing aggressive antineoplastic therapy has objective benefit from nutrition support been clearly documented.[7,8] Thus, controversy remains regarding the clinical efficacy of nutrition support in the tumor-bearing host.

The Development of Cancer Cachexia

Cancer cachexia is a clinical syndrome which consists of anorexia, weight loss, severe tissue wasting, asthenia and organ dysfunction. The relationship between host weight loss and mortality in cancer patients has been recognized.[9] It is evident that the cause of cachexia in cancer patients is multifactorial. Both disease- and treatment-related factors contribute to the syndrome of cachexia which results primarily from distant metabolic effects of the tumor. Thus, cancer cachexia represents a paraneoplastic syndrome which is particularly prevalent in patients with advanced tumors. Although numerous theories have been postulated to explain its etiology, the mechanism of cancer cachexia remains unknown.[10-12]

The relationship of cachexia to tumor burden, stage of disease and tumor histology is inconsistent and does not correlate well with the cachectic state. Recent results suggest that cytokines, which are peptides secreted by host tissues in response to tumors and hormones, play a major role in the development of cachexia. These mediators, such as cachectin (tumor necrosis factor), interferon-γ and the interleukins, exert profound effects on host intermediary metabolism.[12] In the short-term, these mediators promote an acute phase response by rerouting

Nutritional Support in Cancer and Transplant Patients, edited by Rifat Latifi and Ronald C. Merrell. ©2001 Eurekah.com.

nutrients from the periphery to the liver. Long-term effects of this cytokines and hormone responses result in the clinical syndrome of cancer cachexia with marked abnormalities in carbohydrate, protein, lipid and energy metabolism.

Diminished Nutrient Intake

Anorexia accompanies most neoplasms and is a major contributing factor in the development of the cachectic state. Often, loss of appetite is a presenting symptom of an underlying tumor. Both physiologic and mechanical derangements can contribute to the development of anorexia in the cancer patient. Physiologic abnormalities associated with anorexia are common and involve numerous organ systems. Impaired taste perception, such as reduced thresholds for sweet, sour and salty flavors, have been previously demonstrated. For example, DeWys and Walters noted that reduced oral threshold for urea correlates with an aversion for red meat in cancer patients.[13] Deficiencies in zinc and other trace elements are common with advanced malignancy and specific chemotherapy regimens also contribute to altered taste sensation. Patients with hepatic metastases and associated hepatic insufficiency may develop anorexia and nausea from reduced clearance of lactate produced, in part, by anaerobic tumor metabolism of glucose.[1]

The specific metabolic processes that affect nutrient intake in cancer patients are unclear and numerous hypotheses have been proposed to explain this phenomenon. Lucke et al reported a humoral factor that reproduces the metabolic characteristics of cachexia in non-tumor-bearing animals.[14] DeWys et al suggested that tumor peptides acting through neuroendocrine cells and neuroreceptors alter metabolic pathways.[15] Nakahara described a "toxohormone" capable of simulating the cachectic state when injected into the normal animals.[16] Other investigators have shown that blocking antibodies to interferon-γ partially reversed the cachexia in animals with end-stage tumors.[17] Furthermore, recent studies have implicated endogenously produced tumor necrosis factor (TNF) as an important mediator in the development of cachexia in the tumor-bearing host.[12] Circulating levels of TNF are difficult to detect in cancer patients and this cytokine likely works in a paracrine fashion. Finally, as suggested by Krause et al, abnormalities in the central nervous system metabolism of serotonin may be responsible for anorexia associated with the tumor-bearing state.[18]

Local effects of the tumor may also lead to reduced food intake, particularly when the tumor obstructs the upper alimentary tract. Patients with cancer of the oral cavity, pharynx, esophagus or stomach may have reduced nutrient intake because of dysphagia or odynophagia from partial or complete intestinal obstruction. Patients with gastric cancer often have reduced gastric capacity or partial gastric outlet obstruction causing nausea, vomiting, and early satiety. Intestinal tumors and abdominal carcinomatosis can cause partial obstruction or the blind-loop syndrome (associated with intestinal obstruction and bacterial overgrowth) to interfere with nutrient absorption. Pancreatic carcinomas frequently cause exocrine enzyme deficiencies with specific malabsorption syndromes. Finally, psychological factors such as depression, grief or anxiety resulting from the disease or its treatment may lead to poor appetite, abnormal eating behaviors and learned food aversions. These psychological effects may subsequently led to significant metabolic abnormalities, which exacerbate the adverse clinical effects of malnutrition.[19]

Abnormalities of Substrate Metabolism

Extensive abnormalities in energy, carbohydrate, lipid and protein metabolism have been documented in patients with malignancy.[1,2,20] Aberrations in energy expenditure and inefficient energy utilization have been cited as causes of progressive weight loss in the cancer patient.[21,22] Classically, it was believed that all cancer patients were hypermetabolic—i.e., they exhibited increased energy expenditure. With hypermetabolism, it was postulated that cancer patients catabolized endogenous energy stores and excessively consumed exogenous nutrients but were unable to prevent progressive weight loss. However, Young found that resting metabolic rate was not consistently elevated in cancer patients.[21] Leukemia and lymphoma patients were the only groups with commonly elevated resting energy expenditures—increases in metabolic rate paralleled advancing disease in these patients. Other investigators have shown an elevation in energy expenditure in patients with lymphomas, lung cancer, and head and neck cancers.[23,24] Knox et al measured energy expenditure in 200 malnourished patients with gastrointestinal cancer by indirect calorimetry.[25] Only 41% of cancer patients had a normal resting energy expenditure (REE), while decreased and increased REE was observed in 33% and 26% of patients, respectively. Similarly, Heber et al found no clear evidence of hypermetabolism in noncachectic lung cancer patients.[26] Shaw et al concluded that alterations in metabolic rate depended on the type of malignancy.[27] They demonstrated an elevated rate of energy expenditure in sarcoma patients associated with increased Cori cycle activity and glucose turnover, reduced glucose oxidation, and increased protein catabolism. Buzby et al reported a reduction in metabolic rate associated with pancreatic cancer.[28]

Inefficient energy utilization by the tumor-bearing host was studied by Holroyde and Reichard, who reported increased Cori cycle activity in patients with the greatest energy expenditure and weight loss. Another indication of inefficient bioenergetics in cancer patients is an elevated rate of anaerobic glycolysis. Increased anaerobic glycolysis in tumor cells has been known for many years and occurs in host tissues of the cancer patient under certain conditions.[30] Anaerobic glycolysis is an extremely inefficient means of glucose utilization compared to oxidative metabolism. Young suggested that increased rates of protein turnover also result in significant energy losses due to failure of normal mechanisms of adaptation to starvation in the cancer patient.[21] During the first two days of fasting, endogenous glycogen stores of muscle and liver are depleted. Glucose utilization by the brain, leukocytes and other tissues continues, resulting in the breakdown of protein for gluconeogenesis. In non-cancer patients, gluconeogenesis (and associated muscle catabolism) is gradually replaced by fat fuel metabolism in which fatty acids are converted to ketone bodies. Ketone bodies can be used as energy substrates to provide 95% of energy to the brain; this results in decreased glucose utilization with sparing of muscle protein. In cancer patients, the adaptive mechanisms are blunted with continued glucose production fueled by protein catabolism.[3,21,31]

Abnormalities in carbohydrate metabolism include glucose intolerance, impaired whole body insulin sensitivity, decreased glucose oxidation and increased rates of gluconeogenesis and glucose recycling.[20,29] After oral or intravenous administration of glucose, delayed glucose clearance occurs and results in hyperglycemia.[29,32] Despite hyperglycemia, there is impaired insulin release from the pancreas and peripheral tissue insensitivity to circulating insulin.[31,32] Feedback control of glucose production may be impaired because gluconeogenesis and Cori cycle activity are not inhibited by glucose ingestion in the cancer patients. Moley et al and Peacock and Norton studied sarcoma-bearing rats and found that supplemental insulin administration preserves host lean body mass and may influence survival.[33,34] Most other reports indicate that insulin can increase appetite but does not prolong survival of the tumor-bearing host. Shaw and Wolfe recently noted that patients with gastrointestinal tumors had elevated rates of basal hepatic glucose production.[35] They found a direct relationship between tumor burden and the increased rate of gluconeogenesis in this study.

Lipid metabolism is also significantly altered in the cancer patient. Catabolism of body fat stores occurs in patients with malignancy as evidenced by objective changes in host anthropometrics, body composition and increased rate of lipolysis and oxidation of fatty acids.[3] Glycerol and fatty acids, the by-products of lipolysis, serve as substrates for gluconeogenesis and energy production, respectively, during periods of nutrient deprivation. Waterhouse observed that fatty acids are the major substrates utilized in patients with progressive malignant disease.[36]

Increased plasma clearance of endogenous fat stores and exogenously administered fat emulsions occur in cancer patients in both fasting and fed states. Patients with malignancy fail to suppress lipolysis after glucose administration (which is the normal, adaptive response) and continue to oxidize fatty acids.[31] Wilson et al reported that 15 of 21 patients with colorectal cancer demonstrated increased rates of fat clearance.[37] Fat clearance rates were reduced to nearly normal in the 14 patients who were retested 12 weeks after curative resection.

Aberrations in protein metabolism are perhaps the most detrimental to the cancer patient.[38] Since cancer patients commonly exhibit host nitrogen depletion, increased catabolism of muscle protein, decreased muscle protein synthesis and abnormal plasma aminograms, the goal of nutrition support in all patients is incorporation of nitrogen into proteins of host tissues.[1,3] Severe atrophy of lean body mass and negative nitrogen balance frequently occur in the presence of progressive tumor growth and metastasis.[39,40] This selective depletion of host protein at the expense of tumor growth has led to the concept of the tumor as a "nitrogen trap".[3,38] Autonomy of tumor growth and the creation of a hormonal and cytokine milieu which is catabolic to host tissues promotes tumor-versus-host growth.

Norton et al found that sarcoma-bearing limbs released less than 50% of the amount of amino acids released from tumor-free limbs.[41] Studies of whole-body protein kinetics with N^{15}-labeled glycine indicates that cancer patients have increased whole-body protein turnover, a process that contributes to increased energy expenditure.[41,42] Severe wasting of host muscles can occur with depletion of visceral and circulating proteins due primarily to increased protein breakdown. While tumor-induced increases in muscle and visceral protein catabolism contribute to host cachexia, evidence is emerging that tumor tissues may regulate their own protein degradation. Tayek et al compared the rates of host muscle and liver protein synthesis and degradation in rats bearing either syngeneic sarcoma or hepatoma.[43] Eighteen days after tumor implantation, synthesis of rat muscle protein was decreased, but liver protein synthesis increased, with a net decrease in whole body protein synthesis. The metabolic cost to the host of increased protein flux may be substantial and can contribute to the development of cachexia.

Etiology of Cancer Cachexia

The etiology of cancer cachexia remains controversial but is undoubtedly multifactorial. Cachexia is not simply a local effect of the tumor but is caused by systemic factors which elicit a paraneoplastic response to the tumor. Current theories hypothesize that tumors do not directly produce mediators of cachexia. Two prominent classes of cachexia mediators believed to be fundamental to the development of cancer cachexia are cytokines and regulatory hormones.

Cytokines are soluble proteins which are secreted by host tissues in response to various stimuli—including cancer, sepsis, inflammation, and other pathophysiologic insults. Cytokines function by several mechanisms including autocrine, paracrine or circulating/systemic routes. Tumor necrosis factor, interleukin-1, and interleukin-6 are specific cytokines which have been implicated in the development of cancer cachexia by recent experimental evidence.[12,17] Tumor necrosis factor (TNF) or cachectin is a 17 kilodalton molecular weight protein secreted by macrophages in response to endotoxin, malignancy and other catabolic stimuli.[12,44] TNF administration to animals can reproduce many, but not all, of the changes characteristic of cancer cachexia. Anorexia, weight loss, depletion of fat stores, loss of skeletal muscle, hypoproteinemia and increased total body water have been documented in TNF-treated animals. Thus, TNF may cause some of the adverse effects of cancer cachexia, but this cytokine is certainly not the sole mediator. Furthermore, it has been difficult to detect circulating levels of TNF in cancer patients even with severe degrees of cachexia. TNF function primarily by the paracrine mechanism of action.[44]

Interleukin-1 (IL-1) is a cytokine secreted by macrophages in response to endotoxin. This inflammatory cytokine causes anorexia, pyrexia, hypotension, decreased systemic vascular resistance and increased cardiac output.[45] Gene amplification of the IL-1 locus has been found in one cachectic animal model of malignancy and both IL-1 and TNF produce alterations in hepatic protein synthesis similar to the tumor-bearing state.[46] Interleukin-6 (IL-6) is secreted by macrophages stimulated by endotoxin and by fibroblasts in response to TNF or IL-1.[47] This cytokine is also called β_2-interferon, hepatocyte-stimulating factor and hybridoma growth factor and has many activities similar to TNF and IL-1. Elevated levels of IL-6 have been found in tumor-bearing animals and correlate with the hepatic acute phase response to cancer. Although the precise roles of TNF, IL-1 and IL-6 remain to be defined, these cytokines, in part, are involved in the development of cancer cachexia.

Abnormalities in regulatory hormones may also play an important role in the development of cancer cachexia. Cachexia from cancer and other catabolic states has been associated with decreased insulin and increased glucagon levels.[48] The reduced insulin:glucagon ratio establishes a catabolic hormonal milieu which promotes weight loss, muscle breakdown and depletion of fat stores. Insulin alone is inadequate to reverse these metabolic abnormalities due to the associated glucagon response which ensues. In our laboratory we found that providing the anabolic hormones insulin and growth hormone combined with somatostatin (to suppress glucagon secretion) can reverse cachexia in tumor-bearing animals.[49] This hormonal approach to cachexia has improved host nutritional status without stimulating tumor growth in one animal model. Additional clinical and basic research is required to further elucidate the role of cytokines, hormones and other circulating factors in the development of cancer cachexia and to design effective therapeutic strategies to specifically prevent cachexia.

Clinical Efficacy of Nutritional Support

Sixteen randomized, prospective trials have evaluated the effects of preoperative TPN on clinical outcome in cancer patients. Detsky et al reviewed 14 trials utilizing meta-analysis and concluded that "routine use of perioperative TPN in unselected patients having major surgery is not justified, however, this intervention may be helpful in subgroups of these patients who are at high risk".[50] The author agrees that preoperative TPN should be utilized in select subgroups of patients, including severely malnourished patients and those undergoing aggressive chemo- and radiotherapy regimens that may benefit from nutrition support.

Heatley et al studied 74 patients with upper gastrointestinal cancer who underwent operative resection.[51] Perioperative TPN only reduced wound infections in this study. Other prospective randomized trials have noted no significant decrease in operative morbidity and mortality with provision of TPN.[52,53] One notable exception is the trial by Mueller et al which studied the efficacy of 10 days of preoperative TPN in patients with gastrointestinal cancer.[7] Postoperative morbidity was significantly lower in TPN patients (17%) compared with the control group (32%). The mortality rate was 4% in the TPN group and 16% in the control group. Although less morbidity and mortality occurred in the TPN group, this study is criticized for the high mortality rate observed in control patients. In a review of 244 patients with esophageal cancer, 72 patients given at least 5 days of preoperative TPN lost less weight and had significant reduction in major postoperative complications compared to a concurrent group of control patients.[54]

A multi-institutional prospective randomized trial evaluating TPN in surgical patients revealed similar morbidity and mortality rates in both TPN and control patients.[8] A higher incidence of infectious complications was noted in the TPN group while a higher non-infectious complication rate occurred in controls. In the subgroup of patients classified as severely malnourished, preoperative TPN significantly reduced overall major complications from 46% to 21-26%.

In summary, clinical trials of preoperative TPN in cancer patients have demonstrated improved nutritional status indicated by increased body weight and improved serum protein levels, immune function, and nitrogen balance. Effects on clinical outcome are less dramatic except in the severely malnourished high-risk group. Failure to demonstrate improved outcome with TPN in most trials may accurately reflect the limited value ofTPN. In severely malnourished patients undergoing major operative procedures, preoperative TPN appears to have beneficial role.

Numerous clinical reviews have correlated malnutrition with poor prognosis in patients with metastases being treated with chemotherapy. Therapeutic benefit to "adjunctive" nutrition support was suggested and a series of prospective clinical trials were conducted to test this hypothesis. Endpoints of these trials included duration of survival, tumor response to treatment and treatment toxicity.[2,55] The major toxicities evaluated were gastrointestinal (nausea, vomiting, stomatitis, and diarrhea), infectious and hematologic (leukopenia, anemia, and thrombocytopenia). Regarding gastrointestinal toxicity, no difference between TPN and control groups was noted in three studies, improvement (less nausea and vomiting in the TPN group) was noted in one study, and worse stomatitis in the TPN group was reported in two studies. Only 2 of 11 evaluable studies demonstrated less hematologic toxicity. Finally, two of five trials found that septic complications were more common in the TPN group than control group.

One notable exception to these studies concerning the efficacy of nutrition support during chemotherapy is the report by Weisdorf et al.[56] These authors prospectively randomized pediatric bone marrow transplant recipients to either total parenteral nutrition or control electrolyte solution. Improved overall survival, improved disease-free survival and decreased rates of relapse were documented in patients receiving total parenteral nutrition. This study demonstrates the potential of using nutrition to support patients through extremely aggressive antineoplastic therapy. As chemotherapy regimens become more aggressive, additional indications for nutrition support may be evident.

Use of radiation therapy increases the potential for the development of severe nutritional deficits. The onset and degree of malnutrition relate to the tumor's location, the site irradiated and the extent of radiation therapy administered. In patients requiring abdominal or pelvic radiotherapy, Bothe et al reported decreased mortality in patients receiving extensive abdominal or pelvic radiation therapy in TPN versus non-TPN patients.[57] This is one of only a few studies suggesting that nutrition support can reduce mortality in patients receiving radiation therapy. During radiation treatment, use of TPN allows maintenance of nutritional status and restoration of immunocompetence. Solassol prospectively studied patients with advanced ovarian tumors during radiotherapy and found no difference in survival between TPN and non-TPN groups.[58] However, patients receiving TPN had less malnutrition and fewer interruptions of their planned treatments. Douglas and co-workers treated patients with advanced pancreatic, stomach, and colorectal cancer with radiotherapy and enteral feeding.[59] No survival advantage was achieved in the nutritionally supported group. As noted in the previous studies, completion of the prescribed radiotherapy dose and restoration of immunocompetence was noted in the group receiving nutritional support. A significant increase in body weight is typically observed in nutritionally supported versus control patients receiving radiation therapy.

The results of these and other prospective trials have been reviewed by Donaldson and show that routine adjunctive use of nutritional intervention has not improved tolerance to treatment, local control or survival rates after radiation therapy.[60] An increase in weight gain, improvement in nutritional status and fewer interruptions in radiotherapy treatments are seen with nutrition support.

Future Horizons

Since conventional nutrition support has shown limited success in reducing morbidity and mortality of cancer patients, innovative nutritional and metabolic support regimens are currently being investigated. Specific substrates, anabolic hormones, metabolic blocking agents and pharmacologic agents directed at reversing or preventing specific organ atrophy or host functional deficits are being studied. Arginine, glutamine, specific lipid moieties (including medium-chain triglycerides, structured lipids, ω3 fatty acids, and short-chain fatty acids), nucleotides, anabolic hormones and pharmacologic agents are under clinical and research study. These nutrients are designed to prevent critical organ-specific atrophy and functional deficits without enhancing tumor growth or metastasis in the cancer patient.

Immunomodulation has been studied in critically ill and cancer patients in an attempt to reduce septic complications. The immune stimulatory agents arginine, nucleotides, and ω3 fatty acids have been combined for clinical investigation based on prior basic science research.[61,62] Arginine is an anabolic amino acid required for protein and polyamine synthesis and to transport and excrete nitrogen (via the urea cycle). Arginine also functions as a potent hormonal secretagogue and has widespread immunostimulating properties.[63] In vivo and in vitro studies have demonstrated that arginine supplementation can promote nitrogen retention, improve wound tensile strength, improve delayed hypersensitivity, increase T-cell mitogenesis and interleukin-2 production and reduce thymic involution.[64,65] Inhibition of tumor growth during arginine supplementation has been observed and is believed to occur by stimulating antitumor immunity.[66,67]

Specific lipid substrates may play an important role in the nutritional management of cancer patients in the future. Exogenous lipids are incorporated into both host and tumor cell membranes and can significantly affect cellular physiochemical properties including permeability, fluidity, distensibility and sensitivity to chemo- and radiation therapy.[68] Animal studies have demonstrated markedly different effects on tumor growth and tumor metastasis with long chain triglycerides, medium chain triglycerides and ω3 fatty acids.[69,70] In general, long chain triglycerides, which are used in conventional total parenteral nutrient regimens, are immunosuppressive and have been found to stimulate primary tumor growth and metastasis.[69] In contrast, medium chain triglycerides and ω3 fatty acids may inhibit tumor growth and metastasis.[69,70] Structured lipids are composed of a glycerol backbone with a medium chain fatty acid and ω3 fatty acid attached as sidechains.[71] These synthetic lipids have been studied in tumor-bearing animals in an attempt to combine the properties of medium chain and ω3 fatty acids in a single lipid molecule. Studies in sarcoma-bearing rats have demonstrated decreased primary tumor growth with increased tumor protein catabolism and improved hepatic protein fractional synthetic rates.[72] Parenteral infusion of structured lipids maintains nutritional status comparable to conventional nutrient regimens as measured by classic nutritional parameters.[71,72] Clinical investigation is required to evaluate the effect of these and other lipid moieties on tumor growth, metastasis and host metabolism in cancer patient.

ω3 fatty acids, or fish oils, exhibit properties that are dramatically different from ω6 fatty acids, or vegetable oils. Although the cyclo-oxygenase and lipoxygenase enzyme systems metabolize both lipid moieties, the resulting prostaglandin and leukotriene end products have markedly different activities. While ω6 fatty acids suppress T-cell mitogenesis and inhibit host immunity, ω3 fatty acids increase both humeral and cellular immune function.[61,73] Conventional TPN regimens contain ω6 fatty acids which block the reticuloendothelial system and, in part, account for the increased incidence of infection observed in patients receiving parenteral nutrition.[74,75]

Nucleotides may also play a role in disease-specific nutrient regimens of the future. Nucleotides are the integral components of DNA and RNA, the nucleic acids required for cellular proliferation and protein synthesis. Dietary nucleotides are required for normal development of the immune system including NK activity, macrophage activation, and interleukin production.[76,77] Animals maintained on nucleotide-free diets exhibit reduced circulating immune cells and widespread immune functional deficits.[78] Nucleotide supplements can prevent the development of immune dysfunction under certain conditions.

The efficacy of enteral nutrition supplemented with arginine, ω3 fatty acids and nucleotides has been demonstrated in several clinical studies. In a pilot study of critically ill patients, Cerra et al analyzed metabolic and immune function after 7-10 days of enteral nutrient therapy in a randomized, prospective trial.[62] Improved immune function (increased concanavalin A and phytohemagglutinin stimulation indices) was observed in patients receiving the supplemented diet compared to control patients. Furthermore, the supplemented diet was equivalent to the control diet for maintaining nutritional status as measured by classic nutritional parameters. Similar immune findings were reported by Daly et al in a large clinical trial of patients undergoing major elective upper abdominal surgery.[61] Significant improvements in immune function were noted after 4-7 days of nutrient therapy in the supplemented group in this postoperative patient population. The incidence of infectious complications was reduced by 3.5-fold in supplemented patients indicating cost as well as medical efficacy.

Glutamine is a critical amino acid involved in many cellular and systemic metabolic processes. Glutamine is the most abundant circulating amino acid and comprises over 60% of the free intracellular amino acid pool.[79] Glutamine acts as a nitrogen shuttle for protein synthesis, renal ammoniagenesis and nucleotide biosynthesis,[80,81] and is a key fuel for rapidly proliferating cells such as gastrointestinal, immune and endothelial cells.[82] Glutamine supplementation increases macrophage generation of secretory proteins (e.g., tumor necrosis factor, interleukins) to antigenic challenge in tumor-bearing animals.[79] Of particular interest in the cancer patient, glutamine has been shown to maintain intestinal integrity and may prevent bacterial translocation following chemotherapy or radiation therapy.[83,84] By maintaining intestinal integrity and reducing bacterial translocation through the intestinal tract, reduced septic morbidity and mortality has been demonstrated in tumor-bearing animals.[84] Although the clinical relevance of these findings remains controversial, several clinical studies are currently underway to evaluate the efficacy of glutamine and glutamine-containing dipeptides in enteral and parenteral nutrient formulas.

Pharmacologic agents have met limited success in treating cachexia of the cancer patient. Corticosteroids, such as prednisone and methylprednisolone, show minimal objective effects in the cachectic cancer patient.[85,86] Short-term, but no long-term, increased "sense of well-being" has been shown in cancer patients treated with corticosteroids.[71] However, no significant improvement in nutritional status or clinical outcome has been found with corticosteroid therapy in cancer patients.[85] Similarly, progestins cause an increase in appetite and increased weight gain in postmenopausal breast cancer patients.[87] The progestins medroxyprogesterone acetate and megestrol acetate (Megace) do not reduce disease- or treatment-related morbidity or mortality or improve nutritional status.[88]

Hydrazine sulfate therapy is controversial in the cancer patient but presents an important theoretical concept with potential therapeutic efficacy. Hydrazine sulfate inhibits the enzyme phosphoenolpyruvate carboxykinase which catalyzes a rate-limiting reaction in the process of gluconeogenesis.[89] Increased gluconeogenesis has been demonstrated in the cancer patient and is a particularly detrimental metabolic process due to its catabolism of skeletal muscle. Preliminary studies of hydrazine sulfate therapy in cancer patients have documented increased appetite, body weight maintenance and increased serum albumin levels.[90,91] In patients with advanced non-small cell lung cancer receiving combination chemotherapy, hydrazine sulfate maintained host nutritional status and improved patient survival.[91] Further studies are clearly required to investigate hydrazine sulfate therapy, other metabolic blockers, specific cytokines and cytokine blocking agents which can inhibit metabolic pathways critical to the development of cancer cachexia.

Finally, manipulation of regulatory hormones may provide an effective means to selectively treat cancer cachexia in the future. Although single hormone therapy has shown limited success in the treatment of cancer cachexia, combined hormone therapy is promising. Insulin alone moderately improves appetite in the tumor-bearing host.[33] Recent studies conducted by Brennan and colleagues demonstrate that euglycemic insulin infusion can decrease whole body protein breakdown and reduce forearm branched-chain amino acid release in cancer patients—indicating a beneficial effect on host protein metabolism despite persistent insulin resistance to carbohydrate utilization.[92,93] In tumor-bearing animals, insulin combined with somatostatin and growth hormone has been found to selectively support host metabolism without stimulating tumor growth. Somatostatin is used to prevent the compensatory rise in the catabolic hormone glucagon in response to elevated levels of insulin. The result is a marked increase in the anabolic insulin:glucagon ratio.[49] Provision of the potent anabolic agent growth hormone promotes nitrogen retention. Until further research is conducted, growth hormone must be used cautiously in the cancer patient because of the potential to stimulate growth of tumors expressing growth hormone receptors. Clinical investigation is currently underway to study the effects of growth hormone and other hormonal alterations in the cancer patient.

Summary

Conventional nutrition support has produced limited success in reducing morbidity and mortality of the cancer patient. Our vision of nutrition support must be expanded in the future to include specific nutrient supplements, anabolic hormones, cytokines, metabolic and cytokine blockers, and other substrates capable of altering host nutritional and metabolic pathways. Improved understanding of the etiology of cancer cachexia and innovative approaches to the metabolic management of the cancer patient must be designed to successively treat cancer cachexia. Growth hormone can significantly reverse catabolism but must be used with caution in the cancer patient because of its potential to stimulate growth of tumors expressing growth hormone receptors. The combination of growth hormone, insulin and somatostatin in tumor-bearing animals has been shown to selectively treat cancer cachexia by increasing the anabolic insulin:glucagon ratio augmented by the potent anabolic agent growth hormone. Studies manipulating these regulatory hormones as well as cytokines, cytokine antagonists and metabolic blocking agents require clinical investigation in the future.

References

1. Torosian MH, Daly JM. Nutritional support in the cancer-bearing host. Cancer 1986; 58:1915-1929.
2. Brennan MF. Total parenteral nutrition in the cancer patient. N Engl J Med 1981; 305:375-381.
3. Lundholm K, Edstron S, Ekman L. Metabolism in peripheral tissues in cancer patients. Cancer Treat Rep 1981; 65 (Suppl):79-83.
4. Studley HO. Percentage of weight loss: A basic indicator of surgical risk in patients with chronic peptic ulcer. JAMA 1936; 106:458.
5. Mullen JL. Consequences of malnutrition in the surgical patient. Surg Clin North Am 1981; 61:465-473.
6. Smale BF, Mullen JL, Buzby GP et al. The efficacy of nutritional assessment and support in cancer surgery. Cancer 1981; 47:2375-2381.
7. Mueller J, Brenne V, Dienst J et al. Perioperative parenteral feeding in patients with gastrointestinal carcinoma. Lancet 1982; 1:68-71.
8. The Veterans Affairs Total Parenteral Nutrition Cooperative Study Group: Perioperative Total Parenteral Nutrition in Surgical Patients. N Engl J Med 1991; 325:525-532.
9. Warren S. The immediate cause of death in cancer. Am J Med Sci 1932; 184:610-615.

10. Theologides A. Cancer cachexia. Cancer 1979; 43:2004-2012.
11. Theologides A. Anorexia-producing intermediary metabolites. Am J Clin Nutr 1976; 29:552-558.
12. Beutler B, Cerami A. Cachectin and tumor necrosis factor as two sides of the same biological coin. Nature 1986; 320:584-588.
13. DeWys WD, Walters K. Abnormalities of taste sensation in cancer patients. Cancer 1975; 36:1888-96.
14. Lucke B, Borwick M, Zeckwer I. Liver catalase activity in parabiotic rats with one partner tumor bearing. J Natl Cancer Inst 1952; 13:681-686.
15. DeWys WE, Begg C, Lavin PT et al. Prognostic effect of weight loss prior to chemotherapy in cancer patients. Am J Med 1980; 69:491-497.
16. Nakahara W. A chemical basis for tumor host relations. J Natl Cancer Inst 1960; 24:77-86.
17. Langstein H, Fraker D, Norton JA. Reversal of cancer cachexia by antibodies to interferon-gamma but not cachectin/tumor necrosis factor. Surg Forum 1989; 40:408-410.
18. Krause R, Humphreys C, von Meyenfeldt M. A central mechanism for anorexia in cancer: A hypothesis. Cancer Treat Rep 1981; 65 (Suppl): 15-21.
19. Daly JM, Torosian MH. Nutritional support. In: Cancer: Principles and Oncology Devita V, Hellman S, Rosenberg S (Eds), 1992.
20. Brennan MH. Uncomplicated starvation versus cancer cachexia. Cancer Res 1977; 37:2359-2364.
21. Young VR. Energy metabolism and requirements in the cancer patient. Cancer Res 1977; 37:2336-2347.
22. Gunderson AH. The basal metabolism in myelogenous leukemia and its relation to the blood findings. Boston Med Surg J 1921; 185:785.
23. Warnold I, Lundholm K, Schersten T. Energy balance and body composition in cancer. Cancer Res 1978; 38:1801-1807.
24. Shike M, Russell D, Detsky A et al. Changes in body composition in patients with small-cell lung cancer: The effect of TPN as an adjunct to chemotherapy. Ann Intern Med 1984; 101:303-309.
25. Knox LS, Crosby LO, Feurer ID et al. Energy expenditure in malnourished cancer patients. Ann Surg 1983; 197:152-162.
26. Heber D, Chlebowski RT, Ishibashi DE et al. Abnormalities in glucose and protein metabolism in non-cachectic lung cancer patients. Cancer Res 1982; 42:4815-19.
27. Shaw JM, Humberstone DM, Wolfe RR. Energy and protein metabolism in sarcoma patients. Ann Surg 1988; 207:283-289.
28. Buzby GP, Mullen JL, Matthews DC et al. Prognostic nutritional index in gastrointestinal surgery. Am J Surg 1980; 139:160-167.
29. Holroyde CP and Reichard A. Carbohydrate metabolism in cancer cachexia. Cancer Treat Rep 1981; 65 (Suppl):55-59.
30. MacBeth RAL, Bekesi JE. Oxygen consumption and anaerobic glycolysis of human malignant and normal tissue. Cancer Res 1962; 22:244-248.
31. Waterhouse C and Kemperman JH. Carbohydrate metabolism in subjects with cancer. Cancer Res 1971; 31:1273-1278.
32. Schein PS, Kesner D, Haller D et al. Cachexia of malignancy: Potential role of insulin in nutritional management. Cancer 1979; 43:2070-2078.
33. Moley JF, Morrison SD, Gornschboty DM et al. Body composition changes in rats with experimental cancer cachexia: Improvement with exogenous insulin. Cancer Res 1988; 48:2784-2787.
34. Peacock JL, Norton JA. Impact of insulin on survival of cachectic tumor-bearing rats. JPEN 1988; 12:260-264.
35. Shaw JHF, Wolfe RR. Whole body protein kinetics in patients with early and advanced gastrointestinal cancer—the response to glucose infusion and total parenteral nutrition. Surgery 1977; 103:148-155.
36. Waterhouse C. Nutritional disorders in neoplastic diseases. J Chron Dis 1963; 16:637-644.
37. Wilson W, Kirk CJC, Goode AW. The effect of weight loss, operation and parenteral nutrition on fat clearance in patients with colorectal cancer. Clin Sci 1987; 73:489-495.
38. Brennan MF, Burt ME. Nitrogen metabolism in cancer patients. Cancer Treat Rep 1981:65 (Suppl):67-78.
39. Stein TP, Oram-Smith JC, Leskiw MJ et al. Tumor-caused changes in host protein synthesis under different dietary situations. Cancer Res 1976; 36:3936-3950.

40. Waterhouse C, Jeanpretre N, Keilson J. Gluconeogenesis from alanine in patients with progressive malignant disease. Cancer Res 197; 39:1968-1972.
41. Norton JA, Stein TP, Brennan MF. Whole-body protein synthesis and turnover in normal and malnourished patients with and without known cancer. Ann Surg 1981; 194:123-128.
42. Kien CL, Carmitta BN. Close association of accelerated rats of whole body protein turnover (synthesis and breakdown) and energy expenditure in children with newly diagnosed acute lymphocytic leukemia. JPEN 1987; 11:129-134.
43. Tayek JA, Blackburn GL, Bistrian BR. Alterations in whole-body, muscle, liver and tumor tissue protein synthesis and degradation in Novikoff hepatoma and Yoshida sarcoma tumor growth studied in vivo. Cancer Res 1988; 48:1554-1558.
44. Beutler B, Cerami A. Cachectin: More than a tumor necrosis factor. N Engl J Med 1987; 316:379-385.
45. Woloski BMRNJ, Fuller GM. Identification and partial characterization of hepatocyte stimulating factor form leukemia cell lines: Comparison with interleukin-1. Proc Natl Acad Sci 1985; 82(USA):1443-1447.
46. Dinarello CA. Interleukin-1 and the pathogenesis of acute phase response. N Engl J Med 1984; 311:1413-1418.
47. Powanda MC, Beisel WR. Hypothesis: Leukocyte endogenous mediator/endogenous pyrogen/lymphocyte activating factor modulates the development of nonspecific and specific immunity and affects nutritional status. Amer J Clin Nutr 1982; 35:762-768.
48. Unger RH, Orci L. Glucagon and the A cell: Physiology and pathophysiology. N Engl J Med 1981; 304:1575-1580.
49 Bartlett DL, Charland S, Torosian MH. Growth hormone, insulin and somatostatin therapy of cancer cachexia. Cancer (In press).
50. Detsky AS, Baker JP, O'Rourke K et al. Perioperative parenteral nutrition: A meta-analysis. Ann Intern Med 1987; 107:195-203.
51. Heatley RV, Williams R, Lewis M. Preoperative intravenous feeding: A controlled trial. Postgrad Med J 1979; 55:541-545.
52. Holter AR, Fisher JE. The effects of perioperative hyperalimentation on complications in patients with carcinoma and weight loss. J Surg Res 1977; 23:31-34.
53. Thompson B, Julian T, Stremple J. Perioperative TPN in patients with gastrointestinal cancer. J Surg Res 1981; 30:497-500.
54. Daly JM, Massar E, Giacco G et al. Parenteral nutrition in esophageal cancer patients. Ann Surg 1982; 196:91-96.
55. Koretz RL. Parenteral nutrition: Is it oncologically logical? J Clin Oncol 1984; 2:534-538.
56. Weisdorf SA, Lysne J, Wind D. Positive effect of prophylactic total parenteral nutrition on long-term outcome of bone marrow transplantation. Transplantation 1987; 43:833.
57. Bothe A Jr, Valerio D, Bistrian BR et al. Randomized control trial of hospital nutritional support during abdominal radiotherapy. (Abstr) JPEN 1979; 3:292.
58. Solassol C, Joyeuz H, Dubois JB. Total parenteral nutrition (TPN) with complete nutritive mixtures: An artificial gut in cancer patients. Nutr Cancer 1979; 1:13-18.
59. Douglas HO, Milliron S, Nava H et al. Elemental diet as an adjuvant for patients with locally advanced gastrointestinal cancer receiving radiation therapy: A prospectively randomized study. JPEN 1978; 2:682-686.
60. Donaldson SS. Nutritional support as an adjunct to radiation therapy. JPEN 1984; 8:302-309.
61. Daly JM, Lieberman MD, Goldfine J, Shou J et al. Enteral nutrition with supplemental arginine, RNA and omega-3 fatty acids in patients after operation: Immunologic, metabolic and clinical outcome. Surgery 1992; 112:56-67.
62. Cerra FB, Lehmann S, Konstantinides N et al. Improvement in immune function in ICU patients by enteral nutrition supplemented with arginine, RNA and menhaden oil independent of nitrogen balance. Nutrition 1991; 7(3):193-199.
63. Barbul A. Arginine: Biochemistry, physiology, and therapeutic implications. J Parent Ent Nutr 1986; 10(2):227-238.
64. Rettura G, Padawer J, Barbul A et al. Supplemental arginine increases thymic cellularity in normal and murine sarcoma virus-inoculated mice and increases the resistance of mice to the murine sarcoma virus tumor. J Parent Ent Nutr 1979; 3:409-416.
65. Barbul A, Sisto DA, Wasserkrug HL et al. Arginine stimulates lymphocyte immune responses in healthy humans. Surgery 1981; 90:244-251.

66. Seifter E, Barbul A, Levenson SM et al. Supplemental arginine increases survival in mice undergoing local tumor excision. J Parent Ent Nutr 1980; 5:589.
67. Takeda Y, Tominaga T, Tai N et al. Inhibitory effects of 1-arginine on growth of rat mammary tumors induced by 7,12-dimethylbenz(a)anthracene. Cancer Res 1975; 35:2390-2393.
68. Spector AA, Burns CP. Biological and therapeutic potential of membrane lipid modification in tumors. Cancer Res 1987; 47:4529-4537.
69. Bartlett D, Charland S, Torosian M. Differential effect of medium- and long-chain triglycerides on tumor growth and metastasis. J Parent Ent Nutr 1992; 16(Suppl):55S.
70. Cave ST Jr. Dietary omega-3 (ω3) polyunsaturated fatty acid effects on animal tumorigenesis. Fed Am Soc Exp Biol 1991; 5:2160-2166.
71. Mascioli EA, Bistrian BR, Babayan VK et al. Medium-chain triglycerides and structured lipids as unique nonglucose energy sources in hyperalimentation. Lipids 1987; 22:421-423.
72. Mendez B, Ling PR, Istfan NW et al. Effects of different lipid sources in total parenteral nutrition on whole body protein kinetics and tumor growth. J Parent Ent Nutr 1992; 16:545-551.
73. Gottschlich M, Jenkins M, Warden G et al. Differential effects of three enteral dietary regimens on selected outcome variables in burn patients. J Parent Ent Nutr 1990; 14:225-236.
74. Alexander JW. Nutrition and infection: New perspectives for an old problem. Arch Surg 1986; 121:966-972.
75. Wan JMF, Teo TC, Babayan VK et al. Invited comment: Lipids and the development of immune dysfunction and infection. J Parent Ent Nutr 1988; 12(Suppl):43-52.
76. Rudolph FB, Kulkarni AD, Schandle VB et al. Involvement of dietary nucleotides in T lymphocyte function. Adv Exp Med Biol 1984; 165B:175-178.
77. Fanslow WC, Kulkarni A, VanBuren CT et al. Effect of nucleotide restriction and supplementation on resistance to experimental murine candidiasis. J Parent Ent Nutr 1988; 12:49-52.
78. Kulkarni AD, Fanslow WC, VanBuren CT et al. Influence of dietary nucleotide restriction on bacterial sepsis and phagocytic cell function in mice. Arch Surg 1986; 121:169-172.
79. Souba WW, Klimberg VS, Plumley DA et al. The role of glutamine in maintaining a healthy gut and supporting the metabolic response to injury and infection. J Surg Res 1990; 48:383-391.
80. Windmueller HG. Enterohepatic aspects of glutamine and metabolism, in Glutamine: Metabolism, Enzymology and Regulation. Palacios R and Mora J, Eds; Academic Press, New York 1980:235.
81. Windmueller HG. Glutamine utilization by the small intestine. Adv Enzymol 1982; 53:202.
82. Souba WW, Smith RJ, Wilmore DW. Effect of glucocorticoids on glutamine metabolism in visceral organs. Metabolism 1985; 34:450.
83. Souba WW, Klimberg VS, Hautamaki RD et al. Oral glutamine reduces bacterial translocation following abdominal radiation. J Surg Res 1990; 48:1-5.
84. Fox AD, Kripke SA, DePaula JA et al. The effects of a glutamine-supplemented enteral diet on methotrexate-induced enterocolits. J Ent Parent Nutr 1988; 12:325-331.
85. Bruera E, Roca E, Cedaro L et al. Methylprednisolone use in patients with cancer. Cancer Treat Rep 1985; 69:751-754.
86. Wilcox JC, Cou J, Shaw J et al. Prednisolone as an appetite stimulant in patients with cancer. Brit Med J 1984; 288:27-31.
87. Lelli G, Angelelli B, Giambiasi ME et al. The anabolic effect of high dose medroxyprogesterone acetate in oncology. Pharm Res Comm 1983; 15:561-568.
88. Tchekmedyian NS, Tait N, Moody M et al. High-dose megestrol acetate: A possible treatment of cachexia. J Amer Med Assoc 1987; 257:1195-1197.
89. Gold J. Proposed treatment of cancer by inhibition of gluconeogenesis. Oncology 1968; 22:185-207.
90. Chlebowski RT, Bulcavage L, Grosvenor M et al. Hydrazine sulfate in cancer patients with weight loss: A placebo-controlled experience. Cancer 1987; 59:406-410.
91. Chlebowski RT, Bucavage L, Grosvenor M et al. Hydrazine sulfate influence on nutritional status and survival in non-small cell lung cancer. J Clin Oncol 1990; 8:9-15.
92. Heslin MJ, Newman E, Wolf RF et al. Effect of system hyperinsulinemia in cancer patients. Cancer Res 1992; 52:3845-3850.
93. Pisters PWT, Cersosimo E, Rogatko A et al. Insulin action on glucose and branched chain amino acid metabolism in cancer cachexia: Differential effects of insulin. Surgery 1992; 111:301-310.

CHAPTER 3

Glutamine and Cancer

Barrie P. Bode, Steve F. Abcouwer, Cheng-Mao Lin and Wiley W. Souba

Much has been written on the accelerated utilization of glutamine by tumors[1] and tumor-derived cells lines.[2] Given the seemingly "glutamine auxotrophic" nature of tumor cells, "antiglutamine" enzymatic and chemotherapies have been formulated and tested in the past,[3-7] but were fraught with toxicity problems. As a result, the current focus has been the converse; namely the provision of pharmacological levels of this "conditionally essential" amino acid[8] to cancer patients. This approach is based on the observation in animal models that exogenous glutamine administration does not seem to promote tumor growth or exhibit toxicity, but exerts a protective effect on susceptible host tissues during radiation or chemotherapy.[9-13] Given the broad scope of the subject at hand, we will limit our text to a brief review of tumor and host glutamine transport and metabolism followed by a retrospective evaluation of the use of glutamine supplemented nutrition in tumor-bearing animals and human clinical trials. We will conclude with an assessment of the current state of glutamine "nutritional pharmacology" for cancer patients and provide suggestions for future research and development in this evolving field of therapy.

Mammalian Glutamine Metabolism

Glutamine, the most abundant free amino acid in the plasma, is classified as a nonessential amino acid in textbooks because mammalian tissues possess the ability to produce it from glutamate, ammonia and ATP via the enzyme glutamine synthetase. A dynamic balance exists in the body between glutamine utilization and release by individual tissues—processes which collectively maintain circulating glutamine concentrations at approximately 0.6 mM in humans. The simultaneous synthesis and catabolism of glutamine was first described by Hans Krebs in tissues from many different species in 1935.[14] Since that time, this amino acid has received a lot of attention given its diverse and essential roles in cellular metabolism for normal and neoplastic tissues and cells. Among the glutamine-utilizing biosynthetic and metabolic pathways are: gluconeogenesis, glutathione synthesis, oxidative energy metabolism (Krebs Cycle, (CO_2)), protein synthesis, ureagenesis, aminosugar biosynthesis, renal ammoniagenesis, NAD biosynthesis, de novo nucleotide biosynthesis, amino acid biosynthesis (transamination pathways), fatty acid/phospholipid biosynthesis and brain neurotransmitter biosynthesis. Figure 3.1 depicts the major pathways for glutamine transport and metabolism in mammalian cells. The disposition of glutamine to any one or a combination of these pathways is dependent on the role of specific cells within a tissue, and may change depending on the nutritional/hormonal status of the individual. For a detailed discussion of glutamine metabolism, the reader is directed to several comprehensive reviews.[15-20]

Nutritional Support in Cancer and Transplant Patients, edited by Rifat Latifi and Ronald C. Merrell.

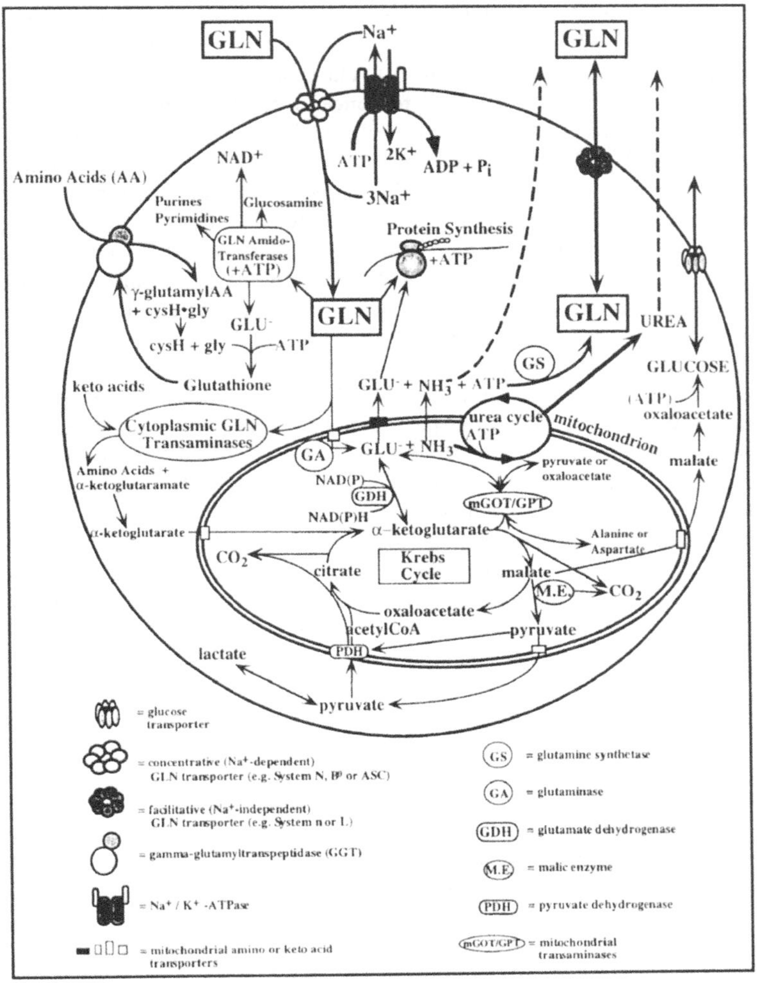

Fig. 3.1. Pathways of glutamine transport and metabolism in mammalian cells. Cytoplasmic glutamine is maintained at levels exceeding those of the plasma by the action of Na^+-dependent transporters which concentrate this amino acid above its equilibrium distribution, while efflux is mediated by nonconcentrative (facilitative) carriers. Combined with the relative activities of glutamine-metabolizing enzymes, the simultaneous operation and relative rates of both types of transporters determine the intracellular free glutamine content. Glutamine is utilized in many energy (ATP) - dependent metabolic pathways including glutathione, urea, glucose aminosugar and nucleotide biosyntheses. Conversely, glutamine may be utilized as an oxidative energy source via the TCA cycle or may be (re)synthesized from glutamate and ammonia via the enzyme glutamine synthetase. Only a subset of the pathways depicted operate in most cells, and some other uses (e.g., neurotransmitter biosynthesis, transglutamination) are not displayed. Dashed lines across the plasma membrane represent diffusion whereas solid lines indicate carrier-mediated processes.

Briefly, major glutamine consumers include rapidly dividing cells such as enterocytes,[21,22] leukocytes[23,24] and fibroblasts,[25,26] while net glutamine producers include skeletal muscle,[27-29] lung[30] and brain.[31] The liver is unique and plays a central role in glutamine homeostasis, as it can switch from an organ of net glutamine balance or consumption to release, depending on the prevailing metabolic demands.[32]

The net uptake or release of glutamine by a tissue or cell is a function of not only metabolism but transport as well. The transport of L-glutamine by mammalian tissues is mediated by a panel of transporters, as shown in Figure 3.1. The expression of Na^+-dependent transporters in the plasma membrane allows a cell to concentrate glutamine well above its normal transmembrane equilibrium distribution by utilizing the energy present in the transmembrane Na^+ electrochemical gradient established by the Na^+/K^+- ATPase. The energetically favorable transport of Na^+ down its gradient is thereby coupled to the "uphill" transport of glutamine against its gradient to make this concentrative process possible. Transporters of this type include Systems N,[33] ASC[34] and under some circumstances, System A.[35,36] Conversely, glutamine release from cells is mediated by facilitative (Na^+-independent) transporters, where the vectorial movement of this amino acid is regulated by its corresponding transmembrane gradient. Examples of such carriers include System n[37] and System L.[35]

In this chapter we will focus on the consumption of glutamine by tumors and the impact of this excess drain on the glutamine economy in the tissues and cells that are profoundly influenced by the pathophysiology of cancer and chemotherapy: muscle, liver, gut and leukocytes. Net glutamine production or utilization by a given tissue is a collective function of transport, anabolic and catabolic rates, which in turn are regulated by hormonal and nutritional factors in the host (see Fig. 3.1). Circulating levels of hormones, cytokines and nutrients are likewise altered during the pathophysiology that ultimately manifests as "cachexia" during tumor growth. Events secondary to cancer such as infection, surgery, radiation and chemotherapy further compromise host tissue glutamine content and provide additional challenges for adequate supplies of this conditionally essential amino acid. Data from both clinical studies as well as animal models of cancer on these alterations will be considered.

Tumor and Tissue Glutamine Utilization

General Considerations

As mentioned above, mammalian glutamine homeostasis is achieved by a harmonious and dynamic balance between net glutamine input (nutrition), production and utilization by individual tissues. The presence of a growing tumor(s), however, represents an additional drain on host glutamine economy. When this additional tax on host glutamine supply exceeds the body's capacity to provide this amino acid, relative decreases in its circulating levels are observed. At this point host tissues—particularly those that consume glutamine such as intestinal and immune cells—are compromised and glutamine thus becomes "conditionally essential".[8,38-40] In many ways a tumor is akin to an additional organ with respect to its heterogeneous cellular composition. One of the disadvantages to the host is that the cells that comprise a tumor are almost all glutamine consumers: tumor cells,[41,42] endothelial cells,[43] fibroblasts,[25,44] macrophages[45-47] and other immune cells.[24] Collectively, these cell types exert a progressively increasing drain on the circulating glutamine supply as the tumor grows. Ironically, the supply of glutamine from the diet may decrease simultaneously as tumor-induced anorexia emerges.[48] As will be discussed later, iatrogenic mechanisms such as surgery, radiation or chemotherapy further exacerbate the effects of tumor-induced host glutamine depletion and tissue compromise.

Hereafter, animal tumor models will be discussed followed by brief reviews of normal, tumor-influenced and transformed host tissue glutamine transport and metabolism and the effects of glutamine supplemented nutrition on these parameters.

Tumor Models

Many animal models have been utilized to study the effects of tumor growth on host tissue metabolism and the progression of cachexia.[49] Most of these models involve the subcutaneous implantation of solid tumors in permissive rodent hosts. With respect to glutamine, the best characterized animal models are the Fisher 344 rat methylcholanthrene (MCA)-induced fibrosarcoma,[11,50-55] the Sprague-Dawley rat Walker 256 carcinosarcoma[56-59] and the murine Ehrlich ascites tumor.[27,60-62] The relevance of the profound tumoral uptake and utilization of glutamine reported in animal models[9,60,63,64] to human cancer may be provided by the depressed plasma levels of glutamine in certain cancer patients.[65-67]

Ascitic vs. solid tumor animal models provide complimentary systems in which to study the impact of tumor growth on host glutamine economy. Whereas solid tumors rely on angiogenesis for adequate oxygen and substrate delivery,[68] ascites tumor cells grow in suspension in the peritoneal cavity. The ascites model therefore provides a system for the study of in vivo host-tumor cell interactions per se, where the animal serves as the "incubator" and the ascitic fluid and plasma serve as the "culture media". In contrast, the implantable solid sarcoma and carcinoma models allow the study of tumor growth and related processes (angiogenesis, inflammatory cell infiltrates, fibrosis, etc.[69]) on host glutamine economy. The early observation by Harry Eagle that glutamine was essential for the proliferation of mammalian cells in tissue culture[70] and the dozens of in vitro studies that have since confirmed that neoplastic cells require glutamine for growth[71-78] seem to hold true in animal tumor models as well. Based on several in vivo studies, the current paradigm is that a growing tumor elicits a mobilization of glutamine from host tissues for its own use, eventually leading to marked declines host cell glutamine content.[79,80]

In the Ehrlich ascites tumor model, plasma glutamine concentrations are persistently higher than those of the ascitic fluid, indicating the net consumption of glutamine by the tumor cells.[60,81] Although there is a rapid transient increase in plasma glutamine levels soon after tumor inoculation, the chronic "sink" created by the ascites tumor cells eventually leads to the impoverishment of glutamine in plasma and host tissues prior to the death of the animals.[27] Similarly, studies with the MCA fibrosarcoma demonstrated marked glutamine consumption and ammonia production by this tumor[82] and an absolute requirement of glutamine for sarcoma cell growth in vitro.[83] Indeed, the implantation of this highly aggressive tumor causes a mobilization of glutamine from skeletal muscle and eventual declines in muscle and arterial glutamine concentrations.[52,84,85] Furthermore, administration of the glutamine analog acivicin to MCA tumor-bearing animals decreased tumor growth,[86] an observation that lends further support to the dependence of this fibrosarcoma on glutamine for growth. Studies with the Walker 256 tumor-bearing rat revealed that this tumor avidly consumed glutamine as well,[56] and caused similar decreases in the glutamine content of skeletal muscle and plasma over time.[58] The increased hepatic availability of glutamine reported in animals bearing this tumor is probably a result of increased efflux from the skeletal muscle.[87] In summary, animal cancer models to date have demonstrated an avid consumption of glutamine by both solid and ascites tumors, with time-dependent host glutamine depletion and compensatory efflux of this amino acid largely from skeletal muscle. It must be noted that in many of the animal models, studies were performed when the tumors were much larger than encountered clinically (i.e., tumor burden from 10-30% of carcass weight). Nonetheless, important information on tumor-host relationships has been obtained.

However, there are notable exceptions to the current tumor/glutamine paradigm. For example, Kallinowski and colleagues examined glutamine utilization in 65 different breast cancer xenografts and found that 80% of them exhibited net glutamine balance or release.[88] There are also reports of "glutamine-independent" tumor cell lines[89,90] including results from our laboratory where some rodent and human cell lines have been observed to grow and survive in the absence of glutamine. Studies with human leukemic cell lines demonstrated that the ability to grow in glutamine-free media correlated with the expression of glutamine synthetase.[74] This observation may be of significance as it has been recently reported that glutamine synthetase is upregulated in human hepatic tumors.[91,92] Although depressed plasma glutamine levels are observed in some cancer patients, these results collectively indicate that similar to normal host tissues, tumors may exhibit a wide range of glutamine dependence.

One general note on the relationship between animal tumor models and human cancer needs to be issued: In the animal models discussed above, the nutritional status and tumor stage are well-documented and controlled. Unfortunately, the same cannot be said for available studies on plasma amino acid profiles of cancer patients, where the effects of nutrition-versus tumor-dependent effects are not easily discriminated. Several studies have been performed on plasma amino acid profiles of cancer patients, and while some indicate diminished glutamine levels, others do not.[93] It was concluded in this retrospective review that many factors influence plasma amino acid profiles in human cancer patients including the type, location and progression of the tumor(s),[94] the absence or presence of weight loss, age and nutritional status. As a result, there is not, and probably will never be a "consensus aminogram" for cancer patients per se. Plasma glutamine levels, even in well controlled animal models, display dynamic temporal profiles, with net increases early after tumor implantation/inoculation followed by normalization and eventual decreases once the tumors become quite large.[79,80] As will be discussed below, plasma glutamine levels alone may not provide a comprehensive index of the glutamine status of the patient, as chronic compensatory processes (increased proteolysis, decreased glutamine uptake and unnecessary energy expenditure) in host tissues that collectively maintain circulating levels near normal during tumor growth may ultimately compromise patient recovery or survival.

In summary, animal tumor models have provided valuable information on tumor-host glutamine relationships, but differences in the models and clinically-encountered oncology patients should be kept in mind. One of the more valuable pieces of information that animal tumor models has yielded is the concept that compensatory changes in host glutamine metabolism occur in response to tumor growth.[15,79,80,95] Although plasma glutamine levels in some cancer patients and some animal tumor models may display little or no alterations, this observation alone may not serve as a good index of host glutamine status per se. It should be kept in mind that the tumor-induced compensatory changes in host glutamine metabolism are probably designed to maintain adequate plasma supplies of this amino acid.[15,79,80,95] As a result, an apparent maintenance of plasma glutamine levels during tumor growth belies the underlying host tissue metabolic alterations responsible for this homeostasis. Ultimately, these chronic processes (e.g., proteolysis, increased energy expenditure and depletion of reducing equivalents) compromise host tissue integrity and the capacity to recover from cancer and the detrimental effects of its treatment.[9-13,20,21,39,96-110] This view of glutamine as a conditionally-essential amino acid for normal host tissue integrity and function during tumor growth is consistent with many animal models of cancer, and will be placed in the context of tumor- and transformation-induced changes in host tissues described hereafter.

Host Tissue Glutamine Metabolism

Liver

With respect to glutamine metabolism the liver is unique in that it can serve as a net consumer or producer, depending on the current metabolic demands. This dynamic ability is afforded by the heterogeneous and position-dependent expression of glutamine-metabolizing enzymes and related transporters along the acinus.[32,111-114] Glutaminase, the primary glutamine-catabolizing enzyme in cells, is expressed in the first few "periportal" hepatocytes containing urea cycle enzymes and surrounding the sinusoidal inflow.[115] The glutaminase isozyme in these cells is liver-specific and linked to the operation of the urea cycle.[115-117] In contrast, the enzyme glutamine synthetase is expressed only in the terminal one or two "perivenous" hepatocytes surrounding the sinusoidal outflow. Net glutamine catabolism in the periportal cells and synthesis in the perivenous cells results in an intercellular glutamine cycle in the liver.[113,117-119] Relative rates of flux through both of these opposing pathways determine the glutamine balance across the liver and can be regulated by the transport of glutamine across liver cell membranes.[120-123]

The transport of glutamine across the hepatocyte plasma membrane in rodents[33] and humans[34] is mediated by System N, a Na^+-dependent carrier with narrow substrate specificity (glutamine and histidine only). The operation of this concentrative carrier allows the maintenance of cytoplasmic glutamine levels of approximately 8 mM, compared to 0.6 mM in the plasma.[124] The efflux of glutamine from the liver appears to be mediated by the Na^+-independent System n,[37] which has a similar substrate profile to System N. In fact, it has been proposed that System n activity is System N operating in the absence of Na^+, a possibility that remains to be determined.

Work from our laboratory with the MCA rat tumor model revealed that the liver—an organ of net glutamine balance under normal postprandial conditions—displays net glutamine output early during tumor growth, followed by a switch to net glutamine consumption later in the course of tumor burden.[52,85,125] The early net release of glutamine has been shown to be associated with increases in hepatic glutamine content, glutamine synthetase activity,[52] and System n and N activities,[122,123,126] with concomitant decreases in hepatic glutaminase activity.[52] As the tumor continues to grow (e.g., at 20% of carcass weight), the hepatic glutamine content decreases, while System N activity remains elevated,[125,126] whereafter it decreases upon excessive tumor burden (>25% of carcass weight) just prior to death (unpublished observations). Similar early increases,[61] and later decreases, in hepatic glutamine content were observed in murine ascites and fibrosarcoma models,[53] respectively. This same general biphasic trend for early diminished glutaminase followed by decreased glutamine synthetase activity later in tumor growth was also seen in the Ehrlich ascites tumor model.[27] Progressive decreases in hepatic glutamine content with tumor growth was also observed in the Walker 256 model.[127] The enhanced activities of both System N and System n allow the liver cells to maintain cytoplasmic levels of this amino acid in the face of progressively diminished plasma levels, and to efficiently export glutamine synthesized de novo, respectively. The effects of tumor growth on the transporters are important as this component of glutamine metabolism has been shown to represent a rate-limiting step under conditions of accelerated intracellular metabolism.[120] Later work demonstrated that the effects on the carriers were attributable to tumor influence, as glutamine transport values returned toward those of control animals five days after tumor resection.[128] Additional studies revealed that the tumor effects on hepatic glutamine transport involved a tumor necrosis factor (TNF)-dependent pathway,[129] possibly via a hepatocyte TNF autocrine mechanism.[130]

Why is hepatic glutamine transport enhanced in response to tumor burden? To the best of our knowledge, there are no "carbon-chasing" studies available to date that quantify the metabolic fates of glutamine in cancer patients or tumor-bearing animals compared to normal subjects. It is well established however, that hepatic energy expenditure is elevated in tumor-bearing animals, eventually resulting in ATP deficit and redox alterations.[127,131-133] ATP-dependent pathways that consume glutamine such as gluconeogenesis[134-138] and protein synthesis[55,139] are elevated in the livers of tumor-bearing animals and cancer patients.[140-146] Increased plasma ammonia levels observed in the MCA rat tumor model[147,148] may also enhance the rate of the energy-dependent urea cycle and stimulate the activity of hepatic glutaminase,[116] a process which is thought to contribute to the net negative nitrogen balance seen in cancer patients. Glutathione, a major redox-regulatory tripeptide in host tissues, is dependent on glutamine for its biosynthesis and appears to be diminished in the liver of AH109A rat hepatoma tumor-bearing rats.[9] Provision of exogenous dietary glutamine, either enterally or parenterally, however, does not seem to affect glutathione levels in the liver, but increases the glutathione content of the intestine.[9,50,106,149,150] Conversely, studies in mice bearing the MCA-105 fibrosarcoma revealed a significant increase in hepatic gamma-glutamyl-cysteine synthetase activity and a decrease in the sulfate level, glutamine/urea ratio, and glutamine/glutamate ratio, suggesting an increased flux through glutathione and urea biosynthetic pathways.[53] Collectively, the acceleration of these glutamine- and energy-dependent processes in the tumor-influenced liver creates an additional drain on the cytoplasmic glutamine pools. System N stimulation may therefore represent a compensatory/adaptive response by the liver to maintain cytoplasmic glutamine levels in the face of accelerated utilization.

Activation of hepatic glutamine transport in the tumor-bearing host may serve a second function in hepatic metabolism as well; to increase the hepatocellular hydration state. The regulation of hepatic metabolism by changes in cell volume has received an increasing amount of attention.[151-157] In general, cell swelling activates anabolic processes in the hepatocyte such as protein synthesis, glycogen synthesis and lipogenesis, whereas cell shrinkage stimulates catabolic processes such as proteolysis, lipolysis and glycogenolysis. System N displays the most rapid rates of substrate (glutamine) transport compared to the other Na^+-dependent amino acid transporters expressed in the hepatocyte.[124,158] As a result, the rapid influx of Na^+ and glutamine via System N results in an increase in the hepatocellular hydration state via mass-action water diffusion across the plasma membrane. Cell swelling likewise causes activation of System N activity[158] and hepatic glutamine uptake[151] which results in a self-activating cycle with respect to hepatic glutamine transport. Of significance is the observation that glutamine exerts its half-maximal effects on hepatocellular volume at normal physiological concentrations of 0.6-0.8 mM, and is maximal at 2 mM.[152,159] The previously described effects of glutamine on carbohydrate, lipid, protein and nitrogen metabolism have now been attributed to its effects on the hepatocellular hydration state.[159] The pathways that have been shown to be accelerated during cancer such as protein synthesis[154,156,160] and gluconeogenesis[161] are regulated by hepatocellular hydration, and could therefore be regulated by glutamine. With respect to glutamine metabolism, cell swelling leads to net glutamine consumption by the liver, whereas cell shrinkage leads to net hepatic glutamine production,[151] changes attributable to altered fluxes through glutaminase and glutamine synthetase. The signal transduction pathways that underlie the volume-induced changes in hepatic metabolism remain poorly understood, but probably involve common kinase/phosphatase networks with branch points specific for individual metabolic pathways and associated enzymes.[161] In summary, accelerated hepatic glutamine transport during tumor growth may serve to regulate metabolism via changes in liver cell volume as well as through provision of intracellular metabolites. With this additional perspective, the role of enhanced hepatic glutamine transport during cancer requires further investigation.

Would glutamine supplementation aid the liver during cancer? Recent studies with Morris hepatoma-bearing rats revealed that tumor burden decreases hepatic glutathione content.[9] Glutamine-supplemented total parenteral nutrition (TPN) failed to rectify this deficit, although it raised glutathione levels in the intestine. Conversely, oral glutamine (1 g/kg) for 2 days in rats bearing the MCA tumor significantly protected host liver reduced glutathione levels (GSH) after a high dose (20 mg/kg) of methotrexate.[162]

Hepatocellular Transformation

While the liver adapts to support tumor-influenced host physiology with respect to glucose and nitrogen metabolism, the changes in glutamine metabolism that occur in this tissue after transformation are perhaps better described than for any other due, in part, to the development of transplantable rat hepatomas in the 1960s.[163] In contrast to liver parenchymal cells which utilize glutamine primarily for urea synthesis and gluconeogenesis, hepatic tumors utilize glutamine primarily as a respiratory fuel and a substrate for nucleotide biosynthesis. Studies showed that both human liver[164] and rat parenchymal cells[43] exhibit a minimal capacity to oxidize glutamine, while human and rat hepatomas displayed prominent oxidation of glutamine for the generation of ATP.[165] Morris rat hepatoma cells displayed a preference for glutamine utilization over glucose for oxidative metabolism,[166] although some rat hepatomas prefer glucose when subjected to physiological mixtures of both substrates. It has been reported by us[167] and others[164] that glutaminase activity in human hepatocellular carcinoma is enhanced approximately 6- to 20-fold when compared to normal human liver. In addition to the increased oxidation of glutamine for ATP production by hepatoma cells, there is also an enhanced utilization of glutamine in de novo nucleotide biosynthetic pathways.[168-172] The level of expression of the enzymes involved in glutamine-dependent nucleotide biosynthesis is commensurate with the growth rates of hepatoma cells. Enhanced rates of glutamine utilization by transformed liver cells are underscored by the observation that intracellular glutamine levels are approximately 10-fold lower than in normal liver tissue.[173]

Collectively, the data suggest that the demand for glutamine increases in hepatocellular transformation and that the metabolic fate of glutamine is altered. Hepatoma cells must therefore possess efficient mechanisms for the extraction and utilization of this amino acid, particularly in poorly vascularized environments of many solid tumors where glutamine levels may be considerably lower than plasma levels. Indeed we have reported that human hepatoma cells transport glutamine at rates 20- to 30-fold faster than isolated human hepatocytes.[34] The basis for this accelerated glutamine uptake is attributable to the expression of a high affinity (Km = 0.050-0.2 mM) glutamine transporter with kinetic and substrate characteristics of System ASC by the hepatoma cells.[34,167] This glutamine transporter is not expressed in normal human hepatocytes, which similar to rats utilize System N for this purpose. Evidence also exists that the expression of intracellular glutamine metabolizing enzymes is altered upon hepatocellular transformation. Early work with transplantable Morris hepatoma cells[174] demonstrated that in spite of retaining varying degrees of liver-specific function, several cell lines expressed the kidney-type glutaminase isozyme (GA^K). The GA^K isozyme possesses a higher affinity for glutamine than liver-type glutaminase (GA^L) with a Km of 3-5 mM.[116] In those studies, the faster growing hepatomas expressed proportionally more GA^K isozyme compared to the GA^L isozyme. Similarly, fetal hepatocytes express exclusively the GA^K isozyme,[116] as well as higher levels of System N activity than adult rat and human hepatocytes;[34,175] only after birth is the GA^L isozyme expressed. Taken together, these observations demonstrate that hepatoma cells utilize more glutamine than normal liver cells—a process made possible by the robust expression of transporters and enzymes with higher affinities for this amino acid. The well-documented consumption of glutamine by rat hepatoma cells may be relevant to the human disease, as patients with hepatomas exhibit decreased plasma levels of this amino acid.[67] Hepatomas may therefore represent an example of a "glutamine-dependent" tumor.

Muscle

It has been estimated that skeletal muscle contains 75% of the glutamine pool in the body, and therefore represents the most significant "bank" for this amino acid.[176] In human skeletal muscle, intracellular glutamine concentrations are approximately 20 mM and represent 60% of the total free amino acids,[177] but these levels can drop as much as 50% during catabolic states.[178] In many animal tumor models[12,27,50,84] and in cancer patients[179,180] the release of amino acids, particularly glutamine, from skeletal muscle is a well-documented process, and is thought to ultimately contribute to the lean muscle wasting observed in cachexia. There remains a debate however, on whether muscle amino acid efflux is elicited by tumor-dependent or nutrition-dependent factors during the pathogenesis of cancer.[181] The complex nature of the etiology and progression of cancer cachexia is underscored by and reviewed in a recent comprehensive volume on this subject to which the reader is referred.[182] While the obvious contribution of hormonal and nutritional factors to the development and progression of cachexia are critical, they are beyond the scope of this chapter, but are nicely summarized in Chapter 2 of this book.[182] We will instead focus on the possible role of glutamine in the manifestations of cachexia and muscle biology and its potential use in the improvement of nitrogen balance in cancer patients.

The role of the muscle during cancer is key to the host with respect to glutamine economy. It is fairly well established that muscle glutamine efflux is accelerated in response to tumor growth, but it is also observed in other catabolic states such as starvation,[183-186] endotoxemia[183,187-189] and after surgery.[190-202] As stated above, it remains unresolved whether tumor growth per se or secondary effects of the cancer (e.g., hypophagia, increased gut permeability, inflammation, etc.) initiate the output of glutamine by the skeletal muscle. Accelerated glutamine efflux during cancer may also be aided by increased blood flow to skeletal muscle, as measured in the MCA tumor-bearing rat.[203] Regardless of the etiology of accelerated skeletal muscle release in cancer, the process helps to maintain circulating levels of this amino acid in the face of enhanced uptake by tissues such as the liver and the tumor.

Why is muscle glutamine important in the host protein economy? Cancer often results in net negative nitrogen balance in the host characterized by decreased muscle protein synthetic rates.[144,204,205] Intracellular glutamine concentrations correlate with protein synthetic rates in this tissue,[185,206] and thus the modulation of glutamine skeletal muscle content may have important implications for host protein economy during cancer.[28,184,207,208] The uptake of glutamine in rat skeletal muscle is mediated by a Na^+-dependent transporter very similar to System N in the liver and termed System N^m.[186,208-214] A similar transporter, although with some notably different characteristics, mediates glutamine uptake in human skeletal muscle.[215,216] The properties of this transporter render it uniquely suited to regulate glutamine import and export as required by host metabolic demands.[29] The relatively high activity of this carrier and its narrow substrate specificity (only glutamine and asparagine) allow the generation of a large transmembrane gradient of glutamine to be established. Decreases in the driving force (transmembrane Na^+ electrochemical gradient) for this carrier therefore allow the efficient efflux of this amino acid when required.[29] The role of carrier-mediated glutamine efflux from the muscle during catabolic states has been reviewed and is implicated in the maintenance of a healthy immune system.[177] For example, increased glutamine efflux from skeletal muscle during catabolic states (including cancer) may be due to decreased System N^m activity secondary to a decrease in the Na^+ electrochemical gradient and an increase in de novo synthesis via glutamine synthetase.[27,84,217,218] Glutamine synthetase (GS) activity is elevated in the muscles of tumor-bearing animals,[27,84] and its important role in the host during cancer is underscored by the observation that methionine sulfoximine, a GS inhibitor, exacerbates anorexia in the tumor-bearing host.[6] Glutamine deprivation in vitro causes increased expression of both System N^m and glutamine synthetase in muscle cells,[213] suggesting that intracellular glutamine levels may regulate the expression of key proteins involved in its net balance in vivo as well.

Similar to System N in the liver, System N^m displays the highest activity of the amino acid transporters expressed in the skeletal muscle.[29] Also similar to liver, the skeletal muscle transporter is activated by cell swelling.[214,219] Activation of the muscle transporter by insulin[214] may also lead to cell swelling and further activation of the carrier. Studies in normal dogs with either glutamine-, glycylglutamine- or alanylglutamine-supplemented total parenteral nutrition displayed higher arterial glutamine concentrations and enhanced extraction across the hindleg than those that received a normal amino acid mixture.[220] These results suggested that glutamine may elicit beneficial anabolic effects by stimulating its own uptake. The subsequent increase in intracellular glutamine may in turn help to promote protein synthesis as described above. This aspect of skeletal muscle glutamine transport regulation may have important implications for possible combination hormonal and nutrient therapy in some cancer patients to help improve nitrogen economy.[221-223]

The effects of glutamine-supplemented nutrition on host muscle protein economy are equivocal and have been reviewed.[180] Many studies suggest that fortification of nutritional formulas (enteral or parenteral) with glutamine or its precursors (α-ketoglutarate), while exerting some beneficial effect is not alone sufficient to rectify the glutamine loss and catabolism in the muscle of tumor-bearing animals[224] or patients[225] especially when the cancer is advanced.[226] Studies in malnourished patients with gastric carcinomas revealed that TPN alone, while preserving muscle glycogen, failed to attenuate the decrease in muscle glutamine and energy status,[193] suggesting that traditional TPN may be inadequate for cancer patients. Results from animal models such as the MCA tumor-bearing rat suggest, however, that both enteral[12] and parenteral[50] nutritional formulas supplemented with glutamine preserve muscle glutamine stores and improve nitrogen balance without stimulating the growth of the tumor. Studies in AH109A hepatoma-bearing rats revealed that glutamine-supplemented TPN improved carcass weight, nitrogen balance and protein synthesis in muscle without stimulating tumor growth.[9,227] Once again, the efficacy of glutamine-supplemented nutrition on muscle metabolism is probably contingent on the nature of the tumor and the stage of the disease.

The above studies examined the differential effects of glutamine-supplemented nutrition on the tumor and host tissues, which provide some insight into the "safety" of these formulations. Historically, there has been a great deal of debate and hesitation on the use of glutamine in cancer patient nutritional formulations for fear that tumor growth might be stimulated by such modalities. As will be discussed below, it appears that this fear, while valid, may be unfounded based on results from some animal models,[9,10,12,50] and in fact may decrease tumor growth. Again, while useful information is gleaned from such studies, the use of large animal tumors may not provide the most ideal model for these types of investigations. As pointed out by others, a more realistic approach is to examine these enriched formulas in models that more closely resemble the clinical scenario.[226] Such models include resection of the primary tumor followed by chemo- or radiation therapy—two procedures that further tax host tissues with respect to glutamine economy.

The first of the two clinical treatments (surgery) is known to decrease muscle glutamine stores and enhance protein catabolism in patients and animals.[15,228] The most continuous body of work on the effects of glutamine or its precursor α-ketoglutarate (in combination with ornithine) on muscle glutamine content, protein synthesis and nitrogen balance in surgical patients has come from Wernerman, Vinnars, Hammarqvist, Petersson and their colleagues.[190,194-200,229,230] The published results are directly relevant to the treatment of patients because they are all clinical studies and support findings in animal models as well.[12,50] Briefly, these authors collectively found that provision of either ornithine/α-ketoglutarate or glutamine (free or as dipeptides) preserves muscle glutamine stores, improves nitrogen balance and enhances protein synthesis. Based on a number of studies, the mechanism by which glutamine preserves muscle protein content seems to be through a stimulation of protein synthesis rather than an inhibition of proteolysis. As recovery from surgery can be a protracted process, one significant caveat to these studies is that glutamine dipeptide TPN must be continuously administered for the beneficial effects to persist.[190,229] The increased costs of such treatments, once approved, will have to be carefully weighed against the physical status of the patient. Erosion of lean body mass (cachexia) decreases the ability of people to endure cancer treatment, and as will be discussed below, one of the dangers of "glutamine depletion" in the cancer patient is immunocompromise and decreased gut integrity which, when combined, predispose the patient to infection. which is a significant cause of mortality in cancer patients.

Gut

The intestine is a major glutamine consumer in the body, as first demonstrated by the pioneering work of Windmueller in the 1970s.[231-233] Most of the glutamine consumed by the gut—whether from basolateral or luminal surface—is converted to CO_2, reflecting its primary use as a respiratory fuel.[232,233] The transport of glutamine across the luminal (brush-border) plasma membrane of intestinal epithelia is mediated by a broad-specificity Na^+-dependent transporter formerly known as neutral brush border (NBB), the gene for which has recently been cloned[234] and is now referred to as System B^0. In contrast, the transport of glutamine across the basolateral (blood-facing or serosal) plasma membrane is mediated by Na^+-dependent System A and Na^+-independent System L in several mammalian species.[235-239] The ability to take up glutamine in a concentrative manner from either side (luminal or serosal) of nutrient supply is probably essential to the dividing intestinal epithelia given its central role in the metabolism of these cells. Indeed the importance of glutamine in the maintenance of a healthy gut has been reviewed extensively.[21,98,240-243] In the MCA tumor-bearing rat, the activity of the luminal glutamine transporter in brush-border membrane vesicles (BBMV's) was found to be stimulated compared to the activity in vesicles from pair-fed controls, whereas glucose transport activity was unchanged.[244] These results suggested that capacity for enteral glutamine uptake may be enhanced in response to tumor growth and may provide a rationale for glutamine-supplemented enteral nutrition for certain cancer patients. Of further significance is the observation that intestinal glutamine uptake can be further enhanced by provision of glutamine-supplemented enteral[245] nutritional formulations.

It must be remembered that the "gut" is also a complex tissue bed, composed of not only intestinal epithelia but also smooth muscle and significant areas of immune cells in Peyer's patches and lymphatic tissue (the so-called gut-associated lymphoid tissue or GALT). The role of glutamine in the maintenance of gut immune function and cellularity has been addressed.[21,96,242,246] There appears to be a correlation between secretory IgA (an index of gut immune function) deficiency and compromise of gut barrier function in both human and animal models.[247-249] While animal models[247,250] and human studies[251] demonstrate that standard TPN may eventually compromise intestinal integrity, studies in human clinical trials are less compelling.[252] Skepticism about the utility of glutamine-supplemented nutrition in patients may be based largely on results obtained in studies from otherwise healthy subjects. It must be remembered, however, that cancer patients may suffer from "glutamine depletion" especially after further "iatrogenic" insults such as surgery and chemo- or radiation therapy. Such patients maintained on TPN because they are too sick to eat may be at particular risk for subsequent infection.

In response to the presence of a growing MCA tumor in the rat, intestinal glutamine extraction (from the basolateral surface) decreased by nearly 50% while arterial levels decreased by 30%.[85] Diminished gut plasma glutamine extraction was associated with decreases in morphometric parameters such as mean villous height and intestinal wet weight. This data suggests that the gut is compromised in response to tumor-induced glutamine depletion and that the ability of the intestine to take up its primary nutrient from the blood is also diminished. Interestingly, recent studies reported an increase in net glutamine extraction by the portal-drained viscera in animals with similar MCA tumor burdens but no change in mean villous height or mucosal cell populations.[64] Although gut barrier function was again compromised by the tumor in these studies, it was associated with a decrease in polyamine levels. Based on data from animal models, another detrimental effect that progressive tumor growth may exert on host intestinal tissue is glutathione depletion.[9,50,100,106] In these studies, it was also found that provision of glutamine-supplemented nutrition helps to restore (or maintain) mucosal glutathione content to control values.

The gut-protective role of glutamine has been reviewed[98,241,253] and may involve its ability to maintain immune cellularity in the intestine,[96,246,254] to maintain glutathione levels as mentioned above, to support the proliferation of enterocytes, or a combination of all known effects. It is generally accepted that enteral nutrition offers advantages to the host versus parenteral nutrition,[21,241,255-258] but is often precluded by the condition of the patient. Reviews of enteral nutritional therapy in cancer patients have been published, along with suggestions for improvement in its utility.[257,259] When enteral feeding is not possible, many studies in animals[50,96,227,260-267] and patients[268] suggest that glutamine-supplemented TPN offers advantages over standard TPN with respect to host gut function and integrity.

Similar to changes observed after hepatocellular transformation, malignant cells from the GI tract have been reported to display increased avidity for glutamine. Studies with human colon carcinoma cell lines in vitro suggested that glutamine increases proliferation and the invasive phenotype of these cells, prompting the authors to warn against the use of glutamine nutrition in cancer patients.[269] Likewise, glutamine concentrations were lower and glutamine-utilizing nucleotide biosynthetic enzyme activities were higher in human colorectal carcinomas than in normal colon.[270] In contrast, recent studies in patients show that while gastrointestinal cancer may deplete arterial glutamine levels, the tumor itself does not consume any more glutamine than the normal intestine.[271] These two examples demonstrate the importance of testing data obtained from cell culture and "test tube" models in in vivo systems. In vitro systems provide a well-controlled environment in which to test hypotheses at the cellular level, but the implications of the results obtained should not be taken too far. This theme will be raised again in the last section of the chapter, where the "fear" of stimulating tumor growth with glutamine therapy (based largely on in vitro data) may not be well-founded.

Glutamine and Iatrogenic Intestinal Compromise

There is evidence that provision of extra glutamine in the diet may protect against radiation-induced enteropathy,[10,13,107,109,272,273] but there are also studies that question this finding.[274] Glutamine has also been shown to protect mammalian cells from radiation-induced killing in vitro.[275] Another anecdotal report suggests that glutamine-enriched TPN may have exerted beneficial effects in gut barrier function in a patient subjected to radiotherapy.[276] There are further indications that oral glutamine may decrease diarrhea after radiation therapy.[277] One other report suggests that glutamine TPN does not aid the repair of radiation-induced mucositis,[278] but in contrast to the other studies, was administered only after the radiation treatment. Taken together, these data may suggest that glutamine enriched nutrition should be given both before and after radiotherapy for maximum benefit. Provision of glutamine after standard cancer treatments may be beneficial, however. In a double-blind, randomized controlled trial the effects of glutamine-supplemented TPN in patients with hematologic malignancies in remission after standard high-dose chemotherapy and radiation therapy and bone marrow transplantation were assessed.[279] The incidence of extracellular fluid expansion and infection were reduced in the glutamine TPN group compared to the standard TPN group in this clinical study.

With respect to chemotherapy-induced enteropathy, glutamine has also been shown to exert a protective role in both animals[10,11,106,260] and patients.[280,281] Again, the superiority of enteral-versus-parenteral nutrition, when possible is emphasized,[282,283] and one study suggested that a protein-based elemental diet is more effective than glutamine-based elemental diet in improving the morbidity and mortality rates associated with methotrexate administration.[284] A prospective double-blind study was conducted in hematologic patients undergoing intensive chemotherapy that evaluated the use of glutamine-supplemented TPN on toxic side effects of the cancer treatment.[285] The authors failed to find any significant effect of the glutamine-supplemented TPN on chemotherapeutic gastrointestinal toxicities, although an improvement in weight gain was reported. Although the results from separate studies are sometimes equivocal, the positive effects coupled with the demonstrated safety of glutamine nutritional therapies together warrant further studies in this area.

In summary, the gut utilizes a great deal of glutamine, but is particularly susceptible to loss of integrity during periods of prolonged catabolic stress or iatrogenic insults such as surgery, chemo- or radiation therapy. Provision of exogenous glutamine (preferably enteral), seems to exert a protective role against many of these insults via support of enterocyte growth, glutathione metabolism and maintenance of immune cellularity. A collective evaluation of the studies to date suggest that glutamine-supplemented nutrition support may not necessarily benefit otherwise healthy patients, or alone be sufficient for extremely sick patients, but could provide some benefit to subjects undergoing surgery and postoperative procedures if administered continuously.

Immune System

Cells of the immune system are the other major glutamine consumer in the body, a point raised above with respect to intestinal barrier function. The utilization of glutamine by immune cells has been reviewed[24,286-288] and studied in humans and rodents.[22,23,45,289-294] Lymphocytes express high levels of the enzyme glutaminase which allows this amino acid to be rapidly metabolized even in resting (inactivated) immune cells.[24,295-298] It has been proposed that the high rates of glutaminolysis in lymphocytes allow these cells to respond rapidly to activation signals, as glutamine can be utilized for de novo nucleotide biosynthesis during specific phases of the cell cycle.[71,177,286] This "metabolic control logic", a phrase coined by Newsholme and colleagues, allows specific biosynthetic pathways to tap into glutamine pools or subsequent pools of glutamine metabolites without altering the flux through other metabolic pathways, because glutaminolysis is already quite high, with rates exceeding those of biosynthetic pathways. Unlike enterocytes, lymphocytes exhibit only partial glutamine oxidation, as major products of its metabolism are glutamate, ammonia and aspartate.[23,45,290-292] The rates of glutamine utilization but not the ratios of its metabolites are altered (stimulated) after mitogenic activation of these cells.[299] As is the case in many tissues, glutamine supports a number of critical cellular functions in lymphocytes such as DNA synthesis,[300-302] B-cell maturation,[303] macrophage phagocytosis and IL-1 production[47] and lymphocyte IL-2 production.[299,304] Of significance is that glutamine exerts its progressive effects on the above immune cell functions at concentrations in the physiological range. The transport of glutamine into lymphocytes from rats,[305] bovines[306] and humans[307] is mediated largely by the Na^+-dependent System ASC carrier. This activity is also accelerated after lymphocyte mitogenic stimulation.[307] Tumor growth also alters host immune cell glutamine metabolism in animal models. Lymphocytes from Walker 256 tumor-bearing rats displayed augmented rates of both glucose and glutamine utilization.[56] In contrast, lymphocytes from rats bearing the Morris 7777 hepatoma displayed decreased glucose and glutamine utilization at two different (1.4% and 6%) tumor burdens.[308] The reason for this disparity is unclear, but may involve the differential effects of each tumor on host hormonal levels.[57]

Exogenous glutamine can also modulate immune function. T-cells from surgical patients maintained on glutamine-supplemented TPN exhibit enhanced mitogenic responses compared to those from patients on standard TPN.[100] MCA tumor-bearing rats maintained on a chow diet and gavaged daily with 1 g/kg L-glutamine for 3 weeks displayed a 40% reduction in tumor growth.[110] This glutamine-associated reduction in tumor volume was attributed to an increase in splenic natural killer (NK) cell activity and increased host glutathione levels. In a rat mammary tumor model (MTF-7), the same glutamine gavage regimen for 7 weeks resulted in decreased tumor growth, incidence of metastases, increased plasma glutathione levels and diminished (2.5-fold) prostaglandin E2 (PGE_2) production (a monocyte-derived suppressor of NK cell activity).[309] Rats bearing the Morris 7777 hepatoma and maintained on a glutamine-supplemented (20 g/kg) oral diet for 2 weeks exhibited decreased tumor growth compared to tumor-bearing rats maintained on normal chow.[224] Splenocytes from these glutamine-supplemented animals displayed increased mitogenic responses and a higher percentage of NK cells as well. In contrast, rats bearing the AH109A hepatoma and maintained on glutamine-supplemented TPN displayed higher muscle protein synthesis rates and restoration of tumor-induced gut glutathione depletion compared to those on standard TPN, but did not display a reduction (or increase) in tumor growth.[9] Possible reasons for this discrepancy with the other models are that the treatment period was only 6 days, and the glutamine was administered intravenously rather than enterally. The results when evaluated together suggest that glutamine decreases the growth of at least three experimental tumors through support of pleiotropic host processes (especially glutathione biosynthesis) that strengthen immune system function.[10] This topic (enhancement of immune function via specific nutritional pharmacology) was the focus of a recent international symposium in Nice, France.[310] When evaluated collectively, the results from human trials and animal models suggest that glutamine may exert many of its beneficial effects during catabolic states (e.g., cancer) through fortification of the immune system.

No studies to date have demonstrated that glutamine protects against chemotherapy-induced leukopenia. However, glutamine-fed MCA tumor-bearing rats exhibit higher gut and plasma glutathione levels, and the tumors from these animals exhibit lower glutathione levels. This differential effect of glutamine on tumor and host tissues has been proposed to play a role the greater sensitization to methotrexate and radiation therapy of the fibrosarcoma in glutamine-fed animals.[10] Whether the beneficial effects of glutamine in this animal cancer model will apply in patients can only be determined by future prospective clinical trials.

Glutamine Nutrition in Cancer Patients

Administration of glutamine to patients has been deemed safe, as no significant toxic side-effects have been reported[39] and, as discussed above, enteral feeding is preferred over parenteral feeding for reasons of economy and efficacy. Intravenous formulas do not contain glutamine because of its instability, but this problem has been addressed and circumvented partially by the use of glutamine dipeptides which seem to be equally effective.[100,178,220,262,311] Many commercially available enteral nutritional formulas contain glutamine, but the levels may not be sufficient to benefit patients for whom pharmacological nutrition is indicated. For commercially-available enteral formulations, it has been recently reported that the daily dietary glutamine intake of patients ranges from 1.3-5.6 g per day for peptide-based products and from 6-8 g per day for protein-based products, whereas critically ill patients may require from 10-20 g of glutamine per day.[40] As often quoted in recent reviews, the best study to date on the use of glutamine-supplemented nutrition in people was performed in a randomized, double-blinded clinical trial in bone marrow transplant patients.[312] Patients in this study that received glutamine (0.57 g/kg BW/day) intravenously had decreased incidence of clinical infections, improved nitrogen balance and shortened hospital stays. This study, in combination with results from animal models above, provides the impetus for further clinical trials in cancer patients.

Summary and Suggestions

We have thus far reviewed the metabolism of glutamine by normal, tumor-influenced and cancerous tissue, and how glutamine levels are affected by tumors and cancer treatments. As this series focuses on current research trends in areas of interest, we will now put all of this information in perspective and make suggestions for future research in this evolving field. It is clear that host tissues become compromised after prolonged periods of catabolic stress associated with cancer such as tumor-induced malnutrition. Cancer therapies—surgery followed by radiation or chemotherapy—exert further challenges upon the patient. Animal models and some human studies suggest that hyperalimentation with glutamine may aid the patient in enduring and recovering from such treatments, although supplementation of nutritional formulations with this amino acid alone may be neither sufficient for extremely sick individuals, nor benefit otherwise healthy individuals. Instead, it is likely that combinations of hormonal and nutritional therapies will ultimately yield maximum benefit to the cancer patient.

Recent research trends have redirected the focus of the field from "antiglutamine" therapies (envisaged as potential anti-tumor agents in the 1970s and 1980s) to glutamine therapy for the host. Based on clinical trials with glutamine analogs such as acivicin ((alphaS,5S)-alpha-amino-3-chloro-4,5-dihydro-5-isoxazoleacetic acid)[313,314] and 6-diazo-5-oxo-L-norleucine (DON)[315,316] and glutaminase infusions,[317] it is clear that the "anti-glutamine" therapies are nonspecific (i.e., not tumor-specific) and induce too many intolerable side effects (e.g., somnolence, ataxia, personality changes, hallucinations, myelosuppression, vomiting, diarrhea and mucositis).[313,318,319] This finding is not surprising, given the information on the essential role of glutamine in the host presented above. Given the demonstrated safety, low cost, gut-protective and immune-promoting role of glutamine, its use as an adjunct therapy in cancer patients is gaining more interest. With the exception of one report in the MCA tumor model,[320] glutamine-supplemented nutrition has yet to be observed to promote tumor growth in vivo, and in fact has been shown to have the opposite effect as discussed in the previous section. The caveat from the single report where glutamine feeding was associated with increased tumor weight is that glutamine was only one of four amino acids (along with asparagine, glutamate and aspartate) given in the TPN. Indeed when all of the evidence (from animal models) is weighed, the benefits of glutamine feeding seem to far outweigh the risks. We feel that glutamine hyperalimentation may be particularly useful after surgery and when the patient undergoes radio- or chemotherapy. Here are the reasons why:

1. Maintenance of plasma glutamine levels near normal (0.6 mM) values with glutamine pharmacology will only benefit the (especially depleted) host. As shown in the human studies discussed in the previous sections, provision of extra glutamine in patient nutritional formulations will help to ease the burden on the already taxed skeletal muscle, and also improve immune function. Fears of stimulating tumor growth with "extra" glutamine have yet to be demonstrated. At this level (0.6 mM) any residual tumor that may persist after surgery (e.g., metastases or lymph-associated) will be saturated with glutamine (based on the Km's of tumor cell glutamine transporters which range from 0.050-0.3 mM on average), and the "extra" glutamine should have no additional affect on growth.

2. Even if additional glutamine stimulates tumor growth, this may allow cell cycle-specific therapies to be more effective. The demonstrated effects of glutamine in the physiological range on immune cell function may also outweigh the risks of stimulating tumor growth. Given the lack of success with anti-glutamine therapies and the adaptable (plastic) nature of tumor cells, it is unlikely that "starving" tumors for glutamine will be effective. The toxicities of the anti-glutamine therapies on the host attests to this. Many tumor cells may have lost (or display deviant regulation of) the pathways that lead to scheduled cell death or apoptosis after cellular stress, whereas normal host cells retain these pathways.[321] This possibility may underlie the observation that after attempted therapies, tumors endure while host tissues continue to deteriorate in patients with terminal cancer. Maintenance or promotion of immune function via glutamine pharmacology therefore must be considered an option worth further investigation.
3. Considering the gut-protective and immune-enhancing effects of glutamine as well as its demonstrated effectiveness in animal models and some human trials, its use as a dietary supplement should be considered before, during and after chemo- or radiotherapy. As leukopenia is a common side-effect of cancer treatments, the provision of adequate glutamine supply may help to speed the recovery of the immune system—as the results from the clinical trial by Ziegler and colleagues suggested.[312] Its use before and during therapies is indicated as the studies where it was given only after radiation therapy were one of the only reports to indicate that its use was ineffective.[278]

Suggestions

In spite of our enthusiasm for the use of glutamine as an adjunct cancer therapy, this topic remains quite contentious. Clearly, more work needs to be done in this area of cancer research. Here are some suggestions for directly pertinent areas of research that we feel should (and will) be addressed.

1. *Use of more relevant tumor models*: While animal models to date have yielded some important information, more clinically relevant animal models of cancer need to be developed and utilized by researchers worldwide, as suggested by LeBricon and colleagues.[226] This includes the use and implantation of primary tumors with metastatic potentials into hosts with normal immune systems, followed by resection at reasonable burdens (e.g., when the tumor is less than 5% of body weight), and standard chemo- or radiation therapies. The effects of enteral versus parenteral glutamine-supplemented nutritional formulas should be evaluated for effects on mortality and incidence of metastases as well as immune function and gut integrity.
2. *Continued evaluation of glutamine pharmacology in human clinical trials*: More large-scale randomized double-blind prospective trials are necessary to ultimately test the utility of glutamine hyperalimentation on the outcomes of patients with specific types of cancer. These trials may be especially warranted in patients with a high likelihood of metastases and resulting aggressive chemotherapies, as they will be at higher risk of cachexia-induced death. This area of clinical research will certainly be highly controversial, but if the immune-enhancing and tumor-sensitizing nature of glutamine observed in studies thus far is globally applicable, then these are the patients who would benefit most from its use.
3. *The biochemical basis for the effects of glutamine on host tissues and tumors*: Studies in animal models to date indicate that glutamine feeding exerts no effect upon, or results in decreased tumor growth.[9,10,12,50] Studies by Klimberg's group in particular have yielded some interesting and surprising results. In both the MCA model and the breast cancer model, it is clear that glutamine raises (preserves) glutathione (GSH) levels in

host tissues and plasma. This support of host GSH was associated with augmented immune cell function and decreased tumor growth. Interestingly, administration of methotrexate (MTX) in conjunction with glutamine feeding decreased tumor volume and GSH levels,[106] while host GSH levels were preserved under the same treatment. In subsequent studies, it was shown that the glutamine-gavaged group methotrexate concentrations in the tumor were higher than in the glycine-fed group.[162] The finding that glutamine pharmacology increases the therapeutic index of chemotherapy was unexpected. The authors postulate that the differential effect of glutamine on host and tumor GSH levels (and subsequent ability to excrete MTX) is due to a differential regulation of glutaminase and gammaglutamyl transferase (GGT) by glutamine in tumors and normal cells.[10] This is a testable hypothesis, but to date, no data are given to support it. Therefore, the questions to be addressed on the basic science level are:

a) Does glutamine augment the therapeutic index of chemotherapy agents in different tumors but not normal cells? What about radiation effects?
b) If so, what is the mechanism by which it exerts its sensitizing effects? Its protective effects? How does glutamine feeding influence the activity and expression of multidrug-resistance transporters or glutathione conjugating enzymes? In tumors? In host cells?
c) Is the glutamine effect on tumor volume during chemotherapy in vivo a function of only intracellular glutathione or does immunoenhancement play a role as well? (similar to glutamine-dependent reduction in tumor volume in the absence of chemotherapy).
d) How does the immune cell population in the tumor change as a function of glutamine nutrition? What are the immune cellular functions that are enhanced by glutamine feeding?

Finally, these questions should be addressed in the more clinically relevant model discussed above.

"Nutritional pharmacology" was the catch-phrase of the 1990s and its utility and mechanisms of action are garnering progressively more attention. We hope that the review presented here—while not intended to be comprehensive—provides enough background and interest to spur further investigation into the use of this "conditionally-essential" amino acid in the treatment of cancer.

References

1. Medina MA, Sanchez-Jimenez F, Marquez J et al. I. Relevance of glutamine metabolism to tumor cell growth. Mol Cell Biochem 1992; 113:1-15.
2. Tannock IF, Steele D, Roberts J. Influence of reduced concentration of L-glutamine on growth and viability of cells in monolayer, in spheroids, and in experimental tumors. Brit J Cancer 1986; 54:733-41.
3. Hanka LJ. Introduction: Possibilities for biochemically rational chemotherapy for some malignancies with depleting enzymes and antimetabolites of specific amino acids. Cancer Treatment Reports 1979; 63:1009-1011.
4. Prajda N. Enzyme targets of antiglutamine agents in cancer chemotherapy. Adv Enzyme Reg 1985; 24:207-23.
5. Lynch G, Kemeny N, Casper E. Phase II evaluation of DON (6-diazo-5-oxo-L-norleucine) in patients with advanced colorectal carcinoma. Am J Clin Oncol 1982; 5:541-3.
6. Chance WT, Zhang FS, Fischer JE. Methionine sulfoximine intensifies cancer anorexia. Pharmacol Biochem Behavior 1991; 39:115-8.
7. Baruchel S, Bernstein M, Whitehead VM et al. A phase I study of acivicin in refractory pediatric solid tumors. A Pediatric Oncology Group study. Investigational New Drugs 1995; 13:211-6.
8. Hall JC, Heel K, McCauley R. Glutamine. Brit J Surg 1996; 83:305-12.

9. Yoshida S, Kaibara A, Yamasaki K et al. T. Effect of glutamine supplementation on protein metabolism and glutathione in tumor-bearing rats. JPEN 1995; 19:492-7.
10. Klimberg VS, McClellan JL. Claude H. Organ, Jr. Honorary Lectureship. Glutamine, cancer, and its therapy. Am J Surg 1996; 172:418-24.
11. Klimberg VS, Nwokedi E, Hutchins LF et al. Glutamine facilitates chemotherapy while reducing toxicity. JPEN1992 ; 16:83S-87S.
12. Klimberg VS, Souba WW, Salloum RM et al. Glutamine-enriched diets support muscle glutamine metabolism without stimulating tumor growth. J Surg Res 1990; 48:319-23.
13. Souba WW, Klimberg VS, Copeland EMd. Glutamine nutrition in the management of radiation enteritis. JPEN 1990; 14:106S-108S.
14. Krebs HA. Metabolism of amino acids. IV. The synthesis of glutamine from glutamic acid and ammonia, and the enzymatic hydrolysis of glutamine in animal tissues. Biochem J 1935; 33:1951-1969.
15. Souba WW. Glutamine: Physiology, Biochemistry and Nutrition in Critical Illness. Austin, TX: R.G. Landes Co., 1992.
16. Abcouwer SF, Bode, B.P. and Souba, W.W. Glutamine as a Metabolic Intermediate. In: Fischer JE, ed. Nutrition and Metabolism in the Surgical Patient. 2nd ed. Boston: Little, Brown and Co., 1996:353-384.
17. Kvamme E, ed. Glutamine and Glutamate in Mammals. Boca Raton, FL: CRC Press, Inc., 1988.
18. Meister A, ed. Glutamate, Glutamine, Glutathione and Related Compounds. MethEneymol, vol. 113. New York, NY: Academic Press, Inc., 1985.
19. Curthoys NP, Watford M. Regulation of glutaminase activity and glutamine metabolism. Ann Rev Nutr 1995; 15:133-59.
20. Moskovitz B, Katz Y, Singer P et al. Glutamine metabolism and utilization: relevance to major problems in health care. Pharmacol Res 1994; 30:61-71.
21. van der Hulst RR, von Meyenfeldt MF, Soeters PB. Glutamine: An essential amino acid for the gut. Nutrition 1996; 12:S78-81.
22. Newsholme EA, Carrie AL. Quantitative aspects of glucose and glutamine metabolism by intestinal cells. Gut 1994; 35:S13-7.
23. Ardawi MS. Glutamine and glucose metabolism in human peripheral lymphocytes. Metabolism: Clinical & Experimental 1988; 37:99-103.
24. Newsholme EA, Crabtree B, Ardawi MS. Glutamine metabolism in lymphocytes: its biochemical, physiological and clinical importance. Quarterly J Experim Physiol 1985; 70:473-89.
25. Kajstura J, Korohoda W. Significance of energy metabolism pathways for stimulation of DNA synthesis by cell migration and serum. Euro J Cell Biol 1983; 31:9-14.
26. Glinos AD, Warren KR. Nonproliferating neoplastic cells in culture: behavior of the adenine nucleotides. J Natl Cancer Inst 1979; 62:819-23.
27. Quesada AR, Medina MA, Marquez J, Sanchez-Jimenez FM et al. Contribution by host tissues to circulating glutamine in mice inoculated with Ehrlich ascites tumor cells. Cancer Res 1988; 48:1551-3.
28. Rennie MJ, Tadros L, Khogali S et al. Glutamine transport and its metabolic effects. J Nutr 1994; 124:1503S-1508S.
29. Rennie MJ, Hundal HS, Babij P et al. Characteristics of a glutamine carrier in skeletal muscle have important consequences for nitrogen loss in injury, infection, and chronic disease. Lancet 1986; 2:1008-12.
30. Welbourne TC. Role of the lung in glutamine homeostasis. Contr Nephrol 1988; 63:178-82.
31. Kvamme E, ed. Glutamine and Glutamate in the Central Nervous System. Boca Raton, FL: CRC Press, Inc., 1988.
32. Haussinger D. Liver glutamine metabolism. JPEN 1990; 14:56S-62S.
33. Kilberg MS, Handlogten ME, Christensen HN. Characteristics of an amino acid transport system in rat liver for glutamine, asparagine, histidine, and closely related analogs. J Biol Chem 1980; 255:4011-9.
34. Bode BP, Kaminski DL, Souba WW et al. Glutamine transport in isolated human hepatocytes and transformed liver cells. Hepatology 1995; 21:511-20.
35. Bussolati O, Sala R, Astorri A, Rotoli BM, Dall'Asta V, Gazzola GC. Characterization of amino acid transport in human endothelial cells. Am J Physiol 1993; 265:C1006-14.

36. Handlogten ME, Kilberg MS, Christensen HN. Incomplete correspondence between repressive and substrate action by amino acids on transport systems A and N in monolayered rat hepatocytes. J Biol Chem 1982; 257:345-8.
37. Pacitti AJ, Inoue Y, Souba WW. Characterization of Na(+)-independent glutamine transport in rat liver. Am J Physiol 1993; 265:G90-8.
38. Lacey JM, Wilmore DW. Is glutamine a conditionally essential amino acid? Nutr Rev 1990; 48:297-309.
39. Ziegler TR, Szeszycki EE, Estivariz CF, Puckett AB, Leader LM. Glutamine: From basic science to clinical applications. Nutrition 1996; 12:S68-70.
40. Kuhn KS, Stehle P, Furst P. Glutamine content of protein and peptide-based enteral products. JPEN 1996; 20:292-5.
41. Baggetto LG. Deviant energetic metabolism of glycolytic cancer cells. Biochimie 1992; 74:959-74.
42. Matsuno T. Bioenergetics of tumor cells: Glutamine metabolism in tumor cell mitochondria. Intl J of Biochem 1987; 19:303-7.
43. Spolarics Z, Lang CH, Bagby GJ et al. Glutamine and fatty acid oxidation are the main sources of energy for Kupffer and endothelial cells. Am J Physiol 1991; 261:G185-90.
44. Dall'Asta V, Rossi PA, Bussolati O et al. The transport of L-glutamine into cultured human fibroblasts. Biochimica et Biophysica Acta 1990; 1052:106-12.
45. Newsholme P, Newsholme EA. Rates of utilization of glucose, glutamine and oleate and formation of end-products by mouse peritoneal macrophages in culture. Biochem J 1989; 261:211-8.
46. Newsholme P, Gordon S, Newsholme EA. Rates of utilization and fates of glucose, glutamine, pyruvate, fatty acids and ketone bodies by mouse macrophages. Biochem J 1987; 242:631-6.
47. Wallace C, Keast D. Glutamine and macrophage function. Metabolism: Clinical & Experimental 1992; 41:1016-20.
48. Daly JMaCAC. Nutrition and Neoplasia. In: Fischer JE, ed. Nutrition and Metabolism in the Surgical Patient. second ed. Boston: Little, Brown and Co., 1996:601-621.
49. Harrison LE. Animal Models of Cancer Cachexia. In: Pisters PWTaB, M.F., ed. Protein and Amino Acid Metabolism in Cancer Cachexia. Austin, TX. R.G. Landes Co., 1996:1-19.
50. Austgen TR, Dudrick PS, Sitren H et al. The effects of glutamine-enriched total parenteral nutrition on tumor growth and host tissues. Ann Surg 1992; 215:107-13.
51. Chance WT, Cao LQ, Fischer JE. Response of tumor and host to hyperalimentation and antiglutamine treatments. JPEN 1990; 14:122-8.
52. Chen MK, Salloum RM, Austgen TR et al. Tumor regulation of hepatic glutamine metabolism. JPEN 1991; 15:159-64.
53. Hack V, Gross A, Kinscherf R et al. Abnormal glutathione and sulfate levels after interleukin-6 treatment and in tumor-induced cachexia. FASEB J 1996; 10:1219-26.
54. Popp MB, Morrison SD, Brennan MF. Total parenteral nutrition in a methylcholanthrene-induced rat sarcoma model. Cancer Treatment Reports 1981; 65:137-43.
55. Warren RS, Jeevanandam M, Brennan MF. Protein synthesis in the tumor-influenced hepatocyte. Surgery 1985; 98:275-82.
56. Fernandes LC, Marques-da-Costa MM, Curi R. Metabolism of glucose, glutamine and pyruvate in lymphocytes from Walker 256 tumor-bearing rats. Brazil J Med Biol Res 1994; 27:2539-43.
57. Fernandes LC, Mattozo CA, Machado UF et al. Insulin treatment can abolish changes in glucose and glutamine metabolism of lymphocytes and macrophages caused by the implantation of the Walker 256 tumor. Cell Biochem Func 1996; 14:187-92.
58. Parry-Billings M, Leighton B, Dimitriadis GD et al. The effect of tumor-bearing on skeletal muscle glutamine metabolism. Intl J Biochem 1991; 23:933-7.
59. Rosa LF, de Almeida AF, Safi DA et al. Thioglycollate stimulus modifies lymphocyte metabolism and proliferation. A comparison with lymphocyte activation by Walker 256 tumor implantation. Cell Biochem Func 1993; 11:251-5.
60. Marquez J, Sanchez-Jimenez F, Medina MA et al. I. Nitrogen metabolism in tumor bearing mice. Arch Biochem Biophys 1989; 268:667-75.
61. Marquez J, Nunez de Castro I. Mouse liver free amino acids during the development of Ehrlich ascites tumour. Cancer Letters 1991; 58:221-4.
62. Holcenberg J, Anderson T, Ritch P et al. Intraperitoneal chemotherapy with melphalan plus glutaminase. Cancer Research 1983; 43:1381-1388.

63. Yoshida S, Ishibashi N, Noake T et al. Glutamine and arginine metabolism in tumor-bearing rats receiving total parenteral nutrition. Metabolism: Clinical & Experimental 1997; 46:370-3.
64. De Blaauw I, Deutz NE, van der Hulst RR et al. Glutamine depletion and increased gut permeability in nonanorectic, non-weight-losing tumor-bearing rats. Gastroenterology 1997; 112:118-26.
65. el-Yazigi A, Al-Mefty O. Decreased cerebrospinal fluid glutamine levels in patients with benign brain tumors. J Neurochem 1985; 45:815-8.
66. Kubota A, Meguid MM, Hitch DC. Amino acid profiles correlate diagnostically with organ site in three kinds of malignant tumors. Cancer 1992; 69:2343-8.
67. Hirayama C, Suyama K, Horie Y et al. Plasma amino acid patterns in hepatocellular carcinoma. Biochem Med Metabol Biol 1987; 38:127-33.
68. Folkman J. New perspectives in clinical oncology from angiogenesis research. Euro J Cancer 1996; 32A:2534-9.
69. Jain RK, Koenig GC, Dellian M et al. Leukocyte-endothelial adhesion and angiogenesis in tumors. Cancer & Metastasis Reviews 1996; 15:195-204.
70. Eagle H. Nutritional needs of mammalian cells in tissue culture. Science 1955; 122:501-504.
71. Newsholme EA, Board M. Application of metabolic-control logic to fuel utilization and its significance in tumor cells. Adv Enzyme Reg 1991; 31:225-46.
72. Petronini PG, Urbani S, Alfieri R et al. Cell susceptibility to apoptosis by glutamine deprivation and rescue: Survival and apoptotic death in cultured lymphoma-leukemia cell lines. J Cellular Physiol 1996; 169:175-85.
73. Schumpp B, Schlaeger EJ. Optimization of culture conditions for high cell density proliferation of HL-60 human promyelocytic leukemia cells. J Cell Science 1990; 97:639-47.
74. Kitoh T, Kubota M, Takimoto T et al. Metabolic basis for differential glutamine requirements of human leukemia cell lines. J Cellular Physiol 1990; 143:150-3.
75. Horakova K, Valentova N, Chorvath B et al. Study of metabolism and growth limitation of human leukemia cell lines. Neoplasma 1989; 36:287-95.
76. Ollenschlager G, Simmel A, Roth E. Availability of glutamine from peptides and acetylglutamine for human tumor-cell cultures. Metabolism: Clinical & Experimental 1989; 38:40-2.
77. Wasa M, Bode BP, Abcouwer SF, Collins CL, Tanabe KK, Souba WW. Glutamine as a regulator of DNA and protein biosynthesis in human solid tumor cell lines. Ann Surg 1996; 224:189-97.
78. Goetz IE, Weinstein C, Roberts E. Inhibition of growth of hamster tumor cells in vitro by pyrrolidone carboxylic acid in a glutamine-dependent system. In Vitro 1973; 8:279-82.
79. Medina MA, Sánchez-Jiménez F, Márquez J et al. I. Relevance of glutamine metabolism to tumor cell growth. Molec Cell Biochem 1992; 113:1-15.
80. Souba WW. Glutamine and cancer. Ann Surg 1993; 218:715-28.
81. Medina MA, Sanchez-Jimenez F, Marquez FJ et al. I. Glutamine and glucose as energy substrates for Ehrlich ascites tumour cells. Biochem Intl 1988; 16:339-47.
82. Fischer JE, Chance WT. Total parenteral nutrition, glutamine, and tumor growth. JPEN 1990; 14:86S-89S.
83. Miller TJ, Franco RS, Chance WT et al. Amino acid requirements of a rat sarcoma as determined by a stem cell assay. JPEN 1987; 11:223-8.
84. Chen MK, Espat NJ, Bland KI et al. Influence of progressive tumor growth on glutamine metabolism in skeletal muscle and kidney. Ann Surg 1993; 217:655-66; discussion 666-7.
85. Souba WW, Strebel FR, Bull JM et al. Interorgan glutamine metabolism in the tumor-bearing rat. J Surg Res 1988; 44:720-6.
86. Chance WT, Cao L, Kim MW et al. Reduction of tumor growth following treatment with a glutamine antimetabolite. Life Sciences 1988; 42:87-94.
87. Argiles JM, Lopez-Soriano FJ. The effects of tumour necrosis factor-alpha (cachectin) and tumor growth on hepatic amino acid utilization in the rat. Biochem J 1990; 266:123-6.
88. Kallinowski F, Runkel S, Fortmeyer HP et al. L-glutamine: A major substrate for tumor cells in vivo? Journal of Cancer Research & Clinical Oncology 1987; 113:209-15.
89. Goetz IE, Weinstein C, Roberts E. Properties of a hamster tumor cell line grown in a glutamine-free medium. In Vitro 1973; 9:46-55.
90. Aubert C, Rouge F, Galindo JR. Differentiation and tumorigenicity of human malignant melanocytes in relation to their culture conditions. J Natl Cancer Inst 1984; 72:3-12.
91. Desdouets C, Fabre M, Gauthier F et al. Proliferation and differentiation of a human hepatoblastoma transplanted in the Nude mouse. J Hepatol 1995; 23:569-77.

92. Christa L, Simon MT, Flinois JP et al. Overexpression of glutamine synthetase in human primary liver cancer. Gastroenterology 1994; 106:1312-20.
93. Hochwald S, Heslin M. Plasma Amino Acid Concentrations in Cancer Cachexia. In: Pisters PWT, Brennan MF, eds. Protein and Amino Acid Metabolism in Cancer Cachexia. Austin, TX: R.G. Landes Co., 1996:73-82.
94. Bozzetti F, Migliavacca S, Scotti A et al. Impact of cancer, type, site, stage and treatment on the nutritional status of patients. Ann Surg 1982; 196:170-9.
95. Medina MA, Márquez, J. and Núnez de Castro, I. Interchange of amino acids between tumor and host. Biochem Med Metabol Biol 1992; 48:1-7.
96. Alverdy JA, Aoys E, Weiss-Carrington P et al. The effect of glutamine-enriched TPN on gut immune cellularity. J Surg Res 1992; 52:34-8.
97. Souba WW. Total parenteral nutrition with glutamine in bone marrow transplantation and other clinical applications [editorial; comment]. JPEN 1993; 17:403.
98. Souba WW, Klimberg VS, Plumley DA et al. The role of glutamine in maintaining a healthy gut and supporting the metabolic response to injury and infection. J Surg Res 1990; 48:383-91.
99. Souba WW, Klimberg VS, Hautamaki RD et al. Oral glutamine reduces bacterial translocation following abdominal radiation. J Surg Res 1990; 48:1-5.
100. O'Riordain MG, De Beaux A, Fearon KC. Effect of glutamine on immune function in the surgical patient. Nutrition 1996; 12:S82-4.
101. O'Riordain MG, Fearon KC, Ross JA et al. Glutamine-supplemented total parenteral nutrition enhances T-lymphocyte response in surgical patients undergoing colorectal resection. Ann Surg 1994; 220:212-21.
102. Anonymous. Glutamine in parenteral solutions enhances intestinal mucosal immune function in rats. Nutr Rev 1993; 51:152-5.
103. Heberer M, Babst R, Juretic A et al. Role of glutamine in the immune response in critical illness. Nutrition 1996; 12:S71-2.
104. Rohde T, MacLean DA, Klarlund Pedersen B. Glutamine, lymphocyte proliferation and cytokine production. Scand J Immunol 1996; 44:648-50.
105. Alexander JW. Immunoenhancement via enteral nutrition. Arch Surg 1993; 128:1242-5.
106. Rouse K, Nwokedi E, Woodliff JE, Epstein J, Klimberg VS. Glutamine enhances selectivity of chemotherapy through changes in glutathione metabolism. Ann Surg 1995; 221:420-6.
107. Klimberg VS, Salloum RM, Kasper M et al. Oral glutamine accelerates healing of the small intestine and improves outcome after whole abdominal radiation. Arch Surg 1990; 125:1040-5.
108. Klimberg VS. Possible mechanism(s) of glutamine-mediated host protection [editorial; comment]. Nutrition 1995; 11:175.
109. Jensen JC, Schaefer R, Nwokedi E et al. Prevention of chronic radiation enteropathy by dietary glutamine. Ann Surg Oncol 1994; 1:157-63.
110. Fahr MJ, Kornbluth J, Blossom S et al. Harry M. Vars Research Award. Glutamine enhances immunoregulation of tumor growth. JPEN 1994; 18:471-6.
111. Haussinger D, Gerok W. Hepatocyte heterogeneity in glutamate uptake by isolated perfused rat liver. Euro J Biochem 1983; 136:421-5.
112. Stoll B, McNelly S, Buscher HP et al. Functional hepatocyte heterogeneity in glutamate, aspartate and alpha-ketoglutarate uptake: a histoautoradiographical study. Hepatology 1991; 13:247-53.
113. Haussinger D, Gerok W, Sies H. Regulation of flux through glutaminase and glutamine synthetase in isolated perfused rat liver. Biochimica et Biophysica Acta 1983; 755:272-8.
114. Haussinger D. Shifting of hepatic glutamine balance from uptake to output: Role of liver cell heterogeneity [editorial]. Nutrition 1990; 6:412-3.
115. Moorman AF, de Boer PA, Watford M et al. Hepatic glutaminase mRNA is confined to part of the urea cycle domain in the adult rodent liver lobule. FEBS Letters 1994; 356:76-80.
116. Watford M. Hepatic glutaminase expression: relationship to kidney-type glutaminase and to the urea cycle. FASEB J 1993; 7:1468-74.
117. Haussinger D. Hepatocyte heterogeneity in glutamine and ammonia metabolism and the role of an intercellular glutamine cycle during ureogenesis in perfused rat liver. Euro J Biochem 1983; 133:269-75.
118. Haussinger D, Sies H, Gerok W. Functional hepatocyte heterogeneity in ammonia metabolism. The intercellular glutamine cycle. J Hepatol 1985; 1:3-14.

119. Haussinger D. Regulation of hepatic ammonia metabolism: the intercellular glutamine cycle. Adv Enzyme Reg 1986; 25:159-80.
120. Haussinger D, Soboll S, Meijer AJ et al. Role of plasma membrane transport in hepatic glutamine metabolism. European Journal of Biochemistry 1985; 152:597-603.
121. Lenzen C, Soboll S, Sies H et al. pH control of hepatic glutamine degradation. Role of transport. Euro J Biochem 1987; 166:483-8.
122. Inoue Y, Bode BP, Souba WW. Dietary regulation of the hepatic system n glutamine transporter in tumor-bearing rats. Am J Surg 1995; 169:173-8.
123. Pacitti AJ, Chen MK, Bland KI et al. Mechanisms of accelerated hepatic glutamine efflux in the tumour-bearing rat. Surg Oncol 1992; 1:173-82.
124. Bode B, Tamarappoo BK, Mailliard M et al. Characteristics and regulation of hepatic glutamine transport. JPEN 1990; 14:51S-55S.
125. Dudrick PS, Bland KI, Copeland EM et al. Hepatocyte glutamine transport in advanced malignant disease. Surgical Forum 1992; 43:13-15.
126. Dudrick PS, Inoue Y, Espat NJ et al. Na(+)-dependent glutamine transport in the liver of tumor-bearing rats. Surg Oncol 1993; 2:205-15.
127. Argiles JM, Lopez-Soriano FJ. The energy state of tumor-bearing rats. J Biol Chem 1991; 266:2978-82.
128. Espat NJ, Bode BP, Lind DS et al. Normalization of tumor-induced increases in hepatic amino acid transport after surgical resection. Ann Surg 1995; 221:50-8.
129. Inoue Y, Bode BP, Copeland EM et al. Enhanced hepatic amino acid transport in tumor-bearing rats is partially blocked by antibody to tumor necrosis factor. Cancer Research 1995; 55:3525-30.
130. Fischer CP, Bode BP, Souba WW. A sarcoma-derived protein regulates hepatocyte metabolism via autocrine production of tumor necrosis factor-alpha. Ann Surg 1996; 224:476-85.
131. Tsuburaya A, Blumberg D, Burt M et al. Energy depletion in the liver and in isolated hepatocytes of tumor-bearing animals. J Surg Rese 1995; 59:421-7.
132. Burt ME, Lowry SF, Gorschboth C et al. Metabolic alterations in a noncachectic animal tumor system. Cancer 1981; 47:2138-46.
133. Roh MS, Ekman LG, Jeevanandam M et al. Elevated energy expenditure in hepatocytes from tumor-bearing rats. J Surg Res 1985; 38:407-15.
134. Noguchi Y, Vydelingum NA, Brennan MF. The reversal of increased gluconeogenesis in the tumor-bearing rat by tumor removal and food intake. Surgery 1989; 106:423-30; discussion 430-1.
135. Blumberg D, Burt ME, Kher U et al. Tolbutamide inhibits gluconeogenesis in the tumor-influenced hepatocyte. J Surg Res 1993; 55:565-9.
136. Lowry SF, Norton JA, Gorschboth CM et al. Glucose turnover and gluconeogenesis during hypocaloric glucose infusion in tumor-bearing F344 male rats. J Natl Cancer Inst 1980; 64:291-6.
137. Lowry SF, Norton JA, Brennan MF. Glucose turnover and gluconeogenesis from alanine in tumor-bearing animals. Surgical Forum 1978; 29:140-2.
138. Roh MS, Ekman L, Jeevanandam M et al. Gluconeogenesis in tumor-influenced hepatocytes. Surgery 1984; 96:427-433.
139. Warren RS, Jeevanandam M, Brennan MF. Comparison of hepatic protein synthesis in vivo versus in vitro in the tumor-bearing rat. J Surg Res 1987; 42:43-50.
140. Rossi-Fanelli F, Cascino A, Muscaritoli M. Abnormal substrate metabolism and nutritional strategies in cancer management. JPEN 1991; 15:680-3.
141. Tayek JA, Katz J. Glucose production, recycling, Cori cycle, and gluconeogenesis in humans: relationship to serum cortisol. Am J Physiol 1997; 272:E476-84.
142. Heber D, Byerly LO, Chlebowski RT. Metabolic abnormalities in the cancer patient. Cancer 1985; 55:225-9.
143. Levin L, Gevers W, Jardine L et al. Serum amino acids in weight-losing patients with cancer and tuberculosis. Euro J Cancer Clin Oncol 1983; 19:711-5.
144. Lundholm K, Edstrom S, Ekman L et al. Metabolism in peripheral tissues in cancer patients. Cancer Treatment Reports 1981; 65:79-83.
145. Brennan MF, Burt ME. Nitrogen metabolism in cancer patients. Cancer Treatment Reports 1981; 65:67-78.
146. Clarke EF, Lewis AM, Waterhouse C. Peripheral amino acid levels in patients with cancer. Cancer 1978; 42:2909-13.

147. Chance WT, Cao L, Foley-Nelson T et al. Possible role of ammonia in experimental cancer anorexia. Brain Research 1989; 486:316-24.
148. Chance WT, Cao LQ, Fischer JE. Tumor-induced alterations in brain neurotransmitter and plasma ammonia concentrations are normalized twenty-four hours after tumor resection. Life Sciences 1991; 48:425-32.
149. Denno R, Rounds JD, Faris R et al. Glutamine-enriched total parenteral nutrition enhances plasma glutathione in the resting state. J Surg Res 1996; 61:35-8.
150. Denno R, Rounds JD, Faris R et al. The enhanced effect of parenteral nutrition on hepatotoxicity [see comments]. Nutrition 1996; 12:30-5.
151. Haussinger D, Lang F, Bauers K et al. Interactions between glutamine metabolism and cell-volume regulation in perfused rat liver. Euro J Biochem 1990; 188:689-95.
152. Wettstein M, vom Dahl S, Lang F et al. Cell volume regulatory responses of isolated perfused rat liver. The effect of amino acids. Biol Chem Hoppe-Seyler 1990; 371:493-501.
153. Haussinger D, Lang F. Exposure of perfused liver to hypotonic conditions modifies cellular nitrogen metabolism. J Cell Biochem 1990; 43:355-61.
154. Hallbrucker C, vom Dahl S, Lang F et al. Control of hepatic proteolysis by amino acids. The role of cell volume. Euro J Biochem 1991; 197:717-24.
155. Haussinger D, Stoll B, Morimoto Y et al. Anisoosmostic liver perfusion: Redox shifts and modulation of alpha-ketoisocaproate and glycine metabolism. Biological Chemistry Hoppe-Seyler 1992; 373:723-34.
156. Stoll B, Gerok W, Lang F et al. Liver cell volume and protein synthesis. Biochem J 1992; 287:217-22.
157. Vom Dahl S, Haussinger D. Nutritional state and the swelling-induced inhibition of proteolysis in perfused rat liver. J Nutr 1996; 126:395-402.
158. Bode BP, Kilberg MS. Amino acid-dependent increase in hepatic system N activity is linked to cell swelling. J Biol Chem 1991; 266:7376-81.
159. Haussinger D, Gerok W, Lang F. Cell Volume and Hepatic Metabolism. In: Lang F and Häussinger D, ed. Interaction of Cell Volume and Cell Function. New York: Springer-Verlag, 1993:33-65.
160. Haussinger D, Hallbrucker C, vom Dahl S et al. Cell volume is a major determinant of proteolysis control in liver. FEBS Letters 1991; 283:70-2.
161. Newsome WP, Warskulat U, Noe B et al. Modulation of phosphoenolpyruvate carboxykinase mRNA levels by the hepatocellular hydration state. Biochem J 1994; 304:555-60.
162. Klimberg VS, Pappas AA, Nwokedi E et al. Effect of supplemental dietary glutamine on methotrexate concentrations in tumors. Arch Surg 1992; 127:1317-20.
163. Morris HP, Slaughter LJ. Historical development of transplantable hepatomas. Adv Experi Med Biol 1977; 92:1-20.
164. Matsuno T, Goto I. Glutaminase and glutamine synthetase activities in human cirrhotic liver and hepatocellular carcinoma. Cancer Research 1992; 52:1192-4.
165. Matsuno T. Pathway of glutamate oxidation and its regulation in the HuH13 line of human hepatoma cells. J Cellular Physiol 1991; 148:290-4.
166. Mares-Perlman JAaS, E. Energy substrate utilization in freshly isolated Morris Hepatoma 7777 cells. Cancer Research 1988; 48:602-608.
167. Bode BP, Souba WW. Modulation of cellular proliferation alters glutamine transport and metabolism in human hepatoma cells. Ann Surg 1994; 220:411-22; discussion 422-4.
168. Reardon MA, Weber G. Increased synthesis of carbamoyl-phosphate synthase II (EC 6.3.5.5) in hepatoma 3924A. Cancer Research 1986; 46:3673-6.
169. Prajda N, Katunuma N, Morris HP et al. Imbalance of purine metabolism in hepatomas of different growth rates as expressed in behavior of glutamine-phosphoribosylpyrophosphate amidotransferase (amidophosphoribosyltransferase, EC 2.4.2.14). Cancer Research 1975; 35:3061-8.
170. Elliott WL, Weber G. Proliferation-linked increase in phosphoribosylformylglycinamidine synthetase activity (EC 6.3.5.3). Cancer Research 1984; 44:2430-4.
171. Denton JE, Lui MS, Aoki T et al. Rapid in vivo inactivation by acivicin of CTP synthetase, carbamoyl-phosphate synthetase II, and amidophosphoribosyltransferase in hepatoma. Life Sciences 1982; 30:1073-80.
172. Aoki T, Weber G. Carbamoyl phosphate synthetase (glutamine-hydrolyzing): Increased activity in cancer cells. Science 1981; 212:463-5.

173. Sebolt JS, Weber G. Negative correlation of L-glutamine concentration with proliferation rate in rat hepatomas. Life Sciences 1984; 34:301-6.
174. Linder-Horowitz M, Knox WE, Morris HP. Glutaminase activities and growth rates of rat hepatomas. Cancer Research 1969; 29:1195-1199.
175. Vadgama JVaC HN. Comparison of System N in fetal hepatocytes and related cell lines. J Biol Chem 1983; 258:6422-6429.
176. Bergstrom J, Furst P, Noree LO et al. Intracellular free amino acid concentration in human muscle tissue. J Appl Physiol 1974; 36:693-7.
177. Newsholme EAaP-B, M. Properties of glutamine release from muscle and its importance for the immune system. JPEN 1990; 14 (Supplement):63S-67S.
178. Furst P, Albers S. Stehle P. Stress-induced intracellular glutamine depletion. The potential use of glutamine-containing dipeptides in parenteral nutrition. Contr Infusion Therapy Clin Nutr 1987; 17:117-136.
179. Kinscherf R, Hack V, Fischbach T et al. Low plasma glutamine in combination with high glutamate levels indicate risk for loss of body cell mass in healthy individuals: the effect of N-acetyl-cysteine. J Molecular Med 1996; 74:393-400.
180. Pisters PW, Pearlstone DB. Protein and amino acid metabolism in cancer cachexia: Investigative techniques and therapeutic interventions. Crit Rev Clin Lab Sci 1993; 30:223-72.
181. Bennegard K, Lindmark L, Eden E et al. Flux of amino acids across the leg in weight-losing cancer patients. Cancer Research 1984; 44:386-93.
182. Pisters PWT, Brennan MF, eds. Protein and Amino Acid Metabolism in Cancer Cachexia. Austin, TX: R.G. Landes Co., 1996.
183. Karner J, Roth E, Ollenschlager G et al. Glutamine-containing dipeptides as infusion substrates in the septic state. Surgery 1989; 106:893-900.
184. Millward DJ, Jepson MM, Omer A. Muscle glutamine concentration and protein turnover in vivo in malnutrition and in endotoxemia. Metabolism: Clinical & Experimental 1989; 38:6-13.
185. Jepson MM, Bates PC, Broadbent P et al. Relationship between glutamine concentration and protein synthesis in rat skeletal muscle. Am J Physiol 1988; 255:E166-72.
186. Rennie MJ, MacLennan PA, Hundal HS et al. Skeletal muscle glutamine transport, intramuscular glutamine concentration, and muscle-protein turnover. Metabolism: Clinical & Experimental 1989; 38:47-51.
187. Michie HR. Metabolism of sepsis and multiple organ failure. World J Surg 1996; 20:460-4.
188. James JH, Hasselgren PO, King JK et al. Intracellular glutamine concentration does not decrease in all muscles during sepsis. J Surg Res 1993; 54:558-64.
189. Parry-Billings M, Leighton B, Dimitriadis G et al. Skeletal muscle glutamine metabolism during sepsis in the rat. Intl J Biochem 1989; 21:419-23.
190. Petersson B, Waller SO, Vinnars E et al. Long-term effect of glycyl-glutamine after elective surgery on free amino acids in muscle. JPEN 1994; 18:320-5.
191. Essen P, Wernerman J, Sonnenfeld T et al. Free amino acids in plasma and muscle during 24 hours post-operatively—a descriptive study. Clin Physiol 1992; 12:163-77.
192. Vinnars E, Bergstom J, Furst P. Influence of the postoperative state on the intracellular free amino acids in human muscle tissue. Ann Surg 1975; 182:665-71.
193. Sandstedt S, Lennmarken C, Symreng T et al. The effect of pre-operative total parenteral nutrition on energy-rich phosphates, electrolytes and free amino acids in skeletal muscle of malnourished patients with gastric carcinoma. Brit J Surg 1985; 72:920-4.
194. Hammarqvist F, Wernerman J, Ali R et al. Addition of glutamine to total parenteral nutrition after elective abdominal surgery spares free glutamine in muscle, counteracts the fall in muscle protein synthesis, and improves nitrogen balance. Ann Surg 1989; 209:455-61.
195. Wernerman J, Hammarkvist F, Ali MR et al. Glutamine and ornithine-alpha-ketoglutarate but not branched-chain amino acids reduce the loss of muscle glutamine after surgical trauma. Metabolism: Clinical & Experimental 1989; 38:63-6.
196. Hammarqvist F, Wernerman J, Ali R et al. Effects of an amino acid solution enriched with either branched chain amino acids or ornithine-alpha-ketoglutarate on the postoperative intracellular amino acid concentration of skeletal muscle. Brit J Surg 1990; 77:214-8.
197. Hammarqvist F, Wernerman J, von der Decken A et al. Alanyl-glutamine counteracts the depletion of free glutamine and the postoperative decline in protein synthesis in skeletal muscle. Ann Surg 1990; 212:637-44.

198. Wernerman J, Hammarqvist F, Vinnars E. Alpha-ketoglutarate and postoperative muscle catabolism. Lancet 1990; 335:701-3.
199. Hammarqvist F, Wernerman J, von der Decken A et al. Alpha-ketoglutarate preserves protein synthesis and free glutamine in skeletal muscle after surgery. Surgery 1991; 109:28-36.
200. Petersson B, Vinnars E, Waller SO et al. Long-term changes in muscle free amino acid levels after elective abdominal surgery. Brit J Surg 1992; 79:212-6.
201. Palmer TE, Griffiths RD, Jones C. Effect of parenteral L-glutamine on muscle in the very severely ill. Nutrition 1996; 12:316-20.
202. Carli F, Webster J, Ramachandra V et al. Aspects of protein metabolism after elective surgery in patients receiving constant nutritional support. Clinical Science 1990; 78:621-8.
203. Van Leeuwen PA, Bading JR, Vydelingum NA et al. Organ blood flow in Fischer-344 rats bearing MCA-induced sarcoma. J Appl Physiol 1991; 71:1674-8.
204. Rennie MJ, Edwards RH, Emery PW et al. Depressed protein synthesis is the dominant characteristic of muscle wasting and cachexia. Clin Physiol 1983; 3:387-98.
205. Emery PW, Lovell L, Rennie MJ. Protein synthesis measured in vivo in muscle and liver of cachectic tumor-bearing mice. Cancer Research 1984; 44:2779-84.
206. MacLennan PA, Brown RA, Rennie MJ. A positive relationship between protein synthetic rate and intracellular glutamine concentration in perfused rat skeletal muscle. FEBS Letters 1987; 215:187-91.
207. Rennie MJ, Ahmed A, Khogali SE et al. Glutamine metabolism and transport in skeletal muscle and heart and their clinical relevance. J Nutr 1996; 126:1142S-9S.
208. Anonymous. Glutamine transport in muscle protein economy. Nutr Rev 1989; 47:215-7.
209. Hundal HS, Rennie MJ, Watt PW. Characteristics of L-glutamine transport in perfused rat skeletal muscle. J Physiol 1987; 393:283-305.
210. Ahmed A, Taylor PM, Rennie MJ. Characteristics of glutamine transport in sarcolemmal vesicles from rat skeletal muscle. Am J Physiol 1990; 259:E284-91.
211. Rennie MJ, Willhoft NM, Taylor PM. Glutamine transport and metabolism in mammalian skeletal muscle [letter]. Biochem J 1992; 285:339-40.
212. Tadros LB, Taylor PM, Rennie MJ. Characteristics of glutamine transport in primary tissue culture of rat skeletal muscle. Am J Physiol 1993; 265:E135-44.
213. Tadros LB, Willhoft NM, Taylor PM et al. Effects of glutamine deprivation on glutamine transport and synthesis in primary culture of rat skeletal muscle. Am J Physiol 1993; 265:E935-42.
214. Low SY, Taylor PM, Rennie MJ. Responses of glutamine transport in cultured rat skeletal muscle to osmotically induced changes in cell volume. J Physiol 1996; 492:877-85.
215. Ahmed A, Gibson JN, Taylor PM et al. Isolation of human skeletal muscle sarcolemmal vesicles for the investigation of glutamine transport. Biochem Soc Trans 1990; 18:1238-9.
216. Ahmed A, Maxwell DL, Taylor PM et al. Glutamine transport in human skeletal muscle. American J Physiol 1993; 264:E993-1000.
217. Hundal HS, Taylor PM, Willhoft NM et al. A role for membrane transport in modulation of intramuscular free glutamine turnover in streptozotocin diabetic rats. Biochimica et Biophysica Acta 1992; 1180:137-46.
218. Hundal HS, Babij P, Taylor PM et al. Effects of corticosteroid on the transport and metabolism of glutamine in rat skeletal muscle. Biochimica et Biophysica Acta 1991; 1092:376-83.
219. Rennie MJ, Khogali SE, Low SY et al. Amino acid transport in heart and skeletal muscle and the functional consequences. Biochem Soci Trans 1996; 24:869-73.
220. Jiang ZM, Wang LJ, Qi Y et al. Comparison of parenteral nutrition supplemented with L-glutamine or glutamine dipeptides. JPEN 1993; 17:134-41.
221. Chance WT, Muggia-Sullam M, Chen MH et al. Reversal of tumor-induced biochemical abnormalities by insulin treatment in rats. J Natl Cancer Inste 1986; 77:497-503.
222. Schein PS, Kisner D, Haller D et al. Cachexia of malignancy: Potential role of insulin in nutritional management. Cancer 1979; 43:2070-6.
223. Pearlstone DB, Wolf RF, Berman RS et al. Effect of systemic insulin on protein kinetics in postoperative cancer patients. Ann Surg Oncol 1994; 1:321-32.
224. Shewchuk LD, Baracos VE, Field CJ. Dietary L-glutamine supplementation reduces the growth of the Morris Hepatoma 7777 in exercise-trained and sedentary rats. J Nutr 1997; 127:158-66.
225. Petersson B, Hultman E, Andersson K et al. Human skeletal muscle protein: effect of malnutrition, elective surgery and total parenteral nutrition. Clin Sci 1995; 88:479-84.

226. Le Bricon T, Cynober L, Field CJ et al. Supplemental nutrition with ornithine alpha-ketoglutarate in rats with cancer-associated cachexia: Surgical treatment of the tumor improves efficacy of nutritional support. J Nutr 1995; 125:2999-3010.
227. Kaibara A, Yoshida S, Yamasaki K et al. Effect of glutamine and chemotherapy on protein metabolism in tumor-bearing rats. J Surg Res 1994; 57:143-9.
228. Souba WW, Austgen TR. Interorgan glutamine flow following surgery and infection. JPEN 1990; 14:90S-93S.
229. Petersson B, von der Decken A, Vinnars E et al. Long-term effects of postoperative total parenteral nutrition supplemented with glycylglutamine on subjective fatigue and muscle protein synthesis. Brit J Surg 1994; 81:1520-3.
230. Blomqvist BI, Hammarqvist F, von der Decken A et al. Glutamine and alpha-ketoglutarate prevent the decrease in muscle free glutamine concentration and influence protein synthesis after total hip replacement. Metabolism: Clinical & Experimental 1995; 44:1215-22.
231. Windmueller HG, Spaeth AE. Uptake and metabolism of plasma glutamine by the small intestine. J Biol Chem 1974; 249:5070-9.
232. Windmueller HG, Spaeth AE. Intestinal metabolism of glutamine and glutamate from the lumen as compared to glutamine from blood. Arch Biochem Biophys 1975; 171:662-72.
233. Windmueller HG, Spaeth AE. Identification of ketone bodies and glutamine as the major respiratory fuels in vivo for postabsorptive rat small intestine. J Biol Chem 1978; 253:69-76.
234. Kekuda R, Prasad PD, Fei YJ et al. Cloning of the sodium-dependent, broad-scope, neutral amino acid transporter Bo from a human placental choriocarcinoma cell line. J Biol Chem 1996; 271:18657-61.
235. al-Mahroos FT, Abumrad N, Ghishan FK. Developmental changes in glutamine transport by rat jejunal basolateral membrane vesicles. Proc Soc Exper Biol Med 1990; 194:186-92.
236. Taylor PM, Egan CJ, Rennie MJ. Transport of glutamine across blood-facing membranes of perfused rat jejunum. Am J Physiol 1989; 256:E550-8.
237. Ghishan FK, Sutter W, Said H et al. Glutamine transport by rat basolateral membrane vesicles. Biochimica et Biophysica Acta 1989; 979:77-81.
238. Ghishan FK, Arab N, Bulus N et al. Glutamine transport by human intestinal basolateral membrane vesicles. Am J Clin Nutr 1990; 51:612-6.
239. Wilde SW, Kilberg MS. Glutamine transport by basolateral plasma-membrane vesicles prepared from rabbit intestine. Biochem J 1991; 277:687-91.
240. Hartmann F, Plauth M. Intestinal glutamine metabolism. Metabolism 1989; 33:18-24.
241. Wilmore DW, Smith RJ, O'Dwyer ST et al. The gut: a central organ after surgical stress. Surgery 1988; 104:917-23.
242. Alverdy JC. Effects of glutamine-supplemented diets on immunology of the gut. JPEN 1990; 14:109S-113S.
243. van der Hulst RR, van Kreel BK, von Meyenfeldt MF et al. Glutamine and the preservation of gut integrity [see comments]. Lancet 1993; 341:1363-5.
244. Salloum RM, Copeland EM, Bland KI et al. Selective stimulation of brush border glutamine transport in the tumor-bearing rat. J Surg Res 1991; 50:391-7.
245. Salloum RM, Souba WW, Fernandez A et al. Dietary modulation of small intestinal glutamine transport in intestinal brush border membrane vesicles of rats. J Surg Res 1990; 48:635-8.
246. Phillips MC, Olson LR. The immunologic role of the gastrointestinal tract. Crit Care Nurs Clin N.A. 1993; 5:107-20.
247. Alverdy JC, Aoys E, Moss GS. Total parenteral nutrition promotes bacterial translocation from the gut. Surgery 1988; 104:185-90.
248. Mestecky J, Russell MW, Jackson S et al. The human IgA system: a reassessment. Clin Immunol Immunopathol 1986; 40:105-14.
249. Alverdy JC, Chi HS, Selivanov V et al. The effect of route of nutrient administration on the secretory immune system. Curr Surg 1985; 42:10-3.
250. Li J, Langkamp-Henken B, Suzuki K et al. Glutamine prevents parenteral nutrition-induced increases in intestinal permeability [see comments]. JPEN 1994; 18:303-7.
251. Buchman AL, Moukarzel AA, Bhuta S et al. Parenteral nutrition is associated with intestinal morphologic and functional changes in humans. JPEN 1995; 19:453-60.
252. Buchman AL. Glutamine: is it a conditionally required nutrient for the human gastrointestinal system?[see comments]. J Am Coll Nutr 1996; 15:199-205.

253. Souba WW, Herskowitz K, Salloum RM et al. Gut glutamine metabolism. JPEN 1990; 14:45S-50S.
254. Alverdy JC. Effects of glutamine-supplemented diets on immunology of the gut. JPEN 1990; 14:109S-113S.
255. Verity S. Nutrition and its importance to intensive care patients. Intens Crit Care Nurs 1996; 12:71-8.
256. Souba WW, Herskowitz K, Austgen TR et al. Glutamine nutrition: theoretical considerations and therapeutic impact. JPEN 1990; 14:237S-243S.
257. Pisters PWT, Brennan MF. Enteral Nutrition in Cancer. In: Pisters PWTaB, M.F., ed. Protein and Amino Acid Metabolism in Cancer Cachexia. Austin, TX: R.G. Landes Co., 1996:133-138.
258. Alverdy J, Chi HS, Sheldon GF. The effect of parenteral nutrition on gastrointestinal immunity. The importance of enteral stimulation. Ann Surg 1985; 202:681-4.
259. Bozzetti F. Is enteral nutrition a primary therapy in cancer patients? Gut 1994; 35:S65-8.
260. Bai MX, Jiang ZM, Liu YW et al. Effects of alanyl-glutamine on gut barrier function. Nutrition 1996; 12:793-6.
261. Haque SM, Chen K, Usui N et al. Alanyl-glutamine dipeptide-supplemented parenteral nutrition improves intestinal metabolism and prevents increased permeability in rats. Ann Surg 1996; 223:334-41.
262. Schroder J, Wardelmann E, Winkler W et al. Glutamine dipeptide-supplemented parenteral nutrition reverses gut atrophy, disaccharidase enzyme activity, and absorption in rats. JPEN 1995; 19:502-6.
263. Okuma T, Kaneko H, Chen K et al. Total parenteral nutrition supplemented with L-alanyl-L-glutamine and gut structure and protein metabolism in septic rats. Nutrition 1994; 10:241-5.
264. Chen K, Okuma T, Okamura K et al. Glutamine-supplemented parenteral nutrition improves gut mucosa integrity and function in endotoxemic rats. JPEN 1994; 18:167-71.
265. Tamada H, Nezu R, Matsuo Y et al. Alanyl glutamine-enriched total parenteral nutrition restores intestinal adaptation after either proximal or distal massive resection in rats. JPEN 1993; 17:236-42.
266. Inoue Y, Grant JP, Snyder PJ. Effect of glutamine-supplemented total parenteral nutrition on recovery of the small intestine after starvation atrophy. JPEN1993; 17:165-70.
267. Burke DJ, Alverdy JC, Aoys E et al. Glutamine-supplemented total parenteral nutrition improves gut immune function. Arch Surg 1989; 124:1396-9.
268. van der Hulst RR, von Meyenfeldt MF, Deutz NE, S et al. The effect of glutamine administration on intestinal glutamine content. J Surg Res 1996; 61:30-4.
269. Turowski GA, Rashid Z, Hong F et al. Glutamine modulates phenotype and stimulates proliferation in human colon cancer cell lines. Cancer Research 1994; 54:5974-80.
270. Natsumeda Y, Lui MS, Emrani J et al. Purine enzymology of human colon carcinomas. Cancer Research 1985; 45:2556-9.
271. van der Hulst RR, von Meyenfeldt MF, Deutz NE et al. Glutamine extraction by the gut is reduced in patients with depleted gastrointestinal cancer. Ann Surg 1997; 225:112-21.
272. Klimberg VS, Souba WW, Dolson DJ et al. Prophylactic glutamine protects the intestinal mucosa from radiation injury. Cancer 1990; 66:62-8.
273. Campos FG, Waitzberg DL, Mucerino DR et al. Protective effects of glutamine enriched diets on acute actinic enteritis. Nutricion Hospitalaria 1996; 11:167-77.
274. McArdle AH. Protection from radiation injury by elemental diet: does added glutamine change the effect? Gut 1994; 35:S60-4.
275. Winters R, Matthews R, Ercal N et al. Glutamine protects Chinese hamster ovary cells from radiation killing. Life Sciences 1994; 55:713-20.
276. Wicke C, Gottwald T, Becker HD. Brief clinical report: Glutamine-enriched total parenteral nutrition in a patient with radiation-induced renal and intestinal fibrosis. Nutrition 1996; 12:S85-6.
277. Richards EW, Long CL, Pinkston JA et al. The role of oral glutamine supplementation in the prevention of radiation-induced enterocolitis in prostate cancer patients (Meeting abstract). FASEB J 1992; 6:A1680.
278. Scott TE, Moellman JR. Intravenous glutamine fails to improve gut morphology after radiation injury. JPEN 1992; 16:440-4.
279. Scheltinga MR, Young LS, Benfell K et al. Glutamine-enriched intravenous feedings attenuate extracellular fluid expansion after a standard stress. Ann Surg 1991; 214:385-95.
280. Skubitz KM, Anderson PM. Oral glutamine to prevent chemotherapy induced stomatitis: A pilot study. J Lab Clin Med 1996; 127:223-8.

281. Muscaritoli M, Conversano L, Petti MC et al. Oral glutamine supplementation and high-dose chemotherapy (HDCT)-induced gastrointestinal toxicity: a pilot study (Meeting abstract). J Invest Med 1995; 43:498A.
282. Ford C, Whitlock JA, Pietsch JB. Glutamine-supplemented tube feedings versus total parenteral nutrition in children receiving intensive chemotherapy. J Pediatr Oncol Nurs 1997; 14:68-72.
283. Muscaritoli M, Micozzi A, Conversano L et al. Oral glutamine in the prevention of chemotherapy-induced gastrointestinal toxicity [letter]. Euro J Cancer 1997; 33:319-20.
284. Shou J, Lieberman MD, Hofmann K et al. Dietary manipulation of methotrexate-induced enterocolitis. JPEN 1991; 15:307-12.
285. van Zaanen HC, van der Lelie H, Timmer JG et al. Parenteral glutamine dipeptide supplementation does not ameliorate chemotherapy-induced toxicity. Cancer 1994; 74:2879-84.
286. Newsholme EA, Crabtree B, Ardawi MS. The role of high rates of glycolysis and glutamine utilization in rapidly dividing cells. Bioscience Reports 1985; 5:393-400.
287. Newsholme EA. The possible role of glutamine in some cells of the immune system and the possible consequence for the whole animal. Experientia 1996; 52:455-9.
288. Newsholme EA, Newsholme P, Curi R. The role of the citric acid cycle in cells of the immune system and its importance in sepsis, trauma and burns. Biochem Soc Symposia 1987; 54:145-62.
289. Brand K, Leibold W, Luppa P et al. Metabolic alterations associated with proliferation of mitogen-activated lymphocytes and of lymphoblastoid cell lines: evaluation of glucose and glutamine metabolism. Immunobiology 1986; 173:23-34.
290. Ardawi MS, Newsholme EA. Glutamine metabolism in lymphocytes of the rat. Biochem J 1983; 212:835-42.
291. O'Rourke AM, Rider CC. Glucose, glutamine and ketone body utilization by resting and concanavalin A activated rat splenic lymphocytes. Biochimica et Biophysica Acta 1989; 1010:342-5.
292. Brand K, Fekl W, von Hintzenstern J et al. Metabolism of glutamine in lymphocytes. Metabolism: Clinical & Experimental 1989; 38:29-33.
293. Wu GY, Field CJ, Marliss EB. Glutamine and glucose metabolism in rat splenocytes and mesenteric lymph node lymphocytes. Am J Physiol 1991; 260:E141-7.
294. Almeida AF, Costa Rosa LF, Curi R. Glutamine metabolism is changed in lymphocytes from aged rats. Physiol Behav 1996; 60:973-8.
295. Ardawi MS, Majzoub MF, Kateilah SM et al. Maximal activity of phosphate-dependent glutaminase and glutamine metabolism in septic rats. J Lab Clin Med 1991; 118:26-32.
296. Newsholme P, Curi R, Gordon S et al. Metabolism of glucose, glutamine, long-chain fatty acids and ketone bodies by murine macrophages. Biochem J1986; 239:121-5.
297. Curi R, Newsholme P, Newsholme EA. Intracellular distribution of some enzymes of the glutamine utilization pathway in rat lymphocytes. Biochem Biophys Res Commun 1986; 138:318-22.
298. Baechtel FS, Gregg DE, Prager MD. The influence of glutamine, its decomposition products, and glutaminase on the transformation of human and mouse lymphocytes. Biochimica et Biophysica Acta 1976; 421:33-43.
299. Calder PC. Glutamine action in cells of the immune system. In: Cynober L, Furst, P. and Lawin, P., ed. Pharmacological Nutrition, Immune Nutrtion. New York: W. Zuckschwerdt Verlag GmbH, 1995:10-20.
300. Szondy Z, Newsholme EA. The effect of glutamine concentration on the activity of carbamoyl-phosphate synthase II and on the incorporation of [3H]thymidine into DNA in rat mesenteric lymphocytes stimulated by phytohaemagglutinin. Biochem J 1989; 261:979-83.
301. Griffiths M, Keast D. The effect of glutamine on murine splenic leukocyte responses to T and B cell mitogens. Immunol Cell Biol 1990; 68:405-8.
302. Chuang JC, Yu CL, Wang SR. Modulation of human lymphocyte proliferation by amino acids. Clin Exper Immunol 1990; 81:173-6.
303. Crawford J, Cohen HJ. The essential role of L-glutamine in lymphocyte differentiation in vitro. J Cell Physiol 1985; 124:275-282.
304. Calder PC, EA N. Glutamine promotes interleukin-2 production by concanavalin A-stimulated lymphocytes [abstract]. Proc Nutr Soc 1992; 41:105A.
305. Ardawi MS, Newsholme EA. The transport of glutamine into rat mesenteric lymphocytes. Biochimica et Biophysica Acta 1986; 856:413-20.
306. Piva TJ, McCabe MG, McEvoy-Bowe E. The kinetics of glutamine transport in bovine lymphocytes. Biochem Intl 1992; 26:309-16.

307. Schroder MT, Schafer G, Schauder P. Characterization of glutamine transport into resting and concanavalin A-stimulated peripheral human lymphocytes. J Cell Physiol 1990; 145:155-61.
308. Shewchuk LD, Baracos VE, Field CJ. Reduced splenocyte metabolism and immune function in rats implanted with the Morris Hepatoma 7777. Metabolism: Clinical & Experimental 1996; 45:848-55.
309. Klimberg VS, Kornbluth J, Cao Y et al. Glutamine suppresses PGE2 synthesis and breast cancer growth. J Surg Res 1996; 63:293-7.
310. Cynober L, Furst P, Lawin P, ed. Pharmacological Nutrition, Immune Nutrition (International Symposium, Nice, France, March 9-11, 1995). New York: W. Zuckschwerdt Verlag, 1995.
311. Babst R, Horig H, Stehle P et al. Glutamine peptide-supplemented long-term total parenteral nutrition: effects on intracellular and extracellular amino acid patterns, nitrogen economy, and tissue morphology in growing rats [published erratum appears in JPEN 1994 Mar-Apr; 18(2):197]. JPEN 1993; 17:566-74.
312. Ziegler TR, Young LS, Benfell K et al. Glutamine-supplemented parenteral nutrition improves nitrogen retention, and reduces hospital mortality versus standard parenteral nutrition following bone marrow transplantation: a randomized, double-blind trial. Ann Int Med 1992; 116:821-828.
313. Poster DS, Bruno S, Penta J et al. Acivicin. An antitumor antibiotic. Cancer Clin Trials 1981; 4:327-30.
314. Ahluwalia GS, Grem JL, Hao Z et al. Metabolism and action of amino acid analog anti-cancer agents. Pharmacol Therap 1990; 46:243-71.
315. Kisner DL, Catane R, Muggia FM. The rediscovery of DON (6-diazo-5-oxo-L-norleucine). Recent Results in Cancer Research 1980; 74:258-63.
316. Catane R, Von Hoff DD, Glaubiger DL et al. Azaserine, DON, and azotomycin: three diazo analogs of L-glutamine with clinical antitumor activity. Cancer Treatment Reports 1979; 63:1033-8.
317. Baskerville A, Hambleton B, Benbough JE Pathologic features of glutaminase toxicity. Brit J Exper Pathol 1980; 61:132-138.
318. Williams MG, Earhart RH, Bailey H, McGovren JP. Prevention of central nervous system toxicity of the antitumor antibiotic acivicin by concomitant infusion of an amino acid mixture. Cancer Research 1990; 50:5475-80.
319. Sullivan MP, Nelson JA, Feldman S et al. Pharmacokinetic and phase I study of intravenous DON (6-Diazo-5-Oxo-L-Norleucine) in children. Cancer Chemother Pharmacol 1988; 21:78-84.
320. Popp MB, Enrione EB, Wagner SC et al. Influence of total nitrogen, asparagine, and glutamine on MCA tumor growth in the Fischer 344 rat. Surgery 1988; 104:152-60.
321. Stewart BW. Mechanisms of apoptosis: Integration of genetic biochemical and cellular indicators. (Review). J Natl Cancer Inst 1994; 86:1286-1296.

Chapter 4

Nutritional Support in Patients with Head and Neck Cancer

Matthew E. Cohen and Rosemarie L. Fisher

Patients with head and neck cancer share many nutritional support issues encountered in most patients with cancer, yet possess unique nutritional challenges due to the location of their cancers in the proximal digestive tract. In this chapter, the following topics are reviewed:

1. risk factors for malnutrition;
2. the relationship between malnutrition and clinical outcome;
3. the impact of enteral and parenteral nutritional support around the time of surgery, radiotherapy or chemotherapy; and
4. methods of delivering enteral nutritional support.

Most studies of nutritional support have been retrospective, and many have suffered from inadequate experimental design, heterogeneous or small groups of patients, or inappropriate endpoints.[1] In addition, many nutritional studies performed in patients with head and neck cancer were descriptive without statistical analyses. Thus, comparisons of studies are, at times, limited.

Risk Factors for Malnutrition

Forty-[2] to sixty-percent[3] of patients with head and neck cancer are malnourished at presentation. There are many possible reasons for this high prevalence of malnutrition, including advanced age, alcohol or tobacco abuse, or dysphagia from the tumor site, as well as psychosocial factors such as concomitant depression or inadequate social support. Data in patients with head and neck cancer are mixed, however, regarding the contribution of these characteristics to the risk of developing malnutrition (Table 4.1).

While some researchers have found a trend toward age being an independent risk factor for malnutrition,[4] others have not.[3,5] Despite the widely held assumption that many patients maintain a marginal nutritional status at baseline because of their unhealthy habits, smoking and alcohol consumption failed to correlate with worse nutritional status in at least two studies.[3,4] A third study found a correlation between pack-years of smoking and better nutritional status, but the implications of this relationship are not known.[5]

Psychosocial factors may play a role in the nutritional status of patients with head and neck cancer. For example, depression may be more common in patients with malnutrition than in those without malnutrition. Westin[6] performed nutritional assessments and psychopathological ratings on 53 patients with various head and neck tumor sites, stages, therapeutic

Nutritional Support in Cancer and Transplant Patients, edited by Rifat Latifi and Ronald C. Merrell. ©2001 Eurekah.com.

Table 4.1. Possible risk factors for malnutrition in patients with head and neck cancer

Variable	Risk Factor?	
Age	Trend[4]	
	No[3, 5]	
Depression	Yes[6]	
Marriage	Yes (if patient >= 60 years old)[5]	
	No (if patient < 60 years old)[5]	
Smoking	No[3-5]	
Tobacco	No[3-5]	
Tumor Site	Yes	Oral cavity/oropharynx/hypopharynx cervical esophagus >larynx/ nasopharynx/paranasal sinuses[2]
	Yes	Oropharynx/hypopharynx > larynx[3]
	No[9]	
Tumor stage	Yes[4]	
	Trend[3]	
	No[9]	

modalities and points in oncologic therapy. While no well-nourished patient was depressed, 30% of the 16 malnourished patients were depressed, which was a statistically significant difference. The five depressed patients had completed their therapies more than one year previously and were admitted because of suspected or known cancer recurrence. The presence or absence of a spouse is another psychosocial variable that may influence the nutritional status of patients with head and neck cancer. In patients less than 60 years old with head and neck cancer, being married decreased the risk of malnutrition, while in those over 60 years of age, being married increased the risk of malnutrition.[5] The source of this discrepancy was not discussed.

Another possible reason for malnutrition in patients with head and neck cancer is the location of the tumor. Oropharyngeal cancer may cause anorexia, nausea, inadequate mastication, xerostomia, dysgeusia, dysphagia or odynophagia.[7] Diminished oral intake and avoidance of firm solids correlated with malnutrition.[4] Maintained oral intake, however, did not prevent weight loss in all cases,[4] perhaps due to tumor-induced metabolic alterations which favor tumor growth at the host's expense.[8] Some studies of patients with head and neck cancer noted that tumors located within the upper digestive tract (but not in the upper respiratory tract) predicted malnutrition (Table 4.1). As with tumor site, tumor stage may correlate with malnutrition (Table 4.1). Other studies, in contrast, found that neither tumor site nor stage predicted nutritional status.[9] Matthews, however, did find a correlation between tumor stage and weight loss, which has predicted poor nutritional status in some[5] (but not other[10]) studies.

Depressed cellular immune response is often attributed at least in part to malnutrition in patients with cancer. However, age (which affects immune response[5]) and nutritional status were not investigated in most studies. In one study which did compare well-nourished to malnourished patients with head and neck cancer, anergy to all seven skin tests and a suppressed in vitro purified lymphocyte response to one of the stimulants (concanavalin A) correlated with malnutrition.[5] In another study, patients with localized head and neck cancer were skin-tested sequentially with DNCB and four common antigens.[11] DNCB reactivity correlated significantly with disease-free survival at six months, one year and four years. In contrast, skin test antigen reactivity did not correlate with clinical course. Any impairment in the immune response was thought to be from deficits other than nutritional, however, since all of the study subjects were outpatients and none had extreme cachexia.[11]

In conclusion, about one-half of patients with head and neck cancer present malnourished, which likely results more from tumor-induced metabolic alterations than from any preexisting malnutrition from tobacco or alcohol abuse. In general, both advanced head and neck cancer and location of tumor within the digestive (versus respiratory) tract appear to increase the risk of malnutrition. Depressed delayed hypersensitivity responses reflect poor nutritional status, but may be as dependent on advanced tumor burden. It is unclear if depression among patients with head and neck cancer causes malnutrition or simply correlates with recurrent disease.

Malnutrition and Clinical Outcome

Some studies of patients with head and neck cancer have found correlations between malnutrition and increased postoperative morbidity,[5,12,13] mortality,[5,13] length of hospitalization,[5] and decreased survival at two years.[2] Others have found no independent correlation between any nutritional parameter and incidence of postoperative complications or death.[9]

Hooley[12] prospectively calculated the Prognostic Nutritional Index (Table 4.2) in 29 patients with head and neck cancer 48 hours before surgery. All patients had completed preoperative high-dose radiotherapy. The more malnourished patients had a greater apparent risk of a major postoperative complications. No consideration was given to the site of the head and neck cancer or to patient comorbidity.

Linn[5] prospectively evaluated 79 men who had surgery for head and neck cancer, using his Protein Energy Malnutrition Scale (Table 4.2).[14] Malnourished elderly patients had the worst surgical outcomes. Potential reservations about the study included the existence of significant differences between the cancer types in the malnourished versus well-nourished groups and uncontrolled preoperative nutritional support (given in 60% of younger malnourished patients and 20% of older malnourished patients).

Goodwin[13] retrospectively studied 50 consecutive patients with stage III, IV or recurrent squamous cell carcinoma of the head and neck, 47 of whom had a variety of treatments, including induction chemotherapy, surgery, and/or radiation. Treatment-related complications in the 14 patients with severe malnutrition based on the Prognostic Nutritional Index were always major and more frequent, compared to the 36 patients with no or mild malnutrition. Statistical analysis was reported only for the 38 patients having surgery, nine of whom were severely malnourished and suffered significantly more morbidity and mortality. There was no attempt, however, to demonstrate a risk of malnutrition independent of tumor or treatment variables.

Brookes[2] prospectively followed 114 patients with untreated squamous cell cancer of the head and neck, and found that a General Nutritional Status (Table 4.2) of less than -10% (undernutrition) correlated with poorer survival, where a life table analysis excluding those patients who received intensive nutritional support showed a 58% survival rate of the adequately nourished patients at two years compared to an 8% survival rate among the undernourished patients. The authors claimed that this correlation was irrespective of tumor site, stage, histology or age, although statistical analyses of these data were not presented.

Matthews[9] prospectively studied 42 patients with newly diagnosed upper aerodigestive squamous cell carcinoma (31 of whom had cancer of the oral cavity, oropharynx, or hypopharynx) who subsequently had surgery with or without radiation therapy. There was a 38% incidence of minor complications, and a 10% incidence of major complications. The study was limited by incomplete compilation of data in the Subjective Global Assessment of Nutritional Status (Table 4.2), varied tumor sites and non-standardized surgical procedures.

Table 4.2. Nutritional indexes used to assess patients with head and neck cancer

Protein Energy Malnutrition Scale[14]

Note that the score of each item ranges along a geometric scale.

Score	**1**	**2**	**4**	**8**
Clinical History				
Inadequate nutrient intake	None	Mild	Moderate	Severe
Excessive nutrient losses	None	Mild	Moderate	Severe
Increased metabolic needs	None	Mild	Moderate	Severe
Anti-nutrient or catabolic medications	None	Mild	Moderate	Severe
Physical Examination				
Cachexia	None	Mild	Moderate	Severe
Hair easily pluckable or nails brittle/ridged	None	Mild	Moderate	Severe
Hepatomegaly or ascites	None	Mild	Moderate	Severe
Muscle atrophy		None	Mild	Moderate
Severe				
Generalized edema	None	Mild	Moderate	Severe
Dry skin, scaling skin, or skin lesions	None	Mild	Moderate	Severe
Anthropometric				
Relation to ideal body weight (%)	>=90	85-89	80-84	<80
Weight loss from usual weight (%)	<=5	6-12	13-19	>=20
Triceps skin fold (mm)				
Men	>=9.0	7.0-8.9	5.0-6.9	<5.0
Women	>=16.0	11.0-15.9	6.0-10.9	<6
Mid-arm muscle circumference (mm)				
Men	>=243	216-242.9	189-215.9	<189
Women	>=192	170-191.9	148-169.9	<148
Laboratory				
Albumin (gm/dL)	>=3.5	2.8-3.4	2.1-2.7	<2.1
Hemoglobin (gm/dL)	>=14.0	13.9-12.0	11.9-10.0	<10.0
Delayed hypersensitivity skin tests (of four)	>=2>5mm	1>5mm	1<5mm	Anergy
Lymphocytes (cells/mL)	>=1500	1200-1500	1000-1200	<1000
Creatinine excretion index (%standard)	>=80	60-79	40-59	<40
Transferrin (mg/dL)	>=200	150-199	100-149	<100
Retinol-binding protein (mg/dL)	>=3.0	2.5-2.9	2.0-2.4	<2.0
Pre-albumin (mg/dl)	>=15	12.5-14.9	10.0-12.4	<10.0
Negative nitrogen balance (g/day)	<=5.0	5.1-10.0	10.1-15.0	>15.0

Prognostic Nutritional Index (PNI)[12]

PNI% = 158% - 16.6(ALB) - 0.78 (TSF) - 0.2(TFN) -5.8(DH)

ALB = albumin (g/dL); TSF = average of three triceps skin fold measurements (mm); TFN = serum transferrin (mg/dL); DH = number of positive delayed hypersensitivity responses measured at 24 and 48 hours after intradermal injection of five antigens (*Candida albicans*, mumps, tuberculin purified protein derivative, *Trichophyton*, and streptokinase-streptodornase). Major post-operative complications occurred in patients with a PNI > 20.[12] In another study, a PNI > 40 indicated a high risk of developing a post-operative infection.[64]

continued on next page

Table 4.2. (cont'd)

General Nutrition Status (GNS)[2]

$$GNS = \frac{P\% + I\% + A}{3}$$

P% = percentage weight change from pre-morbid weight; I% = percentage of ideal weight; A% = percentage arm muscle circumference (AMC) change; AMC = mid-point non-dominant upper arm circumference - (π x triceps skin fold (mm))

GNS	> -10%	Adequate nutrition
	< -10% to 20%	Under-nutrition
	< -20% to 30%	Malnutrition
	<-30%	Severe malnutrition

Subjective Global Assessment of Nutritional Status (SGA)[65]

A routine history and physical examination is utilized for clinical assessment.

History includes inquiry about weight loss	Physical examination includes inspection for
Edema	Cheilosis
Anorexia	Glossitis
Vomiting	Subcutaneous fat loss
Diarrhea	Muscle wasting
Decreased or "unusual" food intake	Edema
Chronic illness	

Based on a global assessment of nutritional status, the examiner classifies a patient as:

A = NORMAL NUTRITIONAL STATUS B = MILD MALNUTRITION C = SEVERE MALNUTRITION

The SGA had a better combination of sensitivity (0.82) and specificity (0.72) in predicting postoperative infection than the PNI or individual measures of creatinine-height index, percentage body fat, serum transferrin, serum albumin, or delayed cutaneous hypersensitivity.[64]

In summary, although data conflict and analyses are imperfect, most studies support the conclusion that malnutrition predicts increased perioperaive morbidity and mortality, and decreased long-term survival.

Surgery, Nutritional Support and Clinical Outcome

Data on the effect of preoperative enteral supplementation on postoperative outcome in malnourished patients with head and neck cancer are mixed.[13,15] Flynn[15] prospectively studied 61 patients with squamous cell cancer of the upper aerodigestive tract who were candidates for operative resection (Fig. 4.1). Malnourished patients receiving "nutritional supplementation" were provided with specific recommendations to meet their individual nutrient requirements and an unspecified number were given nutritional supplements. Additionally, these patients were contacted as necessary during the 10-21 days between counseling and hospital admission to determine their nutritional status and to encourage them to comply with their nutritional programs. The malnourished unsupplemented group received only nutritional counseling, suggestions on ways to cope with eating problems, and no follow-up until hospital admission. Seven of the 25 well-nourished patients (26%) and 23 of the 36 malnourished patients (63%) had received prior radiotherapy. Postoperatively, patients received oral, tube and/or parenteral nutrition. Compared to the malnourished unsupplemented group, the malnourished supplemented patients were younger, had a greater prevalance of advanced disease (68% prevalence versus 35%, respectively), a greater likelihood of prior radiotherapy and a higher rate of extensive resection (26% versus 0%). Yet, the supplemented group had fewer complications and a shorter average hospital stay. The three-day decrease in average hospital stay saved an average of $2,298 per patient. Limitations of the study included small study size, nonstandardized treatment within the supplemented group, differences of stage, prior irradiation and extent of surgery between compared groups, and descriptive nonstatistical analysis.

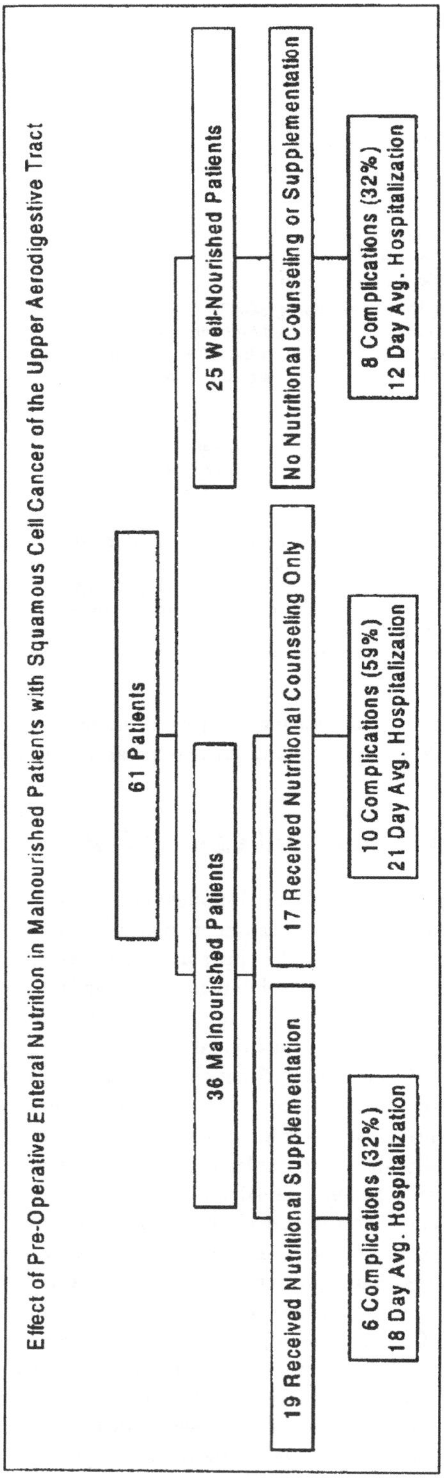

Fig. 4.1. Adapted from Flynn and Leightry.[15]

Goodwin[13] included in their study six patients with severe malnutrition who had an average of 14 days of preoperative enteral and parenteral nutrition, of whom five had major complications. The three severely malnourished patients who underwent surgery without preoperative nutritional support all suffered major complications. Despite the disappointing results, albeit on a small number of patients, Goodwin advocated nearly a decade later one to two weeks of preoperative nutritional support in those patients who have a 15% weight loss, decreased general strength, or low serum proteins.[16]

Data pertaining to altered clinical outcome with perioperative parenteral nutrition are as rare as those addressing the role of enteral nutrition. Shortly after parenteral nutrition became widely available in the 1970s, researchers noted that perioperative parenteral nutrition improved some nutritional parameters. Yet, the majority of patients appeared to possess sufficient nutrient reserves to survive the catabolic and semistarvation recovery period without the need for parenteral nutritional support. Although researchers have argued against its use in the majority of patients due to its high cost,[8] parenteral nutrition has been advocated for selected patients with head and neck cancer, such as those who cannot tolerate enteral nutrition or require rapid repletion of nutritional stores in order to qualify for timely oncologic therapy.[17]

Copeland[17] reported an uncontrolled trial of perioperative (as well as periradiotherapy and convalescent) parenteral nutrition in 23 patients with head and neck cancer who had lost at least 20 pounds and who were severely cachectic or intolerant of nasogastric feeding. Seven of the eight patients who received preoperative parenteral nutrition suffered no postoperative complications. Weight gain, wound healing and recovery were achieved in 20 patients, while only three patients suffered intravenous catheter-related complications. Many patients who previously were intolerant of enteral feeding regained their ability to tolerate enteral nutrition following parenteral nutritional support.

Sako[18] studied 69 patients with head and neck cancer who had no better than a moderate prognosis, stratified them based on nutritional status and prognosis, and randomized them to receive either parenteral or enteral postoperative nutrition (Fig. 4.2). Eight of the 35 patients in the parenteral group received preoperative parenteral nutrition for at least eight days as well. Thirty of the 35 patients received the postoperative parenteral nutrition for at least 12 days. Postoperative wound complications and recurrence of cancer were similar in the two groups. Survival curves were significantly worse in the patients receiving postoperative parenteral versus enteral nutrition in all strata of malnutrition. Parenteral nutrition was superior to enteral nutrition only in maintaining nitrogen balance and weight, and the stable weight was thought to be secondary to fluid retention rather than from preserved tissue mass.

Although the decreased survival among those patients receiving parenteral nutrition in Sako's[18] study was not a result of catheter-related sepsis, this complication is more common in patients treated for head and neck cancer than in patients with any other site of cancer.[19] Contamination of central venous catheters in patients with head and neck cancer may be more frequent due to pharyngostomy or tracheostomy secretions, or from skin compromise secondary to local radiotherapy.[19]

Changes in reimbursement policies may limit flexibility in providing perioperative nutritional support. When 61 patients who were admitted for radical resections of head and neck cancer after the implementation of diagnosis-related group (DRG)-based reimbursement were compared with 59 similar patients admitted before DRGs were used, complications rates had more than doubled in malnourished patients.[20] Comparing the post-DRG with the pre-DRG groups, nutritional status determined by the Protein Energy Malnutrition Scale was similar at admission, but the time from admission to surgery and nutritional status at surgery both decreased in the post-DRG group. Linn[20] concluded that the higher complication rate in the post-DRG patients was attributed to the DRG-driven pressure to limit preoperative hospital stay and nutritional interventions. Although comparisons using historical controls must be

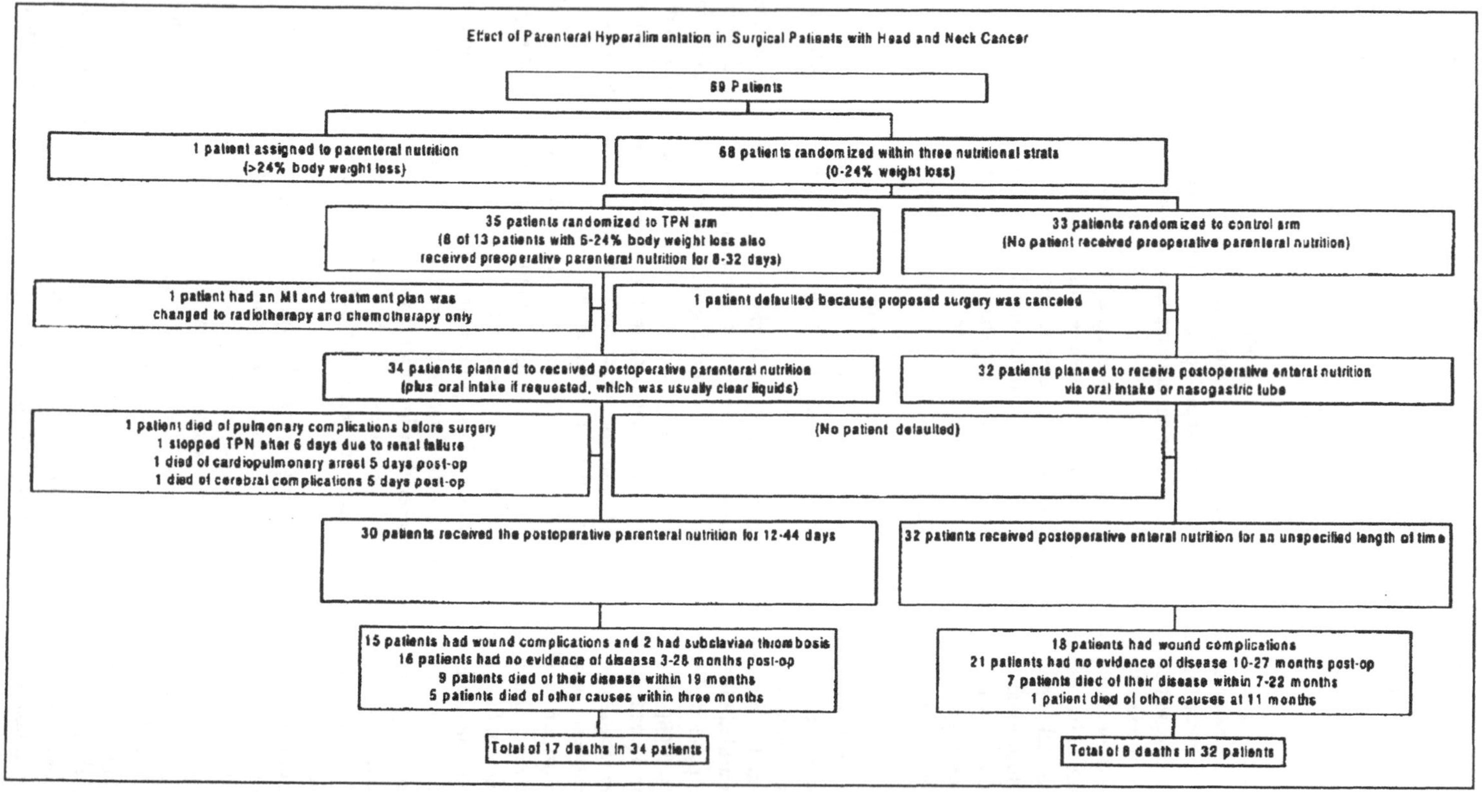

Fig. 4.2. Adapted from Sako and colleagues.[18]

interpreted with caution, the study design usually favors the population treated most recently (not the historical control).

In patients undergoing immediate mandibular reconstruction for oral cavity or oropharyngeal cancer, decreased length of hospital stay was achieved by using a coordinated care plan including oral feeding with speech and swallowing therapy beginning on the first or second postoperative day. The ultimate ability to tolerate regular food was influenced positively by the presence of teeth and tumor originating from the gingiva or retromolar trigone as opposed to the floor of the mouth or the tongue.[21] In a study of 44 patients with dysphagia following head and neck surgery who were enrolled in a comprehensive swallowing rehabilitation program, the severity of impairment was related to the extent of surgical resection, and the number of swallowing phases that were impaired (oral, pharyngeal or esophageal). Severity of the residual swallowing impairment correlated with the magnitude of initial impairment. The ability to compensate for residual impairment varied widely.[22]

Copeland[19] advocated the use of parenteral nutrition during swallowing rehabilitation, after concluding that the lack of a nasogastric tube psychologically facilitated jaw function rehabilitation in five poorly motivated patients and after witnessing a return of competent swallowing function in three patients on parenteral nutrition as general strength returned. Alternative routes of enteral nutrition delivery were not considered, and it is unclear to what degree, if any, the positive outcome in these patients was due to the parenteral nutrition. In the absence of a controlled trial, parenteral nutrition cannot be recommended routinely for swallowing rehabilitation.

Although parenteral nutrition has not been demonstrated to be better than enteral nutrition (and may be worse[18]), relative indications for parenteral nutritional support include a persistent pharyngocutaneous fistula[17] or chylous fistula.[23] Suspending enteral nutrition decreases salivary and mucosal secretions. If the fistula fails to heal spontaneously, surgical closure may be successful after parenteral nutritional support even if a prior attempt at surgical correction was unsuccessful.[19]

It appears reasonable to advocate preoperative coordinated nutritional counseling, enteral nutritional supplementation in severely malnourished patients, and avoidance of preoperative parenteral nutrition. Following surgery, aggressive speech and swallowing therapy should be implemented, as well as enteral tube feeding in those patients with a protracted recovery of deglutition. If one wishes to avoid a nasogastric tube, enteral feeding via a gastrostomy tube (placed prior to surgical resection) is preferable to parenteral nutrition. Parenteral nutrition can be recommended only in patients who cannot tolerate enteral feeding or who have persistent enteric fistulas.

Radiotherapy, Nutritional Support and Clinical Outcome

Radiotherapy is provided commonly to patients with head and neck cancer, but may adversely affect nutritional status by causing mucositis, stomatitis, increased xerostomia, dysgeusia, anorexia or dental defects. Radiotherapy may also affect nutritional status by causing a preference for carbohydrates at the expense of protein,[24,25] perhaps as a result of relative sparing of sweet taste perception.[26] Radiotherapy-induced dental lesions similar to caries may appear as early as one month after the initiation of radiotherapy when the salivary glands are in the radiation field, likely due to alterations in the oral milieu resulting from inadequate salivation.[27] Some degree of anorexia, dysphagia, or dysgeusia may persist for several months.[28]

Of note, zinc may prevent hypogeusia when administered at varying doses prior to the initiation of radiotherapy.[29] Additionally, zinc therapy may improve hypogeusia in patients receiving radiotherapy to the head and neck.[26] One recommended regimen is zinc sulfate 110 mg (elemental zinc 25 mg) orally four times a day during or after meals (to decrease gastrointestinal toxicity).[29]

To address the risk of malnutrition, 31 patients with newly-diagnosed localized head and neck cancer treated with radical external beam radiotherapy for cure were followed for six months.[10] In this descriptive study, weight loss averaging 10% correlated with the size of the radiation field when the oral cavity or oropharynx was included within the field (assuming that the table correlating weight loss with the radiation field only when located outside of the oral cavity was the result of a typographical error). Weight loss did not correlate with pretreatment dietary habits, anthropometric or biochemical measures. Two patients received enteral tube feeding during the last week of four weeks of therapy, and none received parenteral nutrition.

In order to investigate the potential impact of nutritional supplements on nutritional status, Nayel[25] prospectively randomized 11 patients to receive enteral supplementation (Ensure® providing 1,500 to 2,000 kcal/day for 10-31 days) during radiotherapy, and compared them to 12 patients who received radiotherapy without nutritional support. Among the patients receiving enteral supplementation, there were significant improvements in mid-arm circumference, triceps skin-fold thickness and body weight, plus trends toward decreased dysphagia and mucositis. Additionally, there were no interruptions of radiotherapy in the cohort receiving enteral supplements, while 5 of 12 patients (42%) receiving no nutritional support temporarily suspended radiotherapy due to poor performance status or severe mucositis. Although enteral nutritional support improved some anthropometric nutritional parameters and prevented therapy interruption in patients with advanced head and neck cancer having radiotherapy, it was unclear if morbidity was altered, and mortality was not investigated.[25]

Parenteral nutrition has also been used to rehabilitate or maintain patients with head and neck cancer undergoing radiotherapy.[19] Two severely malnourished patients had parenteral nutrition initiated seven to ten days before radiotherapy, while seven patients received it after developing severe stomatitis or pharyngitis which threatened the continuation of their treatment. Parenteral nutrition was delivered for an average of 35 days, and produced weight gain averaging seven pounds. No radiotherapy regimen had to be halted, and all patients completed the protocol except for one who died prematurely of aspiration pneumonia.

Several studies have documented sequelae relevant to nutritional support in patients with head and neck cancer who have had prior radiotherapy. Twenty-four patients with subjective and objective dry mouth at least four months after radiotherapy to the head and neck with curative intent for carcinoma or lymphoma were studied. Energy intake among the patients who had radiotherapy averaged 1,925 kilocalories, versus 2,219 kilocalories in age- and sex-matched controls. This decreased energy intake (nearly statistically significant) was independent of stimulated saliva secretion rate. The patients who had radiotherapy had a significantly lower intake of carbohydrate, sugar, fiber, and most micronutrients compared to controls, despite all but one of the irradiated patients eating food of normal consistency.[30]

In contrast to Bäckström's[30] findings, others have found that a majority of patients with oropharyngeal cancer having radiotherapy alone or radiotherapy combined with surgery required long-term diet restrictions which decreased quality of life somewhat and limited protein and calorie intake. Beeken[31] retrospectively reported 25 disease-free patients who had completed treatment for oropharyngeal cancer at least one year previously. Eighteen patients needed dietary modifications, which limited caloric and protein intake. The four highest ranked side-effects all related to eating (dry mouth, prolonged meals, dysphagia and dysgeusia). Although quality of life scores were high (perhaps reflecting a bias of the retrospective design), all seven patients who scored at or below seven (out of ten) required dietary modifications. As Beeken[31] did not specify how many had surgery, it is impossible to determine if this was the variable that explains why his group, but not Bäckström's,[30] required dietary restrictions.

Harrison[32] analyzed 30 patients with squamous cell cancer of the base of the tongue having primary radiation therapy (local external beam and implant plus external beam to the neck) plus neck dissection if nodes were palpable, compared to ten patients having primary surgery (tumor resection and neck dissection) followed by radiation therapy. Historically, both treatment methods gained local control in more than 80% of patients. Posttreatment performance status in patients having primary radiation was independent of cancer stage, while in patients having primary surgery it was inversely proportional to cancer stage. Subjective performance and quality of life status at least six months after the primary therapy were superior in the patients having primary radiation compared to the those having primary surgery. Nutritional status was not measured, but one criterion of the performance status was normalcy of diet.

To recapitulate, there likely is a role for enteral supplementation during radiotherapy to decrease the risk of intolerable side effects and treatment suspensions. Although performance status may not be jeopardized to the same degree as with surgical resection, radiotherapy-induced side effects also create long-term risk of malnutrition which may be lessened by coordinated nutritional counseling and consideration of enteral supplements. As data are limited regarding the role of parenteral nutrition in patients having radiotherapy, parenteral nutrition cannot be recommended routinely in this population.

Chemotherapy, Nutritional Support and Clinical Outcome

Chemotherapy is less widely used than surgery or radiotherapy for treatment of patients with head and neck cancer. Many of the agents used are emetigenic acutely, and may cause mucositis about one week after being administered. Although chemotherapy may adversely affect nutritional status, there are no data addressing this issue specifically in patients with head and neck cancer. No studies have examined the role of enteral nutrition in this patient population. The same is true for parenteral nutrition, except for one uncontrolled description of 16 severely malnourished patients who were given parenteral nutrition an average of 27 days in order to increase their previously poor chances of tolerating chemotherapy.[19] All the patients tolerated the chemotherapy, and gained an average of 10 pounds during the chemotherapy. Average survival was six months in the five patients who responded to the chemotherapy, compared to one month in the eleven patients who did not respond.

Enteral Nutrition Delivery

Multiple avenues for enteral access are available to patients with head and neck cancer, even for those suffering from relative obstruction of the upper alimentary canal. With advanced planning in patients expected to require preoperative or prolonged postoperative nutritional support, tolerable enteral access can be established. With rare exception, delivery of enteral nutrition is preferable to more costly and potentially less efficacious parenteral nutrition, since patients almost invariably possess a functional gastrointestinal tract distal to the proximal esophagus.

Nasogastric Tubes

Access is usually established using a narrow-bore feeding tube, ideally placed into the jejunum to allow immediate postsurgical feeding,[16] and may be used for many months with long-term complication rates similar to those from percutaneous endoscopic gastrostomy.[33] If nasogastric tubes of sufficient caliber to allow gastric decompression are used, they should remain in place for no more than one to two weeks since they are relatively uncomfortable and predispose the patient to necrosis of the nasal alae, pharyngoesophageal ulceration, postcricoid perichondritis, sinusitis, otitis media and pneumonia due to pooling of secretions.[23,34-36] If the nasogastric tube is dislodged in the early postoperative period, some advocate immediate replacement to take advantage of the initial tensile strength of the suture line, while others prefer to wait several days to allow some healing of the suture line.[34] In one study, 18 of 46 patients needed reinsertion of their nasogastric tubes at least once, and in two of the four patients who accidentally displaced or clogged their tubes at least twice, a tube could not be reintroduced safely.[36] Given the lack of data addressing the timing of nasogastric feeding tube reinsertion, institutional policy is guided by anecdotal experience.

Also based on anecdotal experience, an opinion exists that prolonged nasogastric intubation leads to delayed healing of suture lines, increased risk of fistulization and impaired restoration of normal deglutition.[34] In one study, 10 of 46 patients being fed with nasogastric tubes did develop dysphagia which resolved after removal of the tube, but the incidence of pharyngocutaneous fistulas was similar in patients fed via nasogastric versus percutaneous gastrostomy tubes.[36] Given the discomfort of nasogastric feeding tubes, their frequent suboptimal delivery of nutrition (at least in patients two weeks after acute cerebrovascular accident),[37] cosmetic detraction, and general nuisance in ambulatory patients,[35] alternatives should be considered in patients requiring prolonged enteral nutritional support. Risk factors for needing postoperative nutritional support for more than 30 days are listed in Table 4.3, derived from a descriptive retrospective review of 109 patients with squamous cell carcinoma of the oral cavity, pharynx or larynx treated with surgical resection. Of the 92 who received postoperative enteral feeding, 41 (45%) required prolonged enteral support. Delayed wound healing was the indication for one-half of the patients requiring prolonged enteral feeding.[38]

Cervical Esophagostomy or Pharyngostomy

Percutaneous cervical esophagostomy or pharyngostomy tubes placed via the pyriform sinus may be used as a route for temporary or prolonged enteral nutritional support. If the tube is being placed to improve nutritional status preoperatively, local cutaneous and topical oropharyngeal anesthesia may be sufficient to complete the procedure,[39] although others have preferred general anesthesia.[35] Percutaneous pharyngostomy tubes were placed in 42 patients without significant complication and with much better tolerance than historically witnessed with nasogastric tubes.[35] In another study, however, the complication rate of 60% in the 17 patients with esophagostomy tubes was higher than the 9% complication rate in 21 patients with nasogastric tubes.[38] The most frequent delayed concern is accidental dislodgment of the tube.[39] If the tube has been in place for at least a week prior to being removed, it is usually easy to replace if done so promptly through the established track. If the tube is not replaced, the track has been shown to close spontaneously within four days.[35]

Percutaneous Endoscopic Gastrostomy or Jejunostomy

A percutaneous endoscopic gastrostomy (PEG) or jejunostomy tube is another reasonable alternative in patients who are expected to require protracted enteral supplementation (e.g., at least four weeks), and has been recommended over esophagostomy tubes due to a possible lower risk of major complication.[38] The postoperative course in 43 patients with stage II, III or IV head and neck cancer who received a PEG one day prior to surgery was compared retrospectively to 46 site- and stage-matched patients who received postoperative nutrition via a nasogastric tube.[36] In the patients with the PEG tubes, length of hospital stay was decreased about 60% in patients with cancer of the larynx, pharynx, or tongue base from about 50 days to about 20 days. Patients with cancer of the anterior tongue or floor of mouth (in whom swallowing function is usually relatively preserved) stayed about one month in both groups. When analyzed

Table 4.3. Risk factors for requiring prolonged postoperative enteral nutritional support in patients with head and neck cancer

Individual risk factors	Risk of requiring support > 30 days
Primary pharyngeal tumors	60%
Preoperative weight loss of more than ten pounds	56%
Stage IV cancer	55%
Combined surgery and radiotherapy	50%
Pooled risk factors	
No risk factor	16%
One risk factor	26%
Two risk factors	52%
Three risk factors	56%
All four risk factors	86%

Adapted from Gardine, Kokal, Beatty and colleagues.[38]

by stage, patients with stage III or IV disease who had PEGs also left the hospital earlier. Patients with stage II diseases stayed about the same length of time in both groups. In those patients with PEGs, 63% were discharged with plans for home tube feeding, compared to only 15% of the patients with nasogastric tubes. This discrepancy in outpatient enteral feeding frequencies may explain why PEG tubes allowed more timely discharges.

Patients with head and neck cancer who are likely to require prolonged enteral nutritional support include those who are expected to suffer severe mucositis during radiation therapy preceding surgery. An endoscopic gastrostomy tube placed prior to or at the beginning of radiotherapy may prevent malnutrition from developing. Other patients who may benefit from a prophylactically-placed percutaneous endoscopic gastrostomy tube include those expected to have difficulty establishing safe swallowing postoperatively, such as those with cancer of the tongue or pharyngeal walls.[16] Such patients can have the endoscopic gastrostomy tube placed preoperatively under conscious sedation or in the operating suite after administration of general anesthesia but before resection of the cancer begins.[40]

PEG placement in 114 patients with head and neck cancer was retrospectively compared to PEG placement in 220 patients with neurological impairment.[41] The PEG attempt failed due to pharyngeal or esophageal obstruction in 3% versus 0.5% of patients, respectively. The post-PEG overall complication rate was only 5% in the head and neck cancer compared to 14% in the neurological group. It was unclear if any differences were statistically significant. Of the three patients who received chemotherapy before the PEG and the 12 patients who underwent full-course chemotherapy immediately after PEG placement, only one developed wound breakdown.

Although no mention was made of periprocedure antibiotic use, the PEG site cellulitis or wound breakdown incidences were only 3.5% and 4.5%, respectively,[41] below that observed in patients who received prophylactic doses of antibiotics before PEG placement.[42] In the randomized controlled trial by Jain,[42] the incidence of peristomal wound infection was zero in the 52 patients already on antibiotics, 7% (2 of 27) in patients not already on antibiotics who received cefazolin one gram intravenously 30 minutes prior to the PEG procedure, and 32% (9 of 28) in the control group who received no antibiotics. Based on these findings, the American Society for Gastrointestinal Endoscopy recommended a prophylactic dose of a cephalosporin before PEG procedures.[43]

The reported incidence of respiratory distress during PEG placement in patients with head and neck cancer and an unsecured airway has ranged from 1-10%.[41,44] With judicious use of sedation, airway obstruction was only 0.9% (identical to the control group with neurological impairment), despite inclusion of 19 patients with stage IV pharyngeal cancer and 13 patients with Stage III or IV pyriform cancer.[41] In combining two smaller series (each from the same institutions), however, the risk of respiratory arrest was 14%, occurring in 6 of 44 patients. In 5 of the 6 patients, the airway obstruction occurred after sedation but before endoscopy.[44]

If tumor bulk in head and neck cancer allows passage of the endoscope but is unlikely to allow easy peroral passage of the feeding tube and bumper using either the "push" (Sacks-Vine) technique[45] or "pull" (Ponsky) techniques[46] percutaneous placement under endoscopic guidance can be achieved using the "introducer" (Russell) technique.[47] However, being of smaller diameter and anchored by an inflated balloon rather than a solid bumper, tubes placed by the introducer method are more prone to clogging and premature extrusion. If an endoscopic attempt at gastrostomy placement is halted due to inability to pass even a pediatric fiberoptic endoscope, poor tissue apposition or other technical limitation, an alternative is fluoroscopically-guided percutaneous placement of a gastrostomy tube using the introducer technique, although this method does require passage of an orogastric tube to insufflate the stomach.

When providing nutrition into the stomach, aspiration pneumonia might be decreased by using continuous feeding or slowly delivered intermittent boluses (e.g., 480 mL over one hour).[48,49] Both delivery methods may decrease the risk of inducing gastroesophageal reflux compared to rapidly delivered boluses. Rapid bolus feeding (e.g., 250 mL of formula followed by 100 mL of water, all within 20 seconds) caused marked relaxation of the lower esophageal sphincter on manometry and allowed esophageal reflux to the sternal notch on scintigraphy despite elevation of the head of the bed.[49,50]

Jejunostomy extensions can be added to PEGs and guided through the pylorus endoscopically or fluoroscopically. However, given the extension's risk for clogging, migration into the stomach, and failure to decrease the risk of aspiration (as most aspiration pneumonia appears to result from aspiration of oropharyngeal secretions rather than gastroesophageal reflux[51]), the routine use of jejunostomy extensions cannot be recommended. Rather, direct percutaneous endoscopic jejunostomy[52] should be considered in patients at risk for aspiration of gastric contents who would not be inconvenienced by prolonged pump-driven feedings.

Surgical Gastrostomy or Jejunostomy

Another option is surgical gastrostomy, which is usually a separate procedure but can be performed at the time of cancer resection.[53] When performed in patients under intravenous conscious sedation, open gastrostomy has morbidity, mortality and overall costs comparable to PEG.[54] In a retrospective study comparing laparoscopic to open gastrostomy (performed under general anesthesia in 96% and 67% of the cases, respectively), laparoscopic gastrostomy offered significantly reduced operative time with similar morbidity, mortality, and procedural costs (in the laparoscopic group, additional equipment charges offset reduced room charges).[53] The laparoscopic jejunostomy remains another consideration.

Gastrostomy Site Metastasis

One rare risk of gastrostomy tubes in patients with head and neck cancer is that of gastrostomy site metastases. Most instances occurred after placing PEGs using the pull technique[55-60] (preceded by bougienage of the esophagus only in the first reported case[61]). These and other

Table 4.4. Guidelines for nutritional support in patients with head and neck cancer

1) If the gut works, use it.
2) In severely malnourished patients anticipating major elective surgery, preoperative nutritional support should be provided for ten days.
3) In surgical patients with malnutrition or postoperative complications, postoperative nutritional support should be initiated immediately.
4) In surgical patients who are unable to resume oral feedings by postoperative day ten, postoperative nutritional support should be initiated.
5) In malnourished patients anticipating radiation or chemotherapy, nutritional support may improve toleration of the therapy, but is unlikely to alter morbidity or mortality.
6) In patients requiring nutritional support in whom a nasogastric tube is undesirable, a gastrostomy or jejunostomy is preferable to parenteral nutrition.
7) In patients with a persistent pharyngocutaneous or chylous fistula, parenteral nutritional may improve the likelihood of healing.

reports of pull PEG site metastases suggest that after advancing through the head or neck cancer, the tube may seed the stoma with cancer cells as it emerges from the stomach through the abdominal wall. The absence of reported cases in association with the push technique probably reflects the greater popularity of pull PEGs, not any difference in risk.

There is evidence, however, that the source of PEG site metastases may be from circulating cancer cells rather than those traveling on the feeding tube. In one case, the PEG was placed six weeks after surgical resection of the laryngeal cancer, without evidence for local or regional cancer at the time of PEG placement or 18 months later when metastases were diagnosed in the lung and on the skin at both a prior PEG site and a location several centimeters away.[55] Additionally, a metastasis to the site of an operatively placed gastrostomy tube has been reported.[62] In these cases, development of metastases at the gastrostomy sites presumably was the result of hematogenous inoculation of traumatized tissue having a greater susceptibility to implantation of cancer cells.[63] The incidence of gastrostomy site metastases is unknown, but is presumably low. Also unknown is to what degree, if any, risk of gastrostomy site metastasis is reduced by using introducer or operative placement techniques which avoid contamination of the gastrostomy tube with cancer cells.

Conclusion

Evidence-based guidelines regarding the appropriate nutritional support of surgical, radiotherapy or chemotherapy candidates with head and neck cancer can only be developed after the completion of prospective, randomized studies of sufficient sample size to ensure adequate power.[9] Although such studies in patients with head and neck cancer are lacking, nutritional support guidelines are offered in Table 4.4 based on the available data. These guidelines parallel recent general recommendations for nutritional support.[1]

Malnourished patients with head and neck cancer may improve their nutritional parameters after nutritional support. However, most data suggest that the majority of patients do not improve their outcome with nutritional support. Severely malnourished patients are the exception, in whom enteral nutritional support around the time of therapeutic interventions decreases morbidity and mortality. Several proven methods exist to provide enteral nutrition through a tube placed into the stomach or proximal intestine of appropriate patients. In contrast to enteral nutrition, there is little evidence to support the use of parenteral nutrition in patients with head and neck cancer (the majority of whom can tolerate enteral nutrition).

References

1. Souba W. Nutritional support. N Engl J Med 1997; 336:41-48.
2. Brookes GB. Nutritional status: A prognostic indicator in head and neck cancer. Otolaryngol Head Neck Surg 1985; 93:69-74.
3. Bassett MR, Dobie RA. Patterns of nutritional deficiency in head and neck cancer. Otolaryngol Head Neck Surg 1983; 91:119-125.
4. Guo C, Ma D, Zhang K. Nutritional status of patients with oral and maxillofacial malignancies. J Oral Maxillofac Surg 1994; 52:559-562.
5. Linn BS, Robinson DS, Klimas NG. Effects of age and nutritional status on surgical outcomes in head and neck cancer. Ann Surg 1988; 207:267-273.
6. Westin T, Jansson A, Zenckert C, et al. Mental depression is associated with malnutrition in patients with head and neck cancer. Arch Otol Head Neck Surg 1988; 114:1449-1453.
7. Sobol SM, Conoyer JM, Sessions DG. Enteral and parenteral nutrition in patients with head and neck cancer. Ann Otol 1979; 88:495-501.
8. Sobol SM, Conoyer JM, Zill R, et al. Nutritional concepts in the management of the head and neck cancer patient: I. Basic concepts. Laryngoscope 1979; 89:794-803.
9. Matthews TW, Lampe HB, Dragosz K. Nutritional status in head and neck cancer patients. J Otolaryngol 1995; 24:87-91.
10. Johnston CA, Keane TJ, Prudo SM. Weight loss in patients receiving radical radiation therapy for head and neck cancer: A prospective study. JPEN 1982; 6:399-402.
11. Eilber FR, Morton DL, Ketcham AS. Immunologic abnormalities in head and neck cancer. Am J Surg 1974; 128:534-538.
12. Hooley R, Levine H, Flores TC et al. Predicting postoperative head and neck complications using nutritional assessment: the prognostic nutritional index. Arch Otolaryngol 1983; 109:83-85.
13. Goodwin WJ Jr, Torres J. The value of the prognostic nutritional index in the management of patients with advanced carcinoma of the head and neck. Head Neck Surg 1984; 6:932-937.
14. Linn BS. A protein energy malnutrition scale (PEMS). Ann Surg 1984; 200:747-752.
15. Flynn MB, Leightty FF. Preoperative outpatient nutritional support of patients with squamous cancer of the upper aerodigestive tract. Am J Surg 1987; 154:359-362.
16. Goodwin WJ, Jr, Byers PM. Nutritional management of the head and neck cancer patient. Med Clin N Am 1993; 77:597-610.
17. Copeland EM, MacFadyen BV, MacComb WS, et al. Intravenous hyperalimentation in patients with head and neck cancer. Cancer 1975; 35:606-611.
18. Sako K, Loré JM, Kaufman S, et al. Parenteral hyperalimentation in surgical patients with head and neck cancer: A randomized study. J Surg Oncol 1981; 16:391-402.
19. Copeland EM III, Daly JM, Dudrick SJ. Nutritional concepts in the treatment of head and neck malignancies. Head Neck Surg 1979; 1:350-363.
20. Linn BS, Robinson DS. The possible impact of DRGs on nutritional status of patients having surgery for cancer of the head and neck. JAMA 1988; 260:514-518.
21. Heller KS, Dubner S, Keller A. Long-term evaluation of patients undergoing immediate mandibular reconstruction. Am J Surg 1995; 170:517-520.
22. Aguilar NV, Olson ML, Shedd DP. Rehabilitation of deglutition problems in patients with head and neck cancer. Am J Surg 1979; 138:501-507.
23. Williams EF III, Meguid MM. Nutritional concepts and considerations in head and neck surgery. Head and Neck 1989; 11:393-399.
24. Chencharick JD, Mossman KL. Nutritional consequences of the radiotherapy of head and neck cancer. Cancer 1983; 51:811-815.
25. Nayel H, El-Ghoneimy E, El-Haddad S. Impact of nutritional supplementation on treatment delay and morbidity in patients with head and neck tumors treated with irradiation. Nutrition 1992; 8:13-18.
26. Mossman KL, Henkin RI. Radiation-induced changes in taste acuity in cancer patients. Int J Rad Oncol Biol Phys 1978; 4:663-670.
27. Frank RM, Herdly J, Philippe E. Acquired dental defects and salivary gland lesions after irradiation for carcinoma. J Amer Dent Assn 1965; 70:868-883.
28. Mossman K, Scheer A. Complications of radiotherapy of head and neck cancer. ENT J 1977; 56:90-95.

29. Henkin RI. Prevention and treatment of hypogeusia due to head and neck irradiation (letter). JAMA 1972; 220:870-871.
30. Bäckström I, Funegård U, Andersson I et al. Dietary intake in head and neck irradiated patients with permanent dry mouth symptoms. Eur J Cancer 1995; 31B:253-257.
31. Beeken L, Calman F. A return to "normal eating" after curative treatment for oral cancer: What are the long-term prospects? Eur J Cancer 1994; 30B:387-392.
32. Harrison LB, Zelefsky MJ, Armstrong JG et al. Performance status after treatment for squamous cell cancer of the base of the tongue: A comparison of primary radiation therapy versus primary surgery. Int J Radiat Oncol Biol Phys 1994; 30:953-957.
33. Fay D, Poplausky M, Gruber M et al. Long-term enteral feeding: A retrospective comparison of delivery via percutaneous endoscopic gastrostomy and nasoenteric tubes. Am J Gastroenterol 1991; 86:1604-1609.
34. Sobol SM, Conoyer JM, Zill R et al. Nutritional concepts in the management of the head and neck cancer patient: II. management concepts. Laryngoscope 1979; 89:962-979.
35. Meehan SE, Wood RAB, Cuschieri A. Percutaneous cervical pharyngostomy: A comfortable and convenient alternative to protracted nasogastric intubation. Am J Surg 1984; 148:325-330.
36. Gibson S, Wenig BL. Percutaneous endoscopic gastrostomy in the management of head and neck carcinoma. Laryngoscope 1992; 102:977-980.
37. Norton B, Homer-Ward M, Donnelly MT, et al. A randomized prospective comparison of percutaneous endoscopic gastrostomy and nasogastric tube feeding after acute dysphagic stroke. BMJ 1996; 312:13-16.
38. Gardine RL, Kokal WA, Beatty JD. Predicting the need for prolonged enteral supplementation in the patient with head and neck cancer. Am J Surg 1988; 156:63-65.
39. Noone RB, Graham WP III. Nutritional care after head and neck surgery. Postgrad Med 1973; 53:80-86.
40. Selz PA, Santos PM. Percutaneous endoscopic gastrostomy. A useful tool for the otolaryngologist—head and neck surgeon. Arch Otolaryngol Head Neck Surg 1995; 121:1249-1252.
41. Gibson SE, Wenig BL, Watkins JL. Complications of percutaneous endoscopic gastrostomy in head and neck cancer patients. Ann Otol Rhinol Laryngol 1992; 101:46-50.
42. Jáin NK, Larson DE, Schroeder KW et al. Antibiotic prophylaxis for percutaneous endoscopic gastrostomy: A prospective, randomized, double-blind clinical trial. Ann Intern Med 1987; 107:824-828.
43. ASGE. Antibiotic prophylaxis for gastrointestinal endoscopy. Gastrointest Endosc 1995; 42:630-635.
44. Riley DA, Strauss M. Airway and other complications of percutaneous endoscopic gastrostomy in head and neck cancer patients. Ann Otol Rhinol Laryngol 1992; 101:310-313.
45. Sacks BA, Vine HS, Palestrant AM et al. A non-operative technique for establishment of a gastrostomy in the dog. Invest Radiol 1983; 18:485-489.
46. Ponsky JL, Gauderer MWL. Percutaneous endoscopic gastrostomy: A nonoperative technique for feeding gastrostomy. Gastrointest Endosc 1981; 27:9-11.
47. Russell TR, Brotman M, Forbes N. Percutaneous gastrostomy: A new simplified and cost-effective technique. Am J Surg 1984; 148:132-137.
48. Kocan MJ, Hickisch SM. A comparison of continuous and intermittent enteral nutrition in NICU patients. J Neurosci Nurs 1986; 18:333-337.
49. Hamaoui E. Gastroesophageal reflux during gastrostomy feeding (commentary). JPEN 1995; 19:172-173.
50. Silk DBA, Payne-James JJ. Complications of enteral nutrition. In: Rombeau J, Caldwell M, eds. Clinical Nutrition: Enteral and Tube Feeding. Philadelphia: WB Saunders Co, 1990.
51. Kadakia SC, Sullivan HO, Starnes E. Percutaneous endoscopic gastrostomy or jejunostomy and the incidence of aspiration in 79 patients. Am J Surg 1992; 164:114-118.
52. Shike M, Latkany L, Gerdes H et al. Direct percutaneous endoscopic jejunostomies for enteral feeding. Gastrointest Endosc 1996; 44:536-540.
53. Lydiatt DD, Murayama KM, Hollins RR et al. Laparoscopic gastrostomy versus open gastrostomy in head and neck cancer patients. Laryngoscope 1996; 106:407-410.
54. Stiegmann G, Goff J, Silas D et al. Endoscopic versus operative gastrostomy: Final results of a prospective randomized trial. Gastrointest Endosc 1990; 36:1-5.
55. Bushnell L, White TW, Hunter JG. Metastatic implantation of laryngeal carcinoma at a PEG exit site. Gastrointest Endosc 1991; 37:480-482.

56. Huang DT, Thomas G, Wilson WR. Stomal seeding by percutaneous endoscopic gastrostomy in patients with head and neck cancer. Arch Otolaryngol Head Neck Surg 1992; 118:658-659.
57. Laccourreye O, Chabardes E, Merite-Drancy A et al. Implantation metastasis following percutaneous endoscopic gastrostomy. J Laryngol Otol 1993; 107:946-949.
58. Meurer MF, Kenady DE. Metastatic head and neck carcinoma in a percutaneous gastrostomy site. Head Neck 1993; 15:70-73.
59. Schiano TD, Pfister D, Harrison L et al. Neoplastic seeding as a complication of percutaneous endoscopic gastrostomy. Am J Gastroenterol 1994; 89:131-3.
60. van Erpecum KJ, Akkersdijk WL, Warlam-Rodenhuis CC et al. Metastasis of hypopharyngeal carcinoma into the gastrostomy tract after placement of a percutaneous endoscopic gastrostomy catheter. Endoscopy 1995; 27:124-127.
61. Preyer S, Thul P. Gastric metastasis of squamous cell carcinoma of the head and neck after percutaneous endoscopic gastrostomy: Report of a case. Endoscopy 1989; 21:295.
62. Alagaratnam T, Ong G. Wound implantation: A surgical hazard. Br J Surg 1977; 64:872-875.
63. Murthy SM, Goldschmidt RA, Rao LN et al. The influence of surgical trauma on experimental metastasis. Cancer 1989; 64:2035-2044.

Chapter 5

Nutritional Support in Patients with Gastrointestinal, Pancreatic and Liver Cancer

Matthew E. Cohen

Patients with gastrointestinal cancer who lose weight have poorer survival, with the exception of those with advanced gastric cancer or pancreatic cancer.[1] Malnutrition may compound preexisting immunosuppression, risk of infection, and poor wound healing. Malnutrition may develop secondary to mechanical complications (e.g., obstruction) metabolic derangements (e.g., the catabolic state known as "cancer cachexia"), functional disorders (e.g., postoperative ileus) or psychological reactions (e.g., reactive depression). Although the cause of malnutrition in patients with gastrointestinal cancer may be multifactorial (see Table 5.1), negative energy balance appears to be more closely linked to decreased intake than to increased expenditure.[2] It has been impossible to verify that nutritional status is independent from disease severity.[3] Therefore, it has been difficult to distinguish whether malnutrition associated with gastrointestinal cancer is a cause of, or a result of, the illness.

Multiple studies have demonstrated that nutritional support improves nutritional indices,[4] although anorectic and malnourished patients with advanced gastrointestinal cancer may be an exception.[5] Few studies have demonstrated that improving nutritional parameters translates into improved clinical outcome in cancer patients. One early example is a study of 50 patients with either gastroduodenal or pancreatobiliary malignancy who were unable to maintain adequate enteral nutrition in any form and who had parenteral nutrition for an average of 26 days (range 5-109 days). Discharge with improved physical status and plans for continued therapy were predicted by increasing transferrin levels, total lymphocyte count, and to a lesser extent, arm muscle circumference at two weeks, but not changes in albumin level or skin test reactivity.[6]

Otherwise healthy patients subjected to starvation benefit from nutritional rehabilitation. It is unclear what benefit patients with cancer receive from intensive parenteral or enteral nutritional support, despite studies suggesting that reversing a catabolic state predicted postoperative survival.[7] While basal carbohydrate and fat metabolism in patients with early gastrointestinal cancer (and usually stable weight) parallels healthy people,[8,9] those patients with advanced cancer (and usually weight loss) have elevated rates of protein catabolism,[9] glycolysis,[9] lipolysis,[8] gluconeogenesis[9] (none of which is inhibited by glucose infusion), and lipogenesis,[10] plus impaired free fatty acid oxidation,[8,10] and, in contrast to malnourished patients without cancer, fail to increase muscle strength despite two weeks of parenteral nutrition.[10]

In addition to altered metabolism in gastrointestinal cancer, the physical stress of surgical resection may further contribute to the risk of malnutrition.[11-14] One of the first randomized trials of nutritional support studied its use in the perioperative period of 30 patients with upper gastrointestinal cancer and recent weight loss. Major complications were less in the group that

Nutritional Support in Cancer and Transplant Patients, edited by Rifat Latifi and Ronald C. Merrell.

Table 5.1. Potential causes of weight loss and malnutrition in patients with gastrointestinal cancer

1. Anorexia of "cancer cachexia"
2. Obstruction
3. Malabsorption
   a. Metastatic infiltration of small bowel or mesentery
   b. Pancreatic duct obstruction or insufficiency
   c. Small bowel fistulas
   d. Biliary obstruction or bile salt insufficiency
   e. Bacterial overgrowth
   f. Megaloblastic changes from nutritional deficiencies
4. Fluid and electrolyte imbalance
   a. Hypovolemia from inadequate intake
   b. Emesis from obstruction
   c. Osmotic diarrhea from malabsorption
   d. Secretory diarrhea from hormone-secreting tumors
   e. Fluid loss through fistulas
5. Increased tumor-induced energy expenditure

received parenteral nutrition for three days before and ten days after surgery compared to those who did not, although mortality was the same.[15] In a study of 100 patients undergoing gastrointestinal resection, the majority of whom had cancer, parenteral supplementation of oral diets for at least one week before surgery decreased infectious complications among the malnourished only.[16] In 74 patients given intensive oral feeding and who underwent laparotomy with anticipated surgical resection of esophageal or gastric cancer, those who had 7-10 days of preoperative parenteral nutrition had a reduced incidence of wound infection, despite no improvement in immunological parameters. Of those patients with admission albumin below 3.5 g/dL, 5 of 9 in the control group developed wound infections, while none of the 8 patients who fell in this category from the treated group developed a wound infection. The authors concluded that the limited benefit did not justify the routine use of parenteral nutrition in this population, given the complications from the central venous access and formula, and added expense.[17] In a similar group of 125 patients, those who received 10 days of preoperative parenteral nutrition had fewer anastomotic leaks and reduced mortality (3% versus 11%).[18] With the replacement of suturing by stapling to secure anastomoses, it is quite possible that anastomotic breakdown, and its resultant morbidity and mortality, has become a less significant issue.[19] Additionally, this study has been criticized for failing to stratify for degree of malnutrition and for having been a subgroup analysis, and therefore being at increased risk of type I error (erroneously concluding that a difference exists between similar groups).[20]

In a prospectively study of patients with cancer of the esophagus, stomach, colon, pancreas or biliary system who had lost at least 10 pounds over 3 months, 30 patients were randomized to receive parenteral nutrition for 72 hours prior to surgery and up to 10 days postoperatively until eating at least 1,500 kilocalories a day. Twenty-six patients were randomized to receive no parenteral nutrition. Patient diets were advanced as tolerated. In patients receiving parenteral nutrition, the postoperative albumin level improved significantly over the preoperative value, 53% gained more than 10 pounds, and only 7% lost more than 10 pounds. In contrast, patients who received no parenteral nutrition had no improvement in albumin, none gained more than

10 pounds, and 15% lost more than 10 pounds. Minor complications and mortality (23% and 7%, respectively) were no different between groups. Major complications were 13% in the parenteral nutrition group, and 19% in the unsupplemented group, which was a statistically insignificant difference. The statistical methods were not described.[15]

Other studies have found that despite weight gain in those patients given parenteral nutrition, there were no improvements in morbidity or mortality.[21] A meta-analysis pooling 10 trials (9 of which focused on patients with gastrointestinal cancer) which investigated the impact of preoperative parenteral nutrition on surgical resection of cancer, however, favored parenteral nutrition when assessing endpoints of major complication (95% confidence interval 0.30-0.84) and mortality (95% confidence interval 0.21-0.90).[22] There are fewer data regarding the role of perioperative enteral nutrition. In malnourished patients with gastric or colorectal cancer, preoperative enteral nutrition appeared to protect against infectious complications as well as did parenteral nutrition.[23]

Compared to 16 control patients, 16 patients randomized to receive immediate nasojejunal feeding after small or large bowel resection had improved wound healing, trends toward earlier passage of flatus and feces, and fewer bowel obstructions, despite their failure to meet nutritional requirements until after the introduction of a normal oral diet. Muscle strength, fatigue, and length of stay were similar. Three-quarters of the patients had no problems with the tube or the feedings, and none had diarrhea or complications related to the feeding.[24] Similarly, in a randomized study comparing 14 patients who received immediate jejunal feeding after elective intestinal resection for "quiescent, chronic gastrointestinal disease" with 14 patients who received only intravenous fluids for an average of 6 days, the feeding prevented transient postoperative negative nitrogen balance, attenuated gut permeability, and may have decreased nausea, vomiting, weight loss, and infections (differences between the small groups in these four outcomes did not achieve statistical significance). Baseline nutritional status was not reported.[25]

Other studies have also found benefit to immediate enteral feeding following bowel resection[24,26] which compared favorably to parenteral nutrition at decreased cost.[27] The only randomized study explicitly limited to patients with gastrointestinal cancer (gastric adenocarcinoma) found that postoperative jejunostomy feeding was comparable to parenteral nutrition at one-half the cost, but caused more diarrhea (which was usually controlled by altering the infusion rate and adding loperamide).[28] Parenteral nutrition in those patients with high caloric demands may be appropriate, because patients fed via needle jejunostomy require gradual advancement which delays positive nitrogen balance until the fifth postoperative day, on average.[29] Enteral feeding via jejunostomy has been reported to cause pneumatosis intestinalis and to be associated with small bowel infarction.[30,31] However, in a review of 217 consecutive patients the incidence of the former complication was 1% and the latter complication was not encountered.[32]

In addition to bearing the stress of surgery, patients with gastrointestinal cancer often receive additional therapy which jeopardizes their nutritional status, namely chemotherapy,[33,34] and/or radiation therapy.[33,35] In malnourished patients given parenteral nutrition, up to one-half will improve their indices of immunological function.[36] There has been concern, however, that parenteral nutrition may favor protein synthesis to a greater degree in the tumor than in the host,[14] although this same effect has the potential to increase tumor susceptibility to therapy.[37,38] Most studies have found parenteral nutrition to be associated with more infection,[22,39] poorer tumor response,[22,39] or shorter survival.[39] In a meta-analysis, no protective effect of parenteral nutrition on the gastrointestinal toxicity of chemotherapy was found,[40] although parenteral nutrition may decrease radiation enteritis by suppressing pancreas exocrine function.[41] Reviews have concluded that there is little evidence supporting a role for parenteral nutrition during chemotherapy.[19] Similarly, enteral support has failed to improve nutritional or clinical outcomes following radiation therapy for gastrointestinal malignancies,[42] although a low residue, low-fat diet free of gluten and milk products was reported to prevent acute or delayed radiation enteritis.[35] The American College of Physicians (ACP) concluded that in patients undergoing chemotherapy or combined chemotherapy and radiotherapy, parenteral nutrition was associated with net harm, but conceded that the intervention may be beneficial in patients who are severely malnourished[39] based on a meta-analysis.[43]

Although parenteral nutrition is more expensive than enteral nutrition, is associated with more infectious complications, and may produce limited if any nutritional gains,[5] it may be the only alternative in patients without an intact gastrointestinal tract, or the best alternative in special circumstances. For example, in 25 cancer patients with gastrointestinal fistulas treated with parenteral nutrition, 44% closed spontaneously after an average of one month (including those with cancer involving the fistula) and an additional 28% were closed surgically.[44] (In contrast, patients with enteric fistulas arising in irradiated bowel do not achieve sustained spontaneous closure.) For another example, of eight patients with esophageal anastomotic leaks, only one recovered after emergency surgery. The next eight patients with this complication were treated with parenteral nutrition and fasting, and six recovered.[45]

There may be multiple reasons for the lack of consensus regarding the role of nutritional support around the time of therapy in patients with gastrointestinal cancer. In studies addressing the role of nutritional support in this patient population, methodological shortcomings have included:

1. enrolling patients with heterogeneous sites of disease or stage;
2. failing to stratify patients based on nutritional status;
3. neglecting to account for co-morbid illnesses;
4. providing nutritional support of variable composition, route of delivery, rate or duration;
5. assessing nutritional repletion using unclear criteria;
6. defining post-therapeutic morbidity and mortality inconsistently;
7. neglecting to distinguish malnutrition-related complications from other complications; and
8. using methods to assess malnutrition which are cumbersome and which may not have clear clinical relevance.[46]

No trial has met the ideal of including patients who:

1. share the identical diagnosis;
2. have the same degree of malnutrition;
3. receive consistent nutritional support;
4. undergo a standardized therapeutic intervention performed by equally experienced teams; and
5. are enrolled in sufficient numbers to confirm that quantitative differences between groups are statistically significant.

This chapter reviews the literature addressing the problem of malnutrition and the impact of nutritional support specifically in patients with gastrointestinal cancer, divided into sections on esophageal, gastric, colon, pancreatic, and liver cancer.

Esophageal Cancer

Despite suffering from dysphagia, patients with esophageal cancer may have surprisingly infrequent weight loss (2%, 53% of the time, in one early study[47]). However, patients frequently have decreased indices of cell-mediated immunity, including attenuated reaction to primary and recall antigens, impaired blastogenesis, and decreased T-lymphocyte number, which has correlated with lower survival.[48] This immune dysfunction, however, may be related to the presence of a cancer, per se, rather than to malnutrition, since the changes were independent of albumin level or body weight. Also, three weeks of enteral therapy improved T-lymphocyte numbers, but not anergy or depressed blastogenesis.[49] It is possible that parenteral nutrition would have had an equivalent result. In a study assessing metabolic state, the effects of jejunal feeding and parenteral nutrition were similar, such as the suppression of gluconeogenesis and the conservation of protein stores.[50] In the immediate postoperative period, administering clonidine by continuous infusion (given for prophylaxis against alcohol withdrawal) prevented the negative nitrogen balance seen in nonalcoholic controls.[51] Providing at least 0.2 g N/kg body weight per twenty-four hours can maintain positive nitrogen balance,[52] but it is unknown to what degree, if any, short-term improvements in nitrogen balance influence outcome.

Although surgical resection removes the obstruction and, therefore, at least one barrier to adequate nutrition, it can create new nutritional challenges. In addition to worsening reflux disease due to loss of the lower esophageal sphincter, esophagectomy can cause gastric stasis and isolated fat malabsorption. Both phenomena have been attributed to the effects of vagotomy, although the mechanism for fat malabsorption is unclear.[53] (Substitution of medium-chain triglycerides [which can be absorbed by the small intestine directly] for long-chain fatty acids has led to reduced fecal fat loss.)[54] Anastomotic leaks which are often treated with prolonged parenteral nutrition, are another threat to nutritional repletion. In a study of 617 patients who had esophageal resection and esophagogastric anastomosis, 39 suffered an anastomotic leak (over half of whom died from the complication). Albumin concentration below 3 gm/dL (along with a surgical margin being positive for cancer and use of a cervical anastomosis) was predictive of anastomotic leak.[55] In another study of patients who developed fistulas, those whose leak persisted were more likely to have had either residual tumor after palliative operations or low presurgical albumin levels.[56] Given the poor healing ability of these patients when malnourished, immediate surgical repair of anastomotic leaks should be considered in patients with low preoperative albumin levels.

Not surprisingly, the stage of esophageal cancer has correlated with the degree of negative nitrogen balance and weight loss,[52] and protein-calorie malnutrition has been identified as a risk factor for operative mortality.[57] One group of researchers developed a "Host Defense Index" which included nutritional parameters of arm muscle circumference, albumin, and transferrin to help discriminate between patients who were at high risk from those who were at low risk for perioperative mortality. It was used to identify patients who could benefit from modification of the proposed surgical procedure and more vigilant management of perioperative infections. Prospective implementation of the Host Defense Index may have been among the reasons that fatal complications dropped from 80% to zero.[58]

Despite enteral feedings, those patients who were not able to eat at all after treatment (surgery, radiotherapy, and/or chemotherapy) were less likely to survive, based on univariate analysis. In multivariable analysis, however, the mode of nutrition delivery did not persist as a predictor of survival.[59] Survival differences were better explained by retained variables such as persistent disease, which likely correlated with an inability to eat. For patients with lesions obstructing the esophagus, pyriformostomy tube feeding may maintain enough of an esophageal lumen to allow swallowing of oral secretions,[60] while being more comfortable and cosmetically appealing than nasogastric tubes. In patients who have dysphagia from recurrence of carcinoma after esophagectomy, a feeding tube can be placed percutaneously via direct endoscopic jejunal puncture.[61]

Beneficial effects of parenteral nutrition have been claimed in patients with esophageal cancer as early as 1965 (although in a nonrandomized analysis using historical controls).[62] One early study included 15 patients with esophageal cancer undergoing thoracotomy who were randomized to receive parenteral nutrition for about one week before and one week after surgery. Although patients had similar weight loss compared to the five control patients, wound healing appeared to be better in those who received parenteral nutrition.[63] Another compared 12 patients randomized to 4 weeks of preoperative nutrition via a gastrostomy to 12 patients randomized to parenteral nutrition. The latter group achieved an earlier positive nitrogen balance and greater weight gain, despite the gastrostomy patients receiving a greater nitrogen delivery (although the weight gain may not have been from anabolism—albumin levels were similar between groups). The number of perioperative complications or death was twice as large in the gastrostomy group (14 in 10 patients) compared to the parenteral nutrition group, but the number of observations was too small for any statistical conclusions. The gastrostomy patients were, however, more content with their care. Their hunger was relieved with the feedings, none developed diarrhea, and they were ambulatory. The patients receiving parenteral nutrition, in contrast, were reluctant to walk around despite encouragement for fear of accidents occurring with the intravenous pole, often had hunger for the first week of therapy, and had formulary expenses 15-17 times higher than the enteral formula.[64]

In a study of patients with localized, distal esophageal squamous cell carcinoma who lost more than 20% of their body weight or were unable to swallow liquids, parenteral nutrition administered over two weeks (without any other therapy or interventions) improved nitrogen balance and weight gain better than jejunostomy feedings.[65] However, the improved nitrogen balance may have been solely a reflection of the greater amount of protein delivered in the parenteral product, and the weight gain may have been from the accumulation of fat or water rather than muscle. In one retrospective review of surgical patients treated between 1973 and 1980, malnourished patients who received parenteral nutrition (most beginning two weeks before surgery) lost less weight, had fewer major and minor complications, but had higher perioperative mortality and similar five-year survival compared to well-nourished patients who had no parenteral nutrition.[66]

In a more recent review of 64 patients admitted to the hospital for the first time with cancer of the esophagus, the 37 who received parenteral nutrition had a reduced incidence of weight loss (although not necessarily muscle loss) but an increased incidence of pulmonary sepsis, with a resultant increase in length of hospitalization or death and an average increase of $6,000 in hospital fees (in 1984 dollars).[67] However, a greater proportion of the patients who received parenteral nutrition had surgical resections, and the retrospective design raises the likely possibility of selection bias, where patients considered to be at highest risk for complications were the same patients most likely to be given parenteral nutrition. It has been proposed that in patients with esophageal cancer, nutrition should be delivered enterally whenever possible.[68]

Gastric Cancer

Patients with gastric cancer appear to be particularly susceptible to malnutrition, which may be multifactorial. In one study, 60% of patients with gastric cancer had anorexia, compared with 37% of those with colorectal cancer.[69] Weight loss was seen in 84% of patients with gastric cancer,[70] which was unmatched by patients with esophagus, pancreas or primary liver cancer.[71] Anorexia from functional or mechanical derangements is probably responsible for the majority of malnutrition developing in patients with cancers of the upper versus lower gastrointestinal tract, since energy expenditures appear similar.[72] Patients treated with surgical resection are at risk for esophageal reflux disease and dumping syndrome. In a study of surgical technique, the "pouch and Roux-en-Y" approach for creating an enteric reservoir after total gastrectomy was associated with toleration for greater meal volumes and better weight recovery, when compared to "simple Roux-en-Y" and "pouch and interposition" techniques.[73] Relatively little has been written on the application of enteral or parenteral nutritional support in patients with gastric cancer. Use of postoperative parenteral nutrition in patients with stage III or IV gastric cancer has been claimed to restore cell-mediated immunocompetence, increase tolerance for 5-fluorouracil, and improve three-year survival (54% versus zero). There was no description, however, of how patients were selected to be in the group receiving parenteral nutrition, and hence, any differences could merely have reflected selection bias.[74]

A study of preoperative nutritional assessment of 169 patients with stage IV gastric cancer found that those who were able to undergo gastrectomy had significantly higher albumin, prealbumin, retinol binding protein, transferrin, and vitamin A levels compared to those who received only bypass or exploratory laparotomy. No nutritional parameters in the group were significantly different between those who suffered postoperative complications and those who did not. Thirty-four of the patients received two weeks of preoperative parenteral nutrition for an albumin less than 3.5 g/dl. If albumin rose above 3.5 g/dl, then patients were twice as likely to receive a gastrectomy. In predicting postoperative complications in this subgroup after parenteral nutrition, an albumin of at least 3.0 g/dl, prealbumin of 20 mg/dl, or lymphocyte count of 1,000/mm^3 each possessed specificity in excess of 96%, but being at the terminus of the receiver operating characteristic curves, suffered from low sensitivity. The prealbumin suggested discrete values which possessed both sensitivity and specificity above 80%, but these relevant points on the curve were neither commented upon in the text nor labeled in the figure. Of greater concern, there was no description of the clinical relevance of gastrectomy versus enteral bypass in this population with advanced cancer.[75]

Given the lack of compelling evidence in favor of parenteral nutrition in patients with gastric cancer, enteral access remains the preferred route, although gastric outlet obstruction may require endoscopic stenting, surgical bypass, or decompression gastrostomy and feeding jejunostomy tubes. In patients who are not surgical candidates but in whom nutritional support is desired, inventive routes such as the biliary tree may be used for establishing enteral access.[76] If parenteral nutrition must be used, there is some evidence that specialized formulations may improve nutritional parameters in patients with gastric cancer, although there is no evidence that these compositions improve outcome.

In a randomized multi-center study of 173 patients having surgery for gastric cancer, postoperative parenteral nutrition supplemented with branched-chain amino acids led to decreased 3-methyl-histidine levels (an indication of skeletal muscle catabolism) in patients having subtotal or total gastrectomy, and to improved nitrogen balance in patients having total gastrectomy. There were no differences in albumin or other serum protein levels. There did not appear to be any adverse side effects of higher circulating branched-chain amino acids. Those with complicated postoperative courses, however, were excluded from analysis.[77] In a smaller study, seven patients given methionine-depleting parenteral nutrition and continuous infusion of 5-fluorouracil for seven days before surgery for advanced gastric cancer had marked degeneration of the tumor at the time of resection, compared to little direct impact on the tumor in the seven control patients who received conventional parenteral nutrition with 5-fluorouracil. The proposed mechanism is that the tumor is unable to proliferate without L-methionine, which is essential for methylation in the synthesis of DNA, RNA, and protein. Additionally, the methionine depletion appeared to further increase 5-fluorouracil's inhibition of thymidylate synthase activity.[78] The role for specialized parenteral nutrition in patients with gastric cancer remains to be determined.

Colon Cancer

Patients with cancer of the colon are less likely to have malnutrition compared to patients with cancers of the upper gastrointestinal tract. This observation may be due to less frequent anorexia,[69] nausea, and inanition. Patients with colon cancer may not develop gastrointestinal complaints until late in their disease when they present with colonic obstruction. Colon cancer may also exhibit little effect on energy expenditure. Basal energy expenditure in patients with disease metastatic to the liver did not differ from patients without metastases, and energy expenditure did not change following potentially curative surgical resection.[79] Indicators of poor prognosis on univariate analysis included weight loss, low albumin, and low caloric intake. The best prognostic indicator on multivariate analysis was albumin. Low caloric intake was preserved as a prognostic indicator in the multivariate analysis, but weight loss was discarded (suggesting that it correlated better with albumin or caloric intake than with prognosis).[80] Patients who are malnourished may be more likely to remain malnourished following therapeutic interventions. More than half of 68 patients who were well-nourished before surgery for colorectal cancer established oral intake of at least 60% of their caloric needs by the tenth postoperative day, whereas only one-quarter of the 33 patients who were malnourished had achieved this goal by ten days.[81]

Conventional management of patients after bowel resection includes support with intravenous fluids and nothing by mouth until flatus is passed, heralding the resolution of the postoperative ileus. Although postoperative gastroparesis is common, the small intestine remains functional in the postoperative period.[82] Despite the potential for postoperative gastroparesis, eight consecutive elderly, high-risk patients were allowed immediate regular supplemented meals and cisapride 20 mg twice a day after elective laparoscopic colonic resection for neoplastic disease. Pain was controlled with epidural anesthesia and oral narcotics. Two had mild nausea on one occasion, none vomited, six patients passed feces on postoperative day one, and all were able to be discharged on the second postoperative day. When surveyed one month later, none felt that they had been discharged prematurely.[83]

A less aggressive approach to postoperative enteral nutrition in patients having a bowel resection is to provide immediate jejunal feeding. The clinical value of postoperative maintenance of nitrogen balance and weight is, however, unclear. For example, in a study of parenteral nutrition following major surgery, temporary undernutrition and weight loss in the control group had no impact on recovery.[84] enteral nutrition may decrease gut permeability, but, further study in humans is needed before increased gut permeability can be linked with increased susceptibility to sepsis from bacteria originating in the gut.[85]

Patients receiving chemotherapy or radiation therapy present nutritional challenges as well. Not only do the therapies cause side effects which lead to reduced intake, but the treatment causes increased nutritional losses. For example, patients with Dukes D colon cancer receiving chemotherapy had increased nitrogen losses, likely due to decreased protein synthesis.[86] Although nutritional interventions have allowed some patients to receive therapy who otherwise would have been too depleted to tolerate the intervention, outcomes have remained disappointing. Fifty-one patients randomized to receive oral nutritional support during the first 12 weeks of chemotherapy for colorectal cancer had higher caloric intake than the 33 control patients, but fared no better in weight change, tumor response, tolerance to chemotherapy, time to progression, or survival. Nutritional counseling had no impact on toleration of chemotherapy, progression of tumor, or survival. Enteral tube feedings were refused by the majority of patients who were failing to meet their targeted caloric intake.[80]

Randomized trials of parenteral versus oral nutrition during chemotherapy for colorectal cancer have been disappointing, as well. In a study of 45 patients receiving identical chemotherapy for metastatic colon cancer, 14 days of pretreatment parenteral nutrition continued throughout chemotherapy was well-tolerated, associated with improved mood, and did not stimulate tumor growth, but survival was significantly decreased (79 versus 308 days).[87]

The impact of specialized amino acid formulations and lipids in parenteral nutrition have been investigated in patients with colon cancer. In a study of 12 patients, glutamine-supplemented parenteral nutrition limited negative nitrogen balance and maintained intramuscular glutamine concentrations, but outcomes of the patients were not reported.[88] In addition to benefiting skeletal muscle metabolism, glutamine may be essential for lymphocyte metabolism in times of stress, as well. In a study of 22 patients who had a colorectal resection for a preoperative diagnosis of carcinoma, postoperative glutamine-supplemented parenteral nutrition enhanced in vitro T-lymphocyte response to stimulation, but reduced neither negative nitrogen balance nor infections.[89] Arginine, which also possesses anabolic and immune stimulatory properties, did not enhance mitogen-stimulated lymphocyte stimulation in postoperative colorectal cancer patients treated with parenteral arginine given as the sole protein source.[90] There has been concern that infusion of lipids might cause relative immunosuppression. Parenteral lipids, however, did not affect neutrophil chemotaxis when administered to patients with colon cancer.[91]

The concern for preferential tumor stimulation by parenteral nutrition has been investigated to some extent in patients with colon cancer. In a study of 18 patients with localized colorectal cancer, the nine patients randomized to receive parenteral nutrition for 24 hours before surgery had tumor protein synthesis almost twice as high as those patients who fasted.[92] Potential markers of tumor proliferation include polyamines. Putrescine levels increased significantly after parenteral nutrition in 16 patients with colorectal cancer, while the same nutritional therapy caused no change in levels in control patients without cancer.[93] Others, however, found that such changes probably reflected increased whole-body, rather than tumor-specific, metabolic activity.[94] The more amino acids administered parenterally, regardless of the composition, the more protein synthesis occurred, while the rate of muscle breakdown remained constant.[95] Metabolic expenditures increased when calories administered via parenteral nutrition exceeded basal resting metabolic expenditure.[96] Branched chain amino acid-supplemented parenteral nutrition stimulated in vivo colorectal cancer protein synthesis less than conventional parenteral nutrition, but the effect was not selective. The same trend was seen in skeletal muscle protein synthesis.[97]

Even if parenteral nutrition cannot be recommended routinely, its use must be individualized to patient circumstances. A Jehovah's Witness with an obstructing sigmoid colon cancer had a profound anemia prohibiting surgery. After she failed to respond to oral iron, institution of parenteral nutrition, human erythropoietin and parenteral iron produced enough of a correction in her anemia for her to tolerate surgery.[98] The nutritional support could have improved levels of iron-binding and transport proteins such as ferritin and transferrin, and may play an integral role in treating patients who are profound anemic, unwilling or unable to receive transfusions, and incapable of tolerating enteral nutrition.

Home parenteral nutrition is an option being exercised for many patients with colorectal cancer who have contraindications to enteral nutrition. The OASIS North American Home Nutrition Support Patient Registry followed 1,362 active cancer patients between 1984 and 1989, 20% of whom were those with colorectal cancer, comprising the largest subgroup.[99] The 20% who were able to resume full oral feeding likely were those who survived aggressive therapy which temporarily caused gastrointestinal dysfunction. Home parenteral nutrition for this subgroup of cancer patients appears justified, as it is well tolerated and associated with only a 1% incidence of parenteral nutrition-related mortality. The benefits of home parenteral nutrition remain unclear when being used to extend life a small increment in patients with advanced colorectal cancer. In a retrospective uncontrolled review, patients with advanced colon cancer who received home parenteral nutrition survived longer than those who did not,[100] but the difference could have been due to selection bias.

Home parenteral nutrition is probably inappropriate for the majority of cancer patients who initiate home parenteral nutrition but who are expected to die within six to nine months. Forces encouraging this increasing trend may include a fascination with technology, expanded availability of services, pressure from families, insurers' preference for substituting parenteral nutrition at home for that in the hospital[101] (since despite being $75,000 to $150,000/year [in 1992 dollars] home parenteral nutrition was still one-third the cost of hospital care[102]), Medicare's reimbursement for home parenteral nutrition but not for home parenteral hydration, and the opportunity for hospitals to profit from joint ventures with infusion companies whereas inpatient parenteral nutrition is not profitable.[99]

Pancreatic Cancer

As in patients with esophageal cancer, weight loss may be significant in patients with pancreatic cancer, although it is frequently secondary to anorexia rather than dysphagia. One cause of anorexia may be depression, which was noted more than 60 years ago to be a frequent presenting symptom of patients with pancreatic cancer.[103] In a review of 52 patients with pancreatic cancer reported between 1923 and 1991, 71% had a depression-related disorder, one-third of whom developed the psychiatric symptoms prior to any physical symptoms.[104] Depression appears to be less frequent in patients with other gastrointestinal malignancies. For example, while depression was diagnosed in 50% of patients who ultimately were diagnosed with pancreatic cancer, none of the patients diagnosed with gastric cancer met criteria.[105] It is unclear what psychobiological mechanism causes patients with pancreatic cancer to be at particularly high risk for depression,[104] but it may place patients with pancreatic cancer at particularly high risk for malnutrition. Although a combination of psychotherapy, cognitive-behavioral techniques, and antidepressant medication has been recommended to treat patients with pancreatic cancer and depression,[106] the impact of such treatment on malnutrition remains unknown.

Surgery may relieve the biliary, pancreatic or duodenal obstruction, but may not immediately alleviate signs of gastric outlet obstruction. Following pylorus-preserving pancreaticoduodenectomy (modified Whipple procedure), up to 50% of patients may have delayed gastric emptying and require gastric decompression for a median of 8-14 days. Introduced during the surgical procedure, an apparatus consisting of a 12 French jejunal feeding tube placed through a Y-connector fitted to a modified 32 French malecot catheter can be used both to decompress the stomach (obviating the need for a nasogastric tube) and to provide jejunal feeding. The apparatus can be removed in the outpatient setting when adequate gastric emptying function returns.[107] Another alternative for establishing jejunal feeding is to convert biliary-enteric anastomotic stents to jejunal feeding tubes in the early postoperative period.[108]

Although several studies have identified indicators of malnutrition which predicted postoperative complications, including weight loss greater than 10%, albumin less than 3.0 g/dL, and anergy,[109-111] studies investigating the role of nutritional intervention in the perioperative period have had mixed results. One study investigated preoperative nutrition. Sixty patients with obstructive jaundice (most of whom had pancreatic cancer) who received preoperative enteral or (less often) parenteral nutrition for at least 12 days (and a mean of 20 days) between percutaneous transhepatic biliary drainage and pancreatobiliary surgery reduced their morbidity from 47-18% and their mortality from 13-4%.[112] A study of postoperative nutritional support, however, came to opposite conclusions. In a recent prospective randomized trial of 117 patients undergoing pancreatic resections, the group receiving routine postoperative total parenteral nutrition suffered a statistically significant higher rate of major complications (45% vs. 23%). This relationship remained significant even when complications thought to be reduced by bowel rest were analyzed separately (such as fistulas, abscess, obstruction, and anastomotic leak). Mortality was similar between groups.[113] The authors postulated that the higher rate of infectious complications in the patients receiving parenteral nutrition, in particular abscess formation, may have resulted from increased translocation of bacteria across the intestinal wall occurring in the absence of enteral feeding.[114]

If patients survive for an extended time, their prospects for nutritional repletion are excellent. In 25 patients who were free of recurrent disease at least six months out from either conventional or pylorus-preserving pancreaticoduodenectomy, quality-of-life assessments demonstrated nearly normal well being and little or no impairment in gastrointestinal function. These results were similar to the matched cholecystectomy control group. Compared to the pylorus-preserving pancreaticoduodenectomy group, the group having a conventional pancreaticoduodenectomy were more likely to complain of fullness and restricted food intake, but were less likely to suffer heartburn. Although ten patients required dietary or pharmacological intervention for diabetes, and five patients reported greasy stool, no patients were malnourished and mean weight was greater than mean preoperative weight and ideal weight.[115]

In 23 patients with pancreatic cancer who were not surgical candidates (only one of whom allegedly had disease confined to the pancreas) and who modified their oral intake at least a "moderate extent" toward a macrobiotic diet for at least 3 months, survival averaged 17 months, compared to patients with similar disease and time period in the Surveillance Epidemiology and End Results (SEER) National Tumor Registry who lived only an average of 6 months.[116] The authors pointed out the potential bias in the selection of cases and the limitations of a retrospective study, but their observations remain intriguing, especially when considered with evidence in experimental models that diets high in fat increased the incidence of pancreatic neoplasms, while diets with a 10% reduction of calories protected against neoplasms.[117]

Liver Cancer

In patients with liver disease, nutritional status cannot be determined reliably using traditional methods.[118] For example, a low albumin may reflect limitations in hepatic synthesis rather than depletion of visceral proteins, while protein balance may be overestimated due to impaired urea synthesis and accumulation of ammonia.[119] Edema may cause underestimation of protein or fat loss when determined by mid-arm muscle circumference or skinfold thickness, and, combined with ascites, may produce a falsely reassuring "normal" body weight. Not surprisingly, prognostic nutritional indices have failed to predict complications in patients having liver transplantation.[120]

The majority of patients with primary liver cancer have underlying cirrhosis, which may limit hepatic regenerative capacity or functional reserve, at least one of which is needed to survive liver resection. Adequate nutrition increases liver regeneration in humans[121] and in the rat model.[122] Given that the liver demonstrates a depressed metabolic capacity immediately following liver resection,[123] it is likely that reestablishing functional reserve requires adequate nutrition, as well. Despite having average energy requirements (unless under acute metabolic stress),[124] patients with cirrhosis may have nutritional deficiencies from a combination of poor intake combined with some impairment of digestion, absorption, or metabolism.[125] Digestion, absorption, and metabolism of amino acids from routine protein sources, however, appears to be maintained.[126] Hence, protein should be unrestricted unless there has been prior or current encephalopathy, and patients should be encouraged to eat. When diets were supplemented, patients had improved nutritional intake.[127] Due to depressed hepatic glycogen stores, patients with cirrhosis may sacrifice amino acids for gluconeogenesis even after a short interval of fasting. Including a late evening meal improved nitrogen metabolism in an uncontrolled study.[128] At least in patients preparing for liver transplantation, preoperative ingestion of 120% of the calories calculated by the Harris-Benedict basal level for ideal body weight slightly exceeded resting energy expenditure throughout the perioperative period, and placed more than 40% in a positive nitrogen balance preoperatively.[129] For liver cancer patients with cholestasis and steatorrhea, supplementation of enteral diets with medium chain triglycerides is appropriate, since they are absorbed into the portal circulation without being transformed or transported in chylomicrons.[130] Sixty milliliters of medium chain triglycerides delivered in divided doses in dressings or shakes provide 450 kilocalories per day. If medium chain triglycerides are used as an exclusive fat source, linoleic acid will need to be delivered, as well, to prevent essential fatty acid deficiency.[118]

In patients who are unable to maintain adequate oral intake, enteral feeding via tube may improve clinical condition and outcome.[131] As variceal hemorrhage appears to be a function of the magnitude of portal hypertension rather than of mucosal trauma, enteral feeding via a soft, small-bore feeding tube is acceptable if oral feedings are not feasible.[118] Standard protein delivery has been well tolerated in selected patients.[132] Percutaneous placement of feeding tubes, however, should be avoided due to the risk of hemorrhage from piercing gastric collateral vessels. Ascites is another contraindication to percutaneous technique, although it may not be absolute.[133]

If parenteral nutrition is needed, standard amino acid solutions have been used without precipitating encephalopathy,[134] even in those patients who were previously intolerant of smaller amounts of ingested protein.[118] In some patients with liver disease, choline, cystine and tyrosine may be essential,[135,136] and their inclusion in the formulation of amino acids should be confirmed. Glutamine might also be a beneficial constituent of parenteral nutrition for patients with liver cancer. Free glutamine constitutes 61% of the total intracellular pool of amino acids,[137] and becomes depleted in states of physical stress.[138] While glutamine may be an essential amino acid for intestinal mucosa, it is also a principal fuel for rapidly dividing cancers. Thus, while glutamine-supplemented parenteral nutrition might be more likely to preserve intestinal mucosal integrity and overall protein synthesis, it might also stimulate tumor growth. However, at least in rats inoculated with hepatoma cells, glutamine improved nitrogen balance without enhancing tumor growth,[139] paralleling results in hepatoma-inoculated rats administered standard parenteral nutrition.[140] Additionally, by stimulating the release of glucagon from the pancreas, glutamine-supplemented parenteral nutrition normalized the portal vein insulin to glucagon ratio in rats and protected the liver against steatosis[141] and cholestasis.[142]

Another potentially beneficial intervention when using parenteral nutrition is the inclusion of branched-chain amino acids. They are fuel for skeletal muscle and increase protein synthesis in liver and muscle. They also inhibit muscle protein catabolism, which may be most important, given that the pool of plasma amino acids derived from endogenous protein breakdown is five times greater than the pool derived from the diet in patients with cirrhosis, and 12 times higher in patients with fulminant hepatic failure.[143] In addition to reducing hepatic encephalopathy[144] and possibly improving neurological function even in those patients without encephalopathy,[145] branched-chain amino acids also could treat the protein-calorie malnutrition of some patients with liver cancer. Most trials of branched-chain amino acids focused on encephalopathy and had, at best, inadequate assessments of nutritional status. The only large study with reasonable assessments of nutritional status found improved nitrogen balance compared to casein supplement at three months, but equivalent nitrogen balance at six months.[146] Branched-chain amino acid supplementation did not improve postoperative outcomes in patients with liver dysfunction or cirrhosis having transplantation.[147,148] Since branched-chain amino acid supplements in solution or powder cost at least ten times that of standard amino acids, provide only marginal benefits to nutritional status, and do not improve outcome in liver transplant patients, their routine use in liver cancer patients cannot be supported.[118]

Most investigators have studied nutritional interventions in patients who have hepatic encephalopathy or are having liver transplantation. However, there is a study of 150 patients undergoing potentially curative hepatic resection for hepatocellular carcinoma who were randomized to receive either parenteral nutrition enriched with branched-chain amino acids for one week before and one week after surgery or parenteral crystalloid solution postoperatively. After excluding those patients who had intra-abdominal metastases discovered during surgical exploration, the patients receiving the parenteral nutrition had a statistically significant lower morbidity than the control group (34% vs. 55%), and a trend toward lower mortality (8% vs. 15%).[149] It is possible, however, that the higher "morbidity" in the control group was a result of the liberal definition of pneumonia (positive culture in association with pneumonic or atelectatic changes on chest radiograph) combined with this group's significantly greater incidence of ascites (and thus atelectasis).[150] Duration of hospitalization was similar.

Immediately following liver resection, the ratio of arterial acetoacetate to 3-hydroxybutyrate falls, reflecting a reduced hepatic redox potential of the liver. Because of depressed Krebs cycle activity in this scenario, adenosine triphosphate is generated preferentially by beta oxidation of fatty acids. In the immediate postoperative period, intravenous administration of high concentrations of glucose or doses of insulin should be restricted, since a hyperglycemic and hyperinsulinemic state inhibits fatty acid liberation from adipocytes and hepatic ketone production. In rats, infusion of lipids[151] or monoacetoacetate[152] increased the rate of liver regeneration following liver resection, and lipids remain a safe source of nonprotein calories in patients with liver disease requiring parenteral nutrition.[118]

Extrapolating data from the transplant literature, a randomized study of 24 liver transplantation patients compared parenteral nutrition with nasojejunal feeding started during the first postoperative day. Both groups maintained anthropometric indices of nutritional status, had equivalent incidence of infections (including gut-related infections) and diarrhea, preserved intestinal absorptive capacity and impermeability to macromolecules, and had similar length of stays. By postoperative day ten, 87% of patients had achieved an adequate oral intake. The nasojejunal tube with an integral gastric decompression port was easily positioned in 11 of 14 patients and remained patent in all patients, none of whom suffered pulmonary aspiration. The enteral feeding was one-tenth the cost of parenteral nutrition.[153] Despite the high cost of parenteral nutrition, in another study of liver transplant patients those who received postoperative parenteral nutrition had a mean reduction in hospital costs of $21,000.[147] A similar study in patients with liver cancer has not been performed.

Patients with liver cancer undergoing hepatobiliary surgery or chemotherapy can almost always be fed exclusively, or at least partially, by enteral means.[123] The role of parenteral nutrition in the management of patients with liver cancer remains to be determined.[154] In malnourished patients with gastrointestinal cancer and presumably normal livers, three days of preoperative parenteral nutrition supplementing a hospital diet increased hepatic glycogen content and protein synthesis,[155] although seven days of parenteral nutrition was needed to restore plasma concentrations of several hepatically-synthesized proteins.[156] It is unknown if cirrhotic liver would respond similarly. Also unknown is the clinical impact of this response, when it occurs.

Cost Effectiveness

A cost-effectiveness analysis based on the results of Heatley and colleagues[17] and Müller and colleagues,[18] expressed in 1982 dollars, calculated a net savings of $1,720 per patient given 10 days of preoperative parenteral nutrition.[157] Another cost effectiveness analysis addressing the same question in people having gastrointestinal surgery (for unspecified indications) assumed that when either the risk of postoperative complications or the effectiveness of nutritional support in preventing these complications was high, a strategy of providing nutritional support to all patients was most appropriate. When either variable was lower, the most appropriate strategy was to provide nutritional support only to a high-risk subpopulation, identified using a nutritional assessment technique. For populations in whom the postoperative incidence of nutrition-associated complications is 20%, using the Subjective Global Assessment (SGA) which had the best combination of sensitivity and specificity (82% and 72%, respectively), the incremental cost per complication avoided was $11,515 (in 1980-1981 Canadian dollars).[158]

In the subsequent Veterans Administration Total Parenteral Nutrition Cooperative Study Group conducted in the 1980s (which, granted, did not focus specifically on patients with gastrointestinal cancer), incremental costs in 1992 dollars attributed to perioperative parenteral nutrition were above $3,000, which translated to $13,959 per complication avoided in severely malnourished patients. The costs would have been higher in a major urban teaching hospital.[159] In the mid-1980s, an estimate of the cost of a ten-day course of parenteral nutrition for patients anticipating surgery for gastrointestinal malignancy was $3,340, yielding an increased life expectancy of nine weeks.[19]

There are less data available regarding the costs of enteral nutrition. In a study of 111 postoperative patients enterally supported for at least ten days via needle jejunostomy, net savings totaled $33,000 (in the 1970s) compared to the expenses which would have been incurred using parenteral nutrition.[160] Enteral nutrition via tube feeding, if reducing morbidity and mortality to the same degree assumed for parenteral nutrition, would have saved well over $5,000 per patient using one model, leading the authors to conclude that in patients able to tolerate enteral feeding, the adage of "if the gut works, use it" recognizes principles of both physiology and economics.[157]

Conclusion

Limitation of oral intake in response to illness is a behavior that may have evolved because it protects the host, albeit by an unknown mechanism.[150] Efforts to "force-feed" patients with gastrointestinal cancer, while well-intentioned, may not always be in their best interest. At least one-third of parenteral nutrition administered to patients with cancer is likely inappropriate.[161] Although parenteral nutrition has been shown to improve some indicators of nutritional status in patients with cancer (like body weight, serum proteins, nitrogen balance, and in vitro immune function), its impact on morbidity and mortality has been mixed.[162] It is likely that the effects of parenteral nutrition in patients with gastrointestinal, pancreatic, and liver cancer anticipating surgery parallel those found in the Veterans Affairs Total Parenteral Nutrition Cooperative Study Group. In this landmark study of preoperative parenteral nutrition in malnourished patients scheduled to have major abdominal or noncardiac thoracic surgery, parenteral nutrition was found to be helpful only in the patients who were severely malnourished.[163] A need to reverse the physiological state of starvation in order to effect improved surgical outcomes may explain why randomized trials of two to seven days of preoperative parenteral nutrition in patients with gastrointestinal cancer failed to improve outcome,[15,63] while trials providing 7-10 days of therapy resulted in decreased wound infections[17] and decreased mortality.[18]

Continued research into manipulating the composition of enteral nutrition[164-167] and parenteral nutrition[168] might improve immunologic, metabolic, and clinical outcomes. Some day, nutritional interventions might become adjuvant therapy in the treatment of gastrointestinal cancer. In a study of 44 patients with primary gastrointestinal cancers given 15 days of parenteral nutrition, the cell kinetics of tumors from those who had been on a lipid-based regimen mirrored those from patients in other studies who had received chemotherapy or radiotherapy.[169] In the meantime, the generic ASPEN practice guidelines for cancer and perioperative nutritional therapy (see Table 5.2) and the ACP practice guidelines for chemotherapy and parenteral nutrition (see text) apply equally well to the specific challenge of nutritional support in patients with gastrointestinal, pancreatic and liver cancer.

Table 5.2. ASPEN practice guidelines for cancer and perioperative nutritional support

Cancer

1. Enteral tube feeding and parenteral nutrition support may benefit some severely malnourished patients or those in whom oncologic treatment toxicity is expected to preclude adequate oral nutritional intake for more than one week. Nutritional support should be given in conjunction with the initiation of oncologic therapy.
2. Intensive nutritional support is not routinely indicated for well-nourished or mildly malnourished patients undergoing surgery, chemotherapy, or radiotherapy who are expected to maintain adequate oral intake.
3. Parenteral nutrition is unlikely to benefit patients with advanced cancer unresponsive to chemotherapy or radiation therapy.

Perioperative

1. Preoperative nutritional support may benefit severely malnourished patients undergoing major surgery, when given for seven to ten days.
2. Preoperative nutritional support is not routinely indicated for well-nourished, mildly malnourished, or moderately malnourished patients undergoing major surgery.
3. Preoperative nutritional support should be provided to malnourished patients who are expected to otherwise sustain a prolonged period of starvation while awaiting major surgery.
4. Postoperative nutritional support should be provided to severely malnourished patients as soon as possible. Postoperative nutritional support may be indicated for mildly malnourished patients expected to otherwise sustain a postoperative period of starvation longer than one week. Enteral access should be established at the time of surgery.

Adapted from ASPEN Board of Directors. Clinical Guidelines for the Use of Parenteral and Enteral Nutrition in Adults and Pediatrics, Section IV: Nutrition Support for Adults with Specific Diseases and Conditions.[170]

References

1. Dewys W, Begg C, Lavin P et al. Prognostic effect of weight loss prior to chemotherapy in cancer patients. Am J Med 1980; 69:491-497.
2. Lindmark L, Bennegard K, Eden E et al. Resting energy expenditure in malnourished patients with and without cancer. Gastroenterology 1984; 87:402-8.
3. Buzby G, Willford W, Peterson O et al. A randomized clinical trial of total parenteral nutrition in malnourished surgical patients: the rationale and impact of previous clinical trials and pilot study on protocol design. Am J Clin Nutr 1988; 47(Suppl 2):366-381.
4. Brennan M. Malnutrition in patients with gastrointestinal malignancy: Significance and management. Dig Dis Sci 1986; 31(Suppl):77S-90S.
5. Lindh A, Cedermark B, Blomgren H et al. Enteral and parenteral nutrition in anorectic patients with advanced gastrointestinal cancer. J Surg Oncol 1986; 33:61-65.
6. Eriksson B, Douglass H, Jr. Intravenous hypealimentation: An adjunct to treatment of malignant disease of upper gastrointestinal tract. JAMA 1980; 243:2049-2052.
7. Brandl M, Tonak J, Rotler H. Influence of high caloric parenteral nutrition on catabolism and cellular immune competence in carcinoma patients. Aust NZ J Surg 1982; 52:350-353.
8. Shaw J, Wolfe R. Fatty acid and glycerol kinetics in septic patients and in patients with gastrointestinal cancer: The response to glucose infusion and parenteral feeding. Ann Surg 1987; 205:368-376.
9. Shaw J, Wolfe R. Glucose and urea kinetics in patients with early and advanced gastrointestinal cancer: the response to glucose infusion, parenteral feeding, and surgical resection. Surgery 1987; 101:181-191.

10. Goldstein S, Elwyn D, Askanazi J. Functional and metabolic changes during feeding in gastrointestinal cancer. JACN 1989; 8:530-536.
11. Lawrence W. Nutritional consequences of surgical resection of the gastrointestinal tract for cancer. Cancer Res 1977; 37:2379-2386.
12. Shils M. Effects on nutrition of surgery of the liver, pancreas, and genitourinary tract. Cancer Res 1977; 37:2387-2394.
13. Fredrix E, Soeters P, von Meyenfeldt M et al. Resting energy expenditure in cancer patients before and after gastrointestinal surgery. JPEN 1991; 15:604-607.
14. Stein T, Buzby G, Leskiw M et al. Parenteral nutrition and human gastrointestinal tumor protein metabolism. Cancer 1982; 49:1476-1480.
15. Holter A, Fischer J. The effects of preoperative hyperalimentation on complications in patients with carcinoma and weight loss. J Surg Res 1977; 23:31-34.
16. Bellantone R, Doglietto G, Bossola M et al. Preoperative parenteral nutrition in the high risk surgical patient. JPEN 1988; 12:195-197.
17. Heatley R, RHP W, Lewis M. Preoperative intravenous feeding: A controlled trial. Postgrad Med J 1979; 55:541-545.
18. Müller J, Dienst C, Brenner U et al. Preoperative parenteral feeding in patients with gastrointestinal carcinoma. Lancet 1982; 1:68-71.
19. Koretz R. Nutritional support: How much for how much? Gut 1986; 27 (Suppl 1):85-95.
20. Koretz R. Parenteral nutrition before surgery for gastrointestinal cancer (letter). Lancet 1983;180.
21. Thompson B, Julian T, Stremple J. Perioperative total parenteral nutrition in patients with gastrointestinal cancer. Surg Res 1981; 30:497-500.
22. Klein S, Simes J, Blackburn G. Total parenteral nutrition and cancer clinical trials. Cancer 1986; 58:1378-1386.
23. Meijerink W, von Meyenfeldt M, Rouflart M et al. Efficacy of perioperative nutritional support (letter). Lancet 1992; 340:187-188.
24. Schroeder D, Gillanders L, Mahr K et al. Effects of immediate postoperative enteral nutrition on body composition, muscle function and wound healing. JPEN 1991; 15:376-383.
25. Carr C, Ling K, Boulos P et al. Randomized trial of safety and efficacy of immediate postoperative entral feeding in patients undergoing gastrointestinal resection. BMJ 1996; 312:869-871.
26. Hoover HJ, Ryan J, Anderson E et al. Nutritional benefits of immediate postoperative jejunal feeding of an elemental diet. Am J Surg 1980; 139:153-159.
27. Bower R, Talamini M, Sax H et al. Postoperative enteral vs parenteral nutrition: A randomized controlled trial. Arch Surg 1986; 121:1040-1045.
28. Heylen A, Lybeer M, Penninckx F et al. Parenteral versus needle jejunostomy nutrition after total gastrectomy. Clin Nutr 1987; 6:131-136.
29. Muggia-Sullam M, Bower R, Murphy R et al. Postoperative enteral versus parenteral nutritional support in gastrointestinal surgery: A matched prospective study. Am J Surg 1985; 149:106-112.
30. Gaddy M, Max M, Schwab C. Small bowel ischemia: a consequence of feeding jejunostomy? South Med J 1986; 79:180-182.
31. Smith-Choban P, Max M. Feeding jejunostomy: A small bowel stress test? Am J Surg 1988; 155:112-116 (discussion 116-117).
32. Smith C, Sarr M. Clinically significant pneumatosis intestinalis with postoperative enteral feedings by needle catheter jejunostomy: An unusual complication. JPEN 1991; 15:328-331.
33. Donaldson S, Lenon R. Alterations of nutritional status: Impact of chemotherapy and radiation therapy. Cancer 1979; 43:2036-2052.
34. Ohnuma T, Holland J. Nutritional consequences of cancer chemotherapy and immunotherapy. Cancer Res 1977; 37:2395-2406.
35. Donaldson S. Nutritional consequences of radiotherapy. Cancer Res 1977; 37:2407-2413.
36. Daly J, Dudrick S, Copeland E, III. Intravenous hyperalimentation: Effect on delayed cutaneous hypersensitivity in cancer patients. Ann Surg 1980; 192:587-592.
37. Daly J, Reynolds H, Copeland E et al. Effects of enteral and parenteral nutrition on tumor response to chemotherapy in experimental animals. J Surg Oncol 1981; 16:79-86.
38. Reynolds H, Daly J et al. Effects of nutritional repletion on host and tumor response to chemotherapy. Cancer 1980; 45:3069-3074.
39. McGeer A, Detsky A, O'Rourke K. Parenteral nutrition in patients receiving cancer chemotherapy. Ann Intern Med 1989; 110:734-736.

40. McGeer A, Detsky A, O'Rourke K. Parenteral nutrition in cancer patients undergoing chemotherapy: A meta-analysis. Nutrition 1990; 6:233-240.
41. Copeland EI. Intravenous hyperalimentation as an adjunct to radiation therapy. Cancer 1977; 39:609-616.
42. Douglass H, Milliron S, Nava H et al. Elemental diet as an adjuvant for patients with locally advacned gastrointestinal cancer receiving radiation therapy: A prospectively randomized study. JPEN 1978; 2:682-686.
43. Detsky A, Baker J, O'Rourke K et al. Perioperative parenteral nutrition: A meta-analysis. Ann Intern Med 1987; 107:195-203.
44. Copeland EI. Intravenous hyperalimentation as an adjunct to cancer chemotherapy. CA 1978; 28:322-330.
45. Riboli E, Bertoglio S, Arnulfo G et al. Treatment of esophageal anastomotic leakages after cancer resection: the role of total parenteral nutrition. JPEN 1986; 10:82-85.
46. Campos A, Meguid M. A critical appraisal of the usefulness of perioperative nutritional support. Am J Clin Nutr 1992; 55:117-130 .
47. Lindsay J. Diagnosis of carcinoma of the esophagus. Ann Otol Rhinol Laryngol 1941; 50:675-680.
48. Dellon A, Rotvin C, Chretin P. Thymus dependent lymphocyte levels during radiation therapy for bronchogeneic and esophageal carcinoma: Correlation with clinical course in responders and non-responders. An J Roent Ther Nucl Med 1975; 123:500-511.
49. Advani S, Kutty P, Gopal R et al. Immunity in esophageal carcinoma. J Surg Oncol 1983; 24:268-273.
50. Burt M, Gorschboth C, Brennan M. A controlled, prospective, randomized trial evaluating the metabolic effects of enteral and parenteral nutrition in the cancer patient. Cancer 1982; 49:1092-1105.
51. Mertes N, Goeters C, Kuhmann M et al. Postoperative α2-adrenergic stimulation attenuates protein catabolism. Anesth Analg 1996; 82:258-263.
52. Moghissi K, Teasdale P. Parenteral feeding in patients with carcinoma of the esophagus treated by surgery: energy and nitrogen requirements. JPEN 1980; 4:371-375.
53. Shils M, Gilat T. The effect of esophagectomy on absorption in man: Clinical and "metabolic" observations. Gastroenterology 1966; 50:347-357.
54. Shils M. The esophagus, the vagi and fat absorption (collective review). Surg Gynecol Obstet 1971; 132:709-715.
55. Patil P, Patel S, Mistry R et al. Cancer of the esophagus: Esophagogastric anastomotic leak: A retrospective study of predisposing factors. J Surg Onc 1992; 49:163-167.
56. Fan S, Lau W, Yip W et al. Healing of esophageal fistulas after surgical treatment for carcinoma of the esophagus and the upper part of the stomach. Surg Gyn Obstet 1988; 166:307-310.
57. Saito T, Zeze K, Kuwahara A et al. Correlation between preoperative malnutrition and septic complications of esophageal cancer surgery. Nutrition 1990; 6:303-308.
58. Saito T, Shimoda K, Kinoshita T et al. Prediction of operative mortality based on impairment of host defense systems in patients with esophageal cancer. J Surg Onc 1993; 52:1-8.
59. Kelley D, Wolf R, Shaha A et al. Impact of clinicopathologic parameters on patient survival in carcinoma of the cervical esophagus. Am J Surg 1995; 170:427-431.
60. Nissan S, Bar-Maor J, Lernau O. Piriformostomy in the treatment of malignant tumors obstructing the esophagus. Isr J Med Sci 1980; 16:682-683.
61. de la Torre R, Scott J, Unger S. Percutaneous endoscopic jejunostomy in a patient with previous esophagectomy. Am Surg 1991; 57:269-270.
62. Hadfield J. Preoperative and postoperative intravenous fat therapy. Br J Surg 1965; 52:291-.
63. Moghisi I, Hornshaw J, Teasdale P et al. Parenteral nutrition in carcinoma of the esophagus treated by surgery: nitrogen balance and clinical studies. Br J Surg 1977; 64:125-128.
64. Lim S, Choa R, Lam K et al. Total parenteral nutrition versus gastrostomy in the preoperative preparation of patients with carcinoma of the oesophagus. Br J Surg 1981; 68:69-72.
65. Burt M, Stein T, Brennan M. A controlled, randomized trial evaluating he effects of enteral and parenteral nutrition on protein metabolism in cancer-bearing man. J Surg Res 1983; 34:303-314.
66. Daly J, Massar E, Giacco G et al. Parenteral nutrition in esophageal cancer patients. Ann Surg 1982; 196:203-208.
67. Brister S, Chiu R, Brown R et al. Clinical impact of intravenous hyperalimentation on esophageal carcinoma: Is it worthwhile? Ann Thorac Surg 1984; 38:617-621.

68. Duranceau A. Controversies in esophageal surgery. Canadian J Surgery 1989; 32:415-419.
69. Tchekmedyian N, Zahyna D, Halpert C et al. Clinical staging of nutritional status of cancer patients (abstract). Proc Am Soc Clin Oncol 1992; 11:398.
70. LaDue J, Murison P, McNeer G et al. Symptomatology and diagnosis of gastric cancer. Arch Surg 1950; 60:305-335.
71. Shils M. Nutritional problems associated with gastrointestinal and genitourinary cancer. Cancer Res 1977; 37:2366-2372.
72. Hansell DT, Davies JW, Burns HJ. The effects on resting energy expenditure of different tumor types. Cancer 1986; 58:1739-44.
73. Nakane Y, Okumura S, Akehira K et al. Jejunal pouch reconstruction after total gastrectomy for cancer: a randomized controlled trial. Ann Surg 1995; 222:27-35.
74. Yamada N, Kowyama H, Hioki K et al. Effect of postoperative total parenteral nutrition (TPN) as an adjunct to gastrectomy for advanced gastric cancer. Br J Surg 1983; 70:267-274.
75. Yamanaka H, Nishi M, Kanemaki T et al. Preoperative nutritional assessmant to predict postoperative complication in gastric cancer patients. JPEN 1989; 13:286-291.
76. Haskell L, Gordon R, Salomonowitz E et al. Technical developments and instrumentation: Percutaneous transhepatic feeding jejunostomy. J Surg Oncol 1985; 29:57-58.
77. Okada A, Mori S, Totsuka M et al. Branched-chain amino acids metabolic support in surgical patients: A randomized, controlled trial in patients with subtotal or total gastrectomy in 16 Japanese institutions. JPEN 1988; 12:332-337.
78. Goseki N, Yamazaki S, Shimojyu K et al. Synergistic effect of methionine-depleting total pareneral nutrition with 5-fluorouracil on human gastric cancer: a randomized, prospective clinical trial. Jpn J Cancer Res 1995; 86:484-489.
79. Hansell DT, Davies JW, Burns HJ. Effects of hepatic metastases on resting energy expenditure in patients with colorectal cancer. Br J Surg 1986; 73:659-62.
80. Evans W, Nixon D, Daly J et al. A randomized study of oral nutritional support versus ad lib nutritional intake during chemotherapy for advanced colorectal and non-small-cell lung cancer. J Clin Oncol 1987; 5:113-124.
81. Meguid M, Mughal M, Debonis D et al. Influence of nutritional status on the resumption of adequate food intake in patients recovering from colorectal cancer operations. Surg Clin N Am 1986; 66:1167-1176 .
82. Catchpole B. Smooth muscle and the surgeon. Aust NZ J Surg 1989; 59:199-208.
83. Bardram L, Funch-Jensen P, Jensen P et al. Recovery after laparoscopic colonic surgery with epidural analgesia, and early oral nutrition and mobilization. Lancet 1995; 345:763-764.
84. Sandstrom R, Drott A, Hyltander A et al. The effect of post operative intravenous feeding (TPN) on outcome following major surgery evaluated in a randomized study. Ann Surg 1993; 217:185-.
85. Souba W. Enteral nutrition after surgery: not routinely indicated in well nourished patients (editorial). BMJ 1996; 312:864.
86. Tayer J, Chlebowski R. Metabolic response to chemotherapy in colon cancer patients. JPEN 1992; 16 (Suppl):65S-71S.
87. Nixon D, Moffitt S, Lawson D et al. Total parenteral nutrition as an adjunct to chemotherapy of metastatic colorectal cancer. Cancer Treat Rep 1981; 65(suppl 5):121-128.
88. Stehle P, Zander J, Mertes N et al. Effect of parenteral glutamine peptide supplements on muscle glutamine loss and nitrogen balance after major surgery. Lancet 1989; 1:231-233.
89. O'Riordain M, Fearon K, Ross J et al. Glutamine-supplemented total parenteral nutrition enhances t-lymphocyte response in surgical patients undergoing colorectal resection. Ann Surg 1994; 220:212-221.
90. Sigal R, Shou J, Daly J. Parenteral arginine infusion in humans: nutrient substrate or pharmacologic agent? JPEN 1992; 16:423-428.
91. Escudier E, Escudier B, Henry-Amar M et al. Effects of infused intralipids on neutrophil chemotaxis during total parenteral nutrition. JPEN 1986; 10:596-598.
92. Heys S, Park K, McNurlan M et al. Stimulation of protein synthesis in human tumors by parenteral nutrition: Evidence for modulation of tumor growth. Br J Surg 1991; 78:483-487.
93. Ota D, Nishiok K, Grossie B et al. Erythrocyte polyamine levels during intravenous feeding of patients with colorectal carcinoma. Eur J Clin Oncol 1986; 22:837-842.
94. Pöyhönen M, Takala J, Pitkänen O et al. Polyamine excretion in depleted patients with gastrointestinal malignancy: effect of perioperative nutrition and tumor removal. JPEN 1992; 16:226-231.

95. Neuhäuser M, Bergström J, Chao L et al. Urinary excretion of 3-methylhistidine as an index of muscle protein catabolism in postoperative trauma: The effect of parenteral nutrition. Metabolism 1980; 29:1206-1213.
96. Merrick H, Long C, Grecos G et al. Energy requirements for cancer patients and the effect of total parenteral nutrition. JPEN 1988; 12:8-14.
97. McNurlan M, Heys S, Park K et al. Tumor and host tissue responses to branched-chain amino acid supplementation of patients with cancer. Clin Sci 1994; 86:339-345.
98. Madura J. Use of erythropoietin and parenteral iron dextran in a severely anemic Jehovah's Witness with colon cancer. Arch Surg 1993; 128:1168-1170.
99. Howard L. Home parenteral nutrition in patients with a cancer diagnosis. JPEN 1992; 16 (Suppl):93S-99S.
100. Levin R, Gordon J, Simonich W et al. Phase I clinical trial with floxuridine and high-dose continuous infusion of leucovorin calcium. J Clin Oncol 1991; 9:94-99.
101. Wesley J, Khalidi N, Faubion W et al. Home parenteral nutrition: A hospital-based program with commercial logistic support. JPEN 1984; 8:585-588.
102. Blackburn G, DiScala C, Miller M et al. Preliminary report on collaborative study for home parenteral nutrition patients. In: Johnson I, ed. Advances in Clinical Nutrition. Lancaster, England: MTP Press, Ltd., 1983:433-448.
103. Yaskin J. Nervous symptoms as early manifestations of carcinoma of the pancreas. JAMA 1931; 96:1664-1668.
104. Green A, Austin C. Psychopathology of pancreatic cancer: a psychobiological probe. Psychosomatics 1993; 34:208-221.
105. Joffe R, Rubinow D, Demicoff K, Maher M, Sindelar W. Depression and carcinoma of the pancreas. Gen Hosp Psychiatry 1986; 8:241-245.
106. Passik S, Breitbart W. Depression in patients with pancreatic carcinoma: diagnostic and treatment issues. Cancer 1996; 78:615-626.
107. Hunter J, White T. Gastrostomy and jejunostomy using a transgastric tube for early enteral nutrition after pylorus-preserving pancreaticoduodenectomy. Surg Gyn Obstet 1991; 173:316-318.
108. Burke D, MH T, McLean G et al. Conversion of choledochojejunostomy stents to jejunal feeding tubes for postoperative enteral alimentation. JPEN 1988; 12:225-226.
109. Pitt H, Cameron J, Postier R et al. Factors affecting mortality in biliary tract surgery. Am J Surg 1981; 141:66-71.
110. Dixon J, Armstrong C, Duffy S et al. Factors affecting morbidity and mortality after surgery for obstructive jaundice: A review of 373 patients. Gut 1983; 24:845-852.
111. Halliday A, Benjamin I, Blumgart L. Nutritional risk factors in major hepatobiliary surgery. JPEN 1988; 12:43-48.
112. Foschi D, Cavagna G, Callioni F et al. Hyperalimentation of jaundiced patients on percutaneous transhepatic biliary drainage. Br J Surg 1986; 73:716-719.
113. Brennan M, Pisters P, Posner M et al. A prospective randomized trial of total parenteral nutrition after major pancreatic resection for malignancy. Ann Surg 1994; 220:436-444.
114. Wells C, Rotstein O, Pruett T et al. Intestinal bacteria translocate into experimental intra-abdominal abscesses. Arch Surg 1986; 121:102-107.
115. McLeod R, Taylor B, O'Connor B et al. Quality of life, nutritional status, and gastrointestinal hormone profile following the Whipple procedure. Am J Surg 1995; 169:179-185.
116. Carter J, Saxe G, Newbold V et al. Hypothesis: Dietary management may improve survival from nutritionally linked cancers based on analysis of representative cases. J Am Coll Nutrition 1993; 12:209-226.
117. Watanapa P, Williamson R. Experimental pancreatic hyperplasia and neoplasia: Effects of dietary and surgical manipulation. Br J Cancer 1993; 67:877-884.
118. Muñoz S. Nutritional therapies in liver disease. Sem Liver Dis 1991; 11:278-291.
119. Khatra B, Smith R, Millikan W. Activities of Krebs-Henseleit enzymes in normal and cirrhotic human liver. J Lab Clin Med 1974; 84:708-715.
120. DiCecco S, Wieners E, Weisner R et al. Assessment of nutritional status in patients with end-stage liver disease undergoing liver transplantation. Mayo Clin Proc 1989; 65:95-102.
121. Schmidt G, Tan P. Protein supplementation in a hepatic resection patient. Nutr Clin Pract 1990; 5:251-253.

122. Sato N, Koyama Y, Oyamatsu M et al. Insulin-like growth factor-I (IGF-I) in malnourished rats following major hepatectomy. JPEN 1994; 18:25S.
123. Helton W. Nutritional issues in hepatobiliary surgery. Sem Liver Dis 1994; 14:140-157.
124. Krevsky B, Godley J. Nutritional support in advanced liver disease. Nutr Support Serv 1985; 5:8-17.
125. Silk D, O'Keefe S, Wicks C. Nutritional support in liver disease. Gut 1991; (Suppl):S29-S33.
126. Morgan M, Hawley K, Stambuk D. Amino acid tolerance in cirrhotic patients following oral protein and amino acid loads. Aliment Pharmacol Ther 1990; 4:183-200.
127. Bunout D, Aicardi V, Hirsch S et al. Nutritional support in hospitalized patients with alcoholic liver disease. Eur J Clin Nutr 1989; 43:615-621.
128. Swart G, Zillikens M, van Vuure J et al. Effect of a late evening meal on nitrogen balance in patients with cirrhosis of the liver. Br Med J 1989; 299:1202-1203.
129. Plevak D, DiCecco S, Wiesner R et al. Nutritional support for liver transplantation: Identifying caloric and protein requirements. Mayo Clin Proc 1994; 69:225-230.
130. Greenberger N, Skillman T. Medium-chain triglycerides: physiologic considerations and clinical implications. N Engl J Med 1969; 280:1045-1058.
131. Cabre E, Gonzalez-Huiz F, Abad-Lacruz A et al. Effect of total enteral nutrition on the short-term outcome of severely malnourished cirrhotics. Gastroenterol 1990; 98:715-720.
132. Smith J, Horowitz J, Henderson J et al. Enteral hyperalimentation in undernourished patients with cirrhosis and ascites. Am J Clin Nutr 1982; 35:56-72.
133. Herman L, Hoskins W, Shike M. Percutaneous endoscopic gastrostomy for decompression of the stomach and small bowel. Gastrointest Endosc 1992; 38:314-318.
134. O'Keefe S, Abraham R, Davis M et al. Protein turnover in acute and chronic liver disease. Acta Chir Scand 1980; 507(Suppl):91-101.
135. Chowla R, Wolf D, Kutner M et al. Choline may be an essential nutrient in malnourished patients with cirrhosis. Gastroenterol 1989; 97:1514-1520.
136. Rudman D, Kutner M, Ainsley J et al. Hypotyrosinemia, hypocystinemia, and failure to retain nitrogen during total parenteral nutrition in cirrhotic patients. Gastroenterol 1981; 81:1025-1035.
137. Bergström J, Fürst P, Norée L, Vinnars E. Intracelluar free amino acid concentration in human muscle tissue. J Appl Physiol 1974; 36:693-697.
138. Fürst P, Albers S, Stehle P. Evidence for a nutritional need for glutamine in catabolic patients. Kidney Internat 1989; 36 (Suppl 27):S287-S292.
139. Kaibara A, Yoshida S, Yamasaki K, Ishibashi N, Kakegawa T. Effect of glutamine and chemotherapy on protein metabolism in tumor-bearing rats. J Surg Res 1994; 57:143-149.
140. Daly J, Copeland III E, Dudrick S. Effects of intravenous nutrition on tumor growth and host immunocompetence in malnourished animals. Surgery 1978; 84:655-658.
141. Li S, Nussbaum M, McFadden D et al. Addition of L-glutamine to total parenteral nutrition and its effects on portal insulin and glucagon and the development of hepatic steatosis in rats. J Surg Res 1990; 48:421-426.
142. Li J, Stahlgren L. Glutamine prevents the biliary lithogenic effect of total parenteral nutrition in rats. JPEN 1992; 17:28S.
143. O'Keefe S, Abraham R, Zayadi A et al. Increased plasma tyrosine concentrations in patients with cirrhosis and fulminant hepatic failure associated with increased plasma tyrosine flux and reduced hepatic oxidation capacity. Gastroenterol 1981; 81:1017-1024.
144. Naylor C, O'Rourke K, Detsky A et al. Parenteral nutrition with branched-chain amino acids in hepatic encephalopathy: a meta-analysis. Gastroenterology 1989; 97:1033-1042.
145. Egbergts E, Schomerus H, Hamster W et al. Branched chain amino acids in the treatment of latent portal systemic encephalopathy: a double-blind placebo-controlled crossover study. Gastroenterol 1985; 88:887-895.
146. Marchesini G, Dioguardi F, Bianchi G et al. Long-term oral branched chain amino acid treatment in chronic hepatic encephalopathy: a randomized double blind casein controlled trial. J Hepatol 1990; 11:92-101.
147. Reilly J, Mehta R, Teperman L et al. Nutritional support after liver transplantation: A randomized prospective study. JPEN 1990; 14:386-391.
148. Blackburn G, O'Keefe S. Nutrition in liver failure. Gastroenterol 1989; 97:1049-1051.
149. Fan S, Lo C, Lai E et al. Perioperative nutritional support in patients undergoing hepatectomy for hepatocellular carcinoma. N Engl J Med 1994; 331:1547-1552.
150. Koretz R. Perioperative nutritional support: A tale of two studies. Gastroenterol 1995; 109:628-630.

151. Nishiguchi Y, Sowa M, Birkhahn R. Comparison of effects of long-chain and medium-chain triglyceride emulsions during hepatic regeneration in rats. Nutrition 1991; 7:23-27.
152. Birkahn R, Awad S, Klaunig J et al. Interaction of ketosis and liver regeneration in the rat. J Surg Res 1989; 47:427-432.
153. Wicks C, Somasundaram S, Bjarnason I et al. Comparison of enteral feeding and total parenteral nutrition after liver transplantation. Lancet 1994; 344:837-840.
154. Haupt W, Husemann B, Sailer D. Postoperative parenteral nutrition following segmental liver resection: Are fat emulsions a risk? Infusion 1990; 17:94-98.
155. Zeiderman M, King R, Young G et al. Metabolic changes in human liver associated with preoperative intravenous nutrition. Clin Sci 1989; 77:343-349.
156. Young G, Chem C, Zeiderman M, Thompson M, McMahon M. Influence of preoperative intravenous nutrition upon hepatic protein synthesis and plasma proteins and amino acids. JPEN 1989; 13:596-602.
157. Twomey P, Patching S. Cost-effectiveness of nutritional support. JPEN 1985; 9:3-10.
158. Detsky A, Jeejeebhoy K. Cost-effectiveness of preoperative parenteral nutrition in patients undergoing major gastrointestinal surgery. JPEN 1984; 8:632-637.
159. Eisenberg J, Glick H, Buzby G et al. Does perioperative total parenteral nutrition reduce medical care costs? JPEN 1993; 17:201-209.
160. Page C, Carlton P, Andrassy R et al. Safe cost-effective postoperative nutrition: defined formula diet via needle-catheter jejunostomy. Am J Surg 1979; 138:939-945.
161. Katz S, Oye R. Parenteral nutrition use at a university hospital: factors associated with inappropriate use. West J Med 1990; 152:683-686.
162. Daly J, Redmond H, Gallagher H. Perioperative nutrition in cancer patients. JPEN 1992; 16 (Suppl):100S-105S.
163. Buzby G, Blouin G, Colling C et al. Perioperative total parenteral nutrition in surgical patients. N Engl J Med 1991; 325:525-532.
164. Kemen M, Senkal M, Homann H et al. Early postoperative enteral nutrition with arginine-ω-3 fatty acids and ribonucleic acid-supplemented diet versus placebo in cancer patients: An immunologic evaluation of Impact. Crit Care Med 1995; 23:652-659.
165. Daly J, Lieberman M, Goldfine J et al. Enteral nutrition with supplemented arginine, RNA, and omega-3 fatty acids in patients after operation: Immunologic, metabolic, and clinical outcome. Surgery 1992; 112:56-67.
166. Daly J, Weintraub F, Shou J et al. Enteral nutrition during multimodality therapy in upper gastrointestinal cancer patients. Ann Surg 1995; 221:327-338.
167. Senkal M, Kemen M, Homann H et alV. Modulation of postoperative immune response by enteral nutrition with a diet enriched with arginine, RNA, and omega-3 fatty acids in patients with upper gastrointestinal cancer. Eur J Surg 1995; 161:115-122.
168. Heys S, Park K, Garlick P et al. Nutrition and malignant disease: implications for surgical practice. Br J Surg 1992; 79:614-623.
169. Franchi F, Rossi-Fanelli F, Seminara P et al. Cell kinetics of gastrointestinal tumors after different nutritional regimens: A preliminary report. J Clin Gastroenterol 1991; 13:313-315.
170. ASPEN. Clinical guidelines for the use of parenteral and enteral nutrition in adults and pediatrics: Perioperative therapy. JPEN 1993; 17(Suppl):21SA-22SA.

Chapter 6

Total Parenteral Nutrition in the Perioperative Nutrition Support of Cancer Patients

Rifat Latifi, Ezra Steiger, John Damreis and Ronald C. Merrell

Malnutrition is a common finding in hospitalized patients. Almost 50% of patients in medical as well as surgical services have significant protein malnutrition and manifest various signs and symptoms of nutrient deficiencies. However, malnutrition is even more common in cancer patients. Up to two-thirds of patients dying from cancer manifest some degree of cancer cachexia and in almost one-third of them the primary cause of death is starvation.[1]

The degree of malnutrition depends on the type, site and the stage of malignant involvement.[2] Patients with favorable subtypes of non-Hodgkin's lymphoma, breast cancer, acute nonlymphocytic leukemia and sarcomas manifest less degree of malnutrition and weight loss (48-61%) compared to patients with unfavorable non-Hodgkin lymphoma, colon cancer, prostate cancer and lung cancer. On the other hand, patients with esophageal cancer,[3] pancreatic and gastric cancer exhibit significant weight loss in 83-87% of cases. As the malignant disease develops and progresses, the metabolic consequences and malnutrition become more evident and severe.[3] Patients with most advanced forms of cancer are severely malnourished and frequently malnutrition and cachexia are the initial manifestations of malignancy.[4] Surgical interventions, chemotherapy, radiation therapy and other therapeutic modalities in patients with cancer cachexia are associated with significant morbidity and mortality.[2,5-7] Reversal of malnutrition and specific nutrient deficiencies with aggressive nutritional support can greatly improve nutritional status and, therefore, may improve the prognosis in this group of patients.[8] The most effective means of provision of nutritional support depends on many factors, chief of them being the location and stage of malignancy. Since patients with cancer, especially those with gastrointestinal cancer, often have obstruction of gastrointestinal tract, food aversion, taste abnormalities, alterations of visceral sensing (early satiety), nausea, vomiting and other metabolic and hormonal abnormalities, it is obvious, that they are unable to maintain their nutritional status with oral intake. Furthermore, surgical interventions, chemotherapy, radiation, and immunotherapy may have significant nutritional and metabolic consequences on these patients. When the use of the gastrointestinal tract becomes ill-advised or impossible, the only means to provide all the nutrients and calories required becomes the parenteral technique in order to attempt to prevent and reverse associated abnormalities in energy, protein, carbohydrate, and fat metabolism, as well as deficiencies of trace elements, vitamins, and other micro nutrients.[9]

Nutritional Support in Cancer and Transplant Patients, edited by Rifat Latifi and Ronald C. Merrell.

Determination of malnutrition and specific nutrient deficiencies of patients with malignancies is of utmost importance. This can be achieved with careful assessments, which need to include careful history, physical examination, anthropometric measurements, body composition, and biochemical determinations of immune response and nutritional indices. No single technique has proven to be superior to the other when assessing biochemical status.[10]

Can malnutrition in cancer patients be reversed effectively with TPN and, will these effects transpose into better therapeutic outcomes? When treating patients with cancer, it is difficult to estimate the benefit of nutritional support for many reasons, chief among them being that current conventional TPN formulas do not address the specific cellular metabolic abnormalities and substrate deficiencies.[4] Moreover, the individual response to malignant processes and antimalignant therapy such as surgery, chemotherapy and radiation therapy may vary greatly and it is difficult to quantify.

In the early days of TPN a great deal of enthusiasm for its use in all malnourished cancer patients was present among clinicians. However, the initial success of TPN in cancer patients still remains to be confirmed in large randomized human studies. With the new era of discovery and refinements of enteral feeding techniques and formulas, enteral feedings has become generally a preferred mode of nutritional support whenever possible. There is, however, a great discrepancy between substrate formulations of enteral diets and TPN, and this factor needs to be taken into consideration when comparing immune and other effects of enteral feeding and total parenteral nutrition support, which undoubtedly has saved thousands of patients from death from starvation. In this chapter we will review the current status of total parenteral nutrition support in perioperative period of malnourished cancer patients.

Perioperative Nutrition Support in Cancer Patients

The use of TPN in cancer patients as a means to provide all nutrients is a logical extension of this life saving therapeutic modality in patients that cannot eat, will not eat or should not eat or cannot be fed enterally, as an adjunctive therapy in cancer patients to improve nutritional status.[9] Several studies have reported an improvement in selected nutritional parameters in patients with various cancers, maintained perioperatively on TPN, while receiving chemotherapy or radiation therapy.

In the Eastern Cooperative Oncology Group Study more than 50% of 3000 patients enrolled in 12 chemotherapy protocols, experienced significant weight loss during the previous six months prior to the initiation of chemotherapy.[2] The incidence of weight loss was related to the type of cancer, ranging from 31% in patients with non-Hodgkin's lymphoma to 87% in patients with gastric cancer. The presence and degree of malnutrition at the time of diagnosis was a poor prognostic sign and was associated with a reduced survival rate.[2] In a study of 365 patients with gastric cancer,[11] the presence of malnutrition was highest in patients with Stage III tumors. In a retrospective study of nutritional assessment and support in cancer patients,[12] a linear predictive model, Prognostic Nutritional Index (PNI) was used to categorize patients into low or high risk for postoperative complications. The low risk group suffered an 8% complication rate and zero percent mortality, compared to a 66% complication rate and 21% mortality in the high risk group.[12]

Nutritional support of patients receiving antimalignant therapy (surgery, radiation and chemotherapy) was evaluated retrospectively in 406 patients.[13] In this study parenteral nutritional support proved to be effective in allowing administration and delivery of adequate antineoplastic therapy. In another study a good correlation between adequate nutrition and response to chemotherapy, with an increase in immune competence and improved tolerance to some chemotherapeutic agents (bleomycin, cyclophosphamide, 5-FU, methotrexate and vincristine) was demonstrated in 30 patients with non-oat cell carcinoma.[14] Improvement of immunocompetence, as defined by the conversion of negative skin test to positive skin test, was demonstrated in 13 of 23 patients receiving TPN during an average of 11.4 days. Response to chemotherapy occurred only in those patients that converted their skin test from negative to positive.

Experimentally, an increased tolerance to 5-FU was demonstrated in rats maintained on TPN.[15] Administration of 15 mg/kg/day of 5-FU for seven days killed 80% of rats fed orally, whereas only 30% of rats maintained on TPN died. When this was applied to 26 patients with metastatic colon cancer (16 treatment group and 10 controls), patients treated with TPN and 5-FU had a better response than control.[15] Based on these studies and other clinical experience, TPN has been used extensively in patients with malignancy. Furthermore, when TPN was used perioperatively in 30 of 56 patients with carcinoma and weight loss, fewer complications were seen in TPN group (13.3% versus 19.2%), although this did not have statistical significance.[16] Moreover, in this study marked decrease in postoperative weight loss and increase in serum albumin was observed in the TPN-treated group. There were no significant differences in minor complications or mortality. However, in this randomized study the number of patients was small and the timing of TPN (72 hours before the surgery) was very short. In a similar randomized prospective study, however, 18 days duration of perioperative parenteral nutrition support in patients with gastrointestinal cancer did not alter the rate of major complications or mortality.[17] Yet, when TPN was given preoperatively to a subgroup of patients with PNI > 40%, it was associated with 2.1-fold reduction in all postoperative complications ($P < 0.001$) and 2.9-fold reduction in major sepsis ($P< 0.005$) as well as 2.7-fold reduction in mortality ($P <0.025$).[12] A significant reduction in the rate of postoperative mortality and major complications in patients with carcinoma of the esophagus, stomach, colon, rectum and pancreas ($P <0.05$) was observed when TPN was given for 10 days preoperatively.[18] In a subsequent study of 55 patients with carcinoma of the esophagus and stomach, lower rates of wound infections, pneumonia, major complications and mortality were found in patients receiving preoperative TPN.[19] The major complication rate was 31% in the control group versus 14% in parenteral nutrition group. Operative mortality rate was decreased from 20-8%.

In another study, perioperative TPN significantly lowered the rate of wound infections after gastric resection for carcinoma and ulcer disease.[20] A significant reduction of the local infection rate by 37% was also reported after laryngectomy in patients receiving TPN.[21] A significant reduction in major, infectious and postoperative complications was also demonstrated in patients with esophageal carcinoma treated with 5 days preoperatively.[22] There were no differences in mortality or major organ failure, however.

Patients with esophageal cancer treated with radiation therapy, who received TPN for two weeks, had an improved nitrogen balance, gained weight and total potassium, decreased protein catabolism, and increased protein synthesis.[23,24] In a prospective randomized study[25] patients undergoing radiation therapy for pelvic malignancy who received TPN had significant improvements in weight, serum transferrin levels and response to skin testing. In addition to parenteral nutrition support restoring fat mass in cancer patients,[26,27] several other beneficial effects of TPN were subsequently demonstrated in severely malnourished patients.[28] Severely malnourished patients who received TPN had fewer non-infectious complications (5% versus 43%; $P= 0.03$) and no concomitant increase in infectious complications. When TPN was given to patients undergoing chemotherapy,[29] it was shown that nutritional markers improved significantly. Moreover, in addition to TPN correcting preexisting nutritional alterations and preventing the occurrence of chemotherapy-related worsening of nutritional status, it allows optimal delivery of cytotoxic chemotherapy.[29] Patients with small cell lung cancer undergoing aggressive chemotherapy who received TPN for four weeks gained more weight and body cell mass was improved as well as body fat compared to a group receiving standard oral diet.[27]

Perioperative parenteral nutrition support was also proven beneficial in cirrhotic patients undergoing hepatectomy for hepatocellular carcinoma[30] in which there was a significant reduction in the overall postoperative morbidity (fewer septic complications), better control of ascites with fewer diuretics, less weight loss and less deterioration of liver functions. In this study the TPN formula was fortified with branched chain amino acids. In a most recent prospective randomized study of 117 patients undergoing major pancreatic resection for malignancy, the use of postoperative TPN in which 30% of calories were supplied as fat emulsions, no major measurable benefit was demonstrated. The rate of complications associated with infection was significantly greater in the group receiving TPN.[31] Nonetheless, the review of 18 prospective randomized clinical studies, utilizing a meta-analysis, demonstrated that TPN reduced the risk of major complications by 21% and mortality by 32%.[32]

Radiation and Chemotherapy

The nutritional and metabolic consequences of radiation therapy may compound further the malnutrition in cancer patients. The severity of complications of radiation therapy depends on the location, the site radiated, and the cumulative dose of therapy. Patients with head, neck, and chest cancers when radiated may develop mucositis, dysphasia, xerostomia, trismus, fibrosis, fistula and other complications, which further complicates nutritional status of these patients. Patients with abdominal and pelvic malignancies treated with radiation may develop enteritis and diarrhea with concomitant malabsorption. The use of TPN in these patients who undergo radiation therapy (often in addition to chemotherapy or surgery) becomes the only optimal means of provision of nutrition support.

The effectiveness of TPN in the pediatric population undergoing abdominal and pelvic radiation and chemotherapy was studied in a prospective randomized study.[33] The use of TPN was found to be efficacious in maintaining nutritional status in children during the radiation therapy to abdomen and pelvis, while receiving concurrent chemotherapy. However, the value of parenteral nutritional support for patients undergoing chemotherapy was questioned when 12 randomized studies were pooled and analyzed for odd ratio.[34]

In a study of 31 patients treated with chemotherapy for small cell lung cancer who were randomized to receive four weeks of TPN and four weeks of self-regulated diet, during the TPN therapy body weight, total body fat and body potassium were increased significantly.[27] In addition, although very small increase of body nitrogen occurred, significant functional improvement was reported.[27] In another randomized prospective multi-center study of 119 patients, TPN was given to patients for the first 30 days for small lung cancer.[35] No improvement in survival or response to chemotherapy was demonstrated. In this study, however, 50% of calories were given as fat. The rate of TPN complications in this study group was unacceptably high as well. In addition to significant electrolyte abnormalities, 50% of patients were fluid overloaded and 37% had fever or catheter related infection. Due to TPN, related complications only 39 of 54 patients completed three weeks of TPN.

The most significant positive effects of TPN in the cancer population have been reported in patients undergoing bone marrow transplantation (BMT). (See Chapter on total parenteral nutrition in patient undergoing bone marrow transplantation). Protein-losing entheropathy is a common finding in half of the patients undergoing BMT, resulting in severe nutritional abnormalities. In a randomized prospective trial, 137 patients were randomized to receiving either TPN or standard hydration for four weeks (one before and three after BMT).[36] A prolonged overall survival, decreased incidence of relapse and longer relapse free survival in patients receiving TPN during cytoreductive therapy was demonstrated. Furthermore, multivariate analysis revealed that nutritional support influenced independently the survival and relapse, and in allogenic patients the incidence of graft-versus-host disease.

Metabolic and Tumor Growth Effects of TPN

Intermediary metabolism and its response to TPN in patients with malignancy has been reviewed elegantly elsewhere.[4] In cancer patients adjuvant TPN therapy improves nitrogen balance, decreases the urinary 3-methylhistidine excretion, increases whole body protein turnover and synthesis and significantly decreases the breakdown of protein.[4] While glucose turnover in cancer patients receiving TPN is increased,[23] gluconeogenesis and alanine-glucose conversion is decreased. These effects on alanine and glucose kinetics, however, are similar to the effects of jejunal feeding, as both, enteral and parenteral feedings, are equally efficacious in suppressing gluconeogenesis in cancer patients.[23]

The role of TPN in the complex equation between tumor growth biology and nutritional status is not well defined. Potential concerns that exacerbation of tumor growth with feeding the patient have not been demonstrated clinically. Induced tumorigenesis in animal models has not yet been demonstrated clinically in human studies, although few studies have attempted to address this important problem. A recent study of 19 malnourished patients with biopsy proven gastric cancer evaluated the impact of perioperative TPN on tumor cell proliferation.[37] These investigators evaluated the variations in the S-phase cell fraction, defined as 3-H-thymidine labeling index (TLI), before and after 10 days of TPN, or nonadministration of nutritional support in 19 malnourished gastric cancer patients. The TLI was determined at the time of diagnosis and intraoperatively. Although enhanced tumor cell proliferation was seen in 50% of patients given TPN, the clinical importance of this increase is not clear. In one of the earliest studies of the effect of TPN on cell cycle kinetics, 14 patients with previously untreated squamous cell carcinomas of the head and neck region underwent biopsies of normal and malignant tissues and then were placed either on TPN or served as control group. Eight patients were given TPN. Biopsy taken after 3-17 days of TPN therapy showed an increase in percentage of hyperdiploid cells of cancer tissue. The percentage of hyperdiploid cells after TPN was significantly increased in patients with cancer compared in the control group.[38]

Summary

The ultimate goal in the nutritional management of patients with cancer is the same as with all other patients: provision of optimal nutrition to all patients under all conditions at all times, as long as there is a reasonable chance for curing or improving the quality of life of that patient.[9] TPN is not indicated in most patients undergoing surgery for cancer, but may be of value in certain patients in their postoperative course, as well as during chemotherapy or radiation therapy. TPN, however, should not be used in cancer patients who are completely unresponsive to therapy and in whom extraordinary measures to provide nutrients can serve only to prolong unrelieved suffering and inevitable death. Nutritional support of patients with cancer, nonetheless remains an important aspect of therapy of these patients.[9]

References

1. Waren S. The immediate causes of death in cancer. Am J Med Sci 1932; 184:610615.
2. DeWys WD, Begg C, Lavin PT et al. Prognostic effect of weight loss prior to chemotherapy in cancer patients. Am J Med 1980; 69:491497.
3. Bozzetti F, Migliavacca S, Scott A et al. Impact of cancer, type, site, stage and treatment on nutritional status of patients. Ann Surg 1982; 196:170-179.
4. Harrison L, Brennan MF. The role of total parenteral nutrition in the patient with cancer. Curr Prob Surg 1995; 10:837-917.
5. Kern KA, Norton JA. Cancer cachexia. JPEN 1988; 12:286-298

6. Nixon DW, Heymsfield SB, Cohen AE et al. Protein-calorie undernutrition in hospitalized cancer patients. Am J Med 1980; 68:683-690.
7. Strain AJ. Cancer cachexia in man. A review. Invest Cell Pathol 1979; 2:181-193.
8. Silberman H. The role of preoperative parenteral nutrition in cancer patients. Cancer 1985; 55:254-257.
9. Dudrick SJ, Wilmore DW, Wars HM et al. Can intravenous feeding as the sole means of nutrition support growth in the child and restore weight loss in an adult? An affirmative answer. Ann Surg 1969; 169:974-984.
10. McAddams P, DeChicco RS, Matarese LE et al. Biochemical assessment and monitoring of nutritional status. In: Latifi R, Dudrick SJ, eds.: Current Surgical Nutrition. London, Austin, Chapman & Hall and RG Landes, 1996:1-32.
11. Meguid MM, Meguid V. Preoperative identification of the surgical patient in need for postoperative total parenteral nutrition. Cancer 1985; 55:258-262.
12. Smale BF, Mullen JE, Buzby GP et al. The efficacy of nutritional assessment and support in cancer surgery. Cancer 1981; 47:2375-2381.
13. Copeland EM, Daly JM, Dudrick SJ. Nutritional as an adjunct to cancer treatment in the adults. Cancer Res 1977; 37:2451-2456.
14. Lanzotti VC, Copeland EM, George SL et al. Cancer chemotherapeutic response and intravenous hyperalimentation. Cancer Chemotherapy Rep 1975; 59:437-439.
15. Souchon EA, Copeland EM, Watson P Dudrick SJ. Intravenous hyperalimentation as an adjunct to cancer chemotherapy with fluorouracil. J Surg Res 1975; 18:451-454.
16. Holter AR, Fischer JE. The effects of perioperative hyperalimentation on complications in patients with carcinoma and weight loss. J Surg Res 1977; 23:31-34.
17. Thompson BR, Julian TB, Stremple JF. Perioperative total parenteral nutrition in patients with gastrointestinal cancer. J Surg Res 1981; 30:497-500.
18. Muller JM, Dients C, Brenner U et al. Preoperative parenteral feeding in patients with gastrointestinal carcinoma. Lancet 1982; 1:68-71.
19. Muller JM, Keller HW, Brenner U et al. Indications and effects of preoperative parenteral nutrition. World J Surg 1986; 10:53-63.
20. Williams RHP, Heatly RV, Lewis MLT et al. A randomized controlled trial of preoperative intravenous nutrition in patients with stomach cancer. BR J Surg 1976; 63:667-670.
21. Simm JM, Oliver E, Smith JAR. A study of total parenteral nutrition (TPN) in major gastric and esophageal resection for neoplasia. JPEN 1980; 4:422-429.
22. Daly JM, Massar E, Giacco G et al. Parenteral Nutrition in esophageal cancer patients. Ann Surg 1982; 2:203-208.
23. Burt ME, Goochboth CM, Brennan MF. A controlled prospective randomized trial evaluating metabolic effects of enteral and parenteral nutrition in cancer patient. Cancer 1982; 49:1092-1105.
24. Burt ME, Stein TP, Swade JG et al. Whole-body protein metabolism in cancer-bearing patients. Effect of total parenteral nutrition and associated serum insulin response. Cancer 1984; 53:1246-1252.
25. Kinsella TJ, Malcom AW, Bothe A et al. Prospective study of nutritional support during pelvic irradiation. Int J Rad Onc Biol Phys 1981; 7:543-548.
26. Nixon DW, Moffitt S, Lawson DH et al. Total parenteral nutrition as an adjunct to chemotherapy of metastatic colorectal cancer. Cancer Treat Rep 1981; 65:121-128.
27. Shike M, Russell D, Detsky AS et al. Changes in body composition in patients with small cell lung cancer. The effect of total parenteral nutrition as an adjunct to chemotherapy. Ann Int Med 1984; 101:303-309.
28. The Veterans Affairs Total Parenteral Nutrition Cooperative Study Group. Perioperative total parenteral nutrition in surgical patients. N Eng J Med 1991; 325:525-532.
29. DeCicco M, Panarello G, Fantin D et al. Parenteral nutrition in cancer patients receiving chemotherapy: Effects on toxicity and nutritional status. JPEN 1993; 17:513-518.
30. Fan ST, Lo CM, Lai E et al. Preoperative nutritional support on patients undergoing hepatectomy for hepatocellular carcinoma. N Eng J Med 1994; 331:1547-1552.
31. Brennan MF, Pister PW, Posner M et al. A randomized trial of total parenteral nutrition after pancreatic resection for malignancy. Ann Surg 1994; 220: 436-444.
32. Detsky AS, Baker JP, O'Rouke K, and Goel V. Preoperative parenteral nutrition: A meta-analysis. Ann Int Med 1987; 107: 195-203.

33. Ghavimi F, Shils ME, Scott BF et al. Comparison of morbidity in children requiring abdominal radiation and chemotherapy, with and without total parenteral nutrition. J Ped 1982; 101:530-537.
34. American College of Physicians. Parenteral nutrition in patients receiving cancer therapy. Ann Int Med 1989; 110:734-736.
35. Weiner RS, Barnett SK, Clamon GH et al. Effects of intravenous hyperalimentation during treatment in patients with small-cell lung cancer. J Clin Oncol 1985; 3:949-957.
36. Weisdorf SA, Lysne J, Wind D et al. Positive effects of prophylactic total parenteral nutrition on long-term outcome of bone marrow transplantation. Transplantation 1987; 39:159-165.
37. Bozzeti F, Gavazzi C, Cozzglio C et al. Total parenteral nutrition and tumor growth in malnourished patients with gastric cancer. Tumori 1999; 85:163-166.
38. Baron PL, Lawrence W, Jr, Chan WMY et al. Effect of parenteral nutrition on cell cycle kinetics of head and neck cancer. Arch Surg 1986; 121:1282-1286.

CHAPTER 7

Cell Cycle Kinetics in Cancer Patients Receiving Total Parenteral Nutrition

Michael H. Torosian

A complex interaction exists between nutrition, host metabolism, tumor cell cycle kinetics and response to antineoplastic therapy. Although extensive nutritional and metabolic alterations have been documented in the cancer patient, controversy exists regarding the use of nutrition support in this clinical population. Although retrospective clinical studies have suggested that nutrition support can reduce morbidity and mortality associated with antineoplastic therapy, prospective, randomized trials have, in general, failed to support this concept. Current guidelines suggest that only severely malnourished patients benefit from the use of nutrition support during aggressive antineoplastic therapy. Preoperative and peritreatment nutrition support should be administered to severely malnourished patients undergoing aggressive surgery, chemotherapy or radiation therapy. In mild or moderately malnourished patients undergoing antineoplastic therapy, no significant reduction in complications or increased survival has been found. A potential concern with the use of nutritional support in cancer patients is stimulation of tumor growth. One mechanism of nutrient-induced acceleration of tumor growth is alteration of tumor cell cycle kinetics to increase the growth fraction of tumor cells.

Animal studies have clearly demonstrated that spontaneous tumor development and growth of established tumors can be stimulated during periods of exogenous nutrient administration. Protein/calorie deprivation significantly reduces spontaneous tumorigenesis and carcinogen-induced tumor development in animal models. Refeeding can stimulate spontaneous and carcinogen-induced malignancies in many tumor models. Increased caloric and dietary fat intake are particularly associated with an increased incidence of spontaneous tumorigenesis. Stimulation of growth kinetics of established tumors can occur following enteral or parenteral nutrient administration. Acceleration of tumor growth has been demonstrated by objective increases in tumor volume, tumor mitotic activity, H^3-thymidine labeling index, tumor protein synthesis, and increased S-phase and aneuploid tumor cells. Several recent clinical studies have suggested that the cell cycle kinetics of human tumors can likewise be altered by exogenous nutrient administration. This review will emphasize both laboratory and clinical studies which analyze the effect of nutrition support on tumorigenesis, tumor cell cycle kinetics, primary tumor growth, and tumor metastasis.

Tumorigenesis

Animal studies have demonstrated that protein, calorie and lipid restriction can significantly reduce the incidence of spontaneous tumorigenesis. Furthermore, rates of tumor

Nutritional Support in Cancer and Transplant Patients, edited by Rifat Latifi and Ronald C. Merrell.

establishment and growth of transplanted tumors are significantly inhibited under conditions of protein/calorie restriction. Tannenbaum and Silverstone in 1953 published an extensive review demonstrating that protein and/or calorie deprivation significantly reduced spontaneous tumorigenesis in the majority of rat and mouse tumor systems studied.[1] Ross and Bras in 1971 similarly reported a reduced incidence of spontaneous benign and malignant tumors in COBS rats with underfeeding.[2] This pattern of tumorigenesis was established early in life in this animal model and remained stable with subsequent nutrient alterations. In serially transplanted tumor models, tumor establishment was significantly delayed with protein/calorie deprivation and returned to normal with refeeding.[3]

The interaction of carcinogens with host tissues is also affected by dietary intake and the nutritional status of the host. Protein/calorie deprivation effectively inhibits the development of various carcinogen-induced malignancies.[4] Moore and Tittle documented these effects in DMBA-induced breast tumors in Sprague-Dawley rats.[5] Thus, it is clear in numerous animal models that protein/calorie restriction is associated with a decreased incidence of spontaneous and carcinogen-induced tumors, inhibition of primary tumor growth and decreased establishment of transplanted tumors.

Few clinical trials have been conducted to specifically test this hypothesis in patients. Despite the paucity of objective clinical studies, numerous associations have been made between high caloric intake, high fat intake, obesity and the development of human tumors. In particular, the incidence of hormone-sensitive tumors such as breast carcinoma, ovarian carcinoma and colorectal cancer is increased with increased calorie and fat intake. In contrast, increased fiber intake is associated with reduction of carcinogenesis. Fiber appears to be an independent factor which inhibits the development of colorectal cancer but may be inversely related to dietary intake as a risk factor for breast cancer. This rationale forms the basis for the development of current chemoprevention trials which are designed to determine the effect of external interventions on the incidence of human cancers. Furthermore, these findings have led to the general recommendations of the American Cancer Society and National Cancer Institute that decreased fat intake and increased fiber intake are associated with reduced risk of human cancers.

Primary Tumor Growth: Animal Models, Nutrition Support Studies

Numerous animal studies have demonstrated that tumor growth and metastasis can be significantly stimulated during periods of nutrition support.[6-8] Both oral and intravenous nutrients have been shown to accelerate tumor growth rates. In 1976, Cameron and Pavlat studied growth of the Morris hepatoma compared with Buffalo rats receiving a standard oral diet or total parenteral nutrition (TPN).[6] A two-week period of parenteral nutrition significantly accelerated tumor growth as indicated by increases in tumor volume, tumor mitotic index and tumor:host weight ratio in this model. Using this same tumor system, Cameron in 1981 demonstrated a dramatic increase in H^3-thymidine labeling index of tumor cells after starting total parenteral nutrition.[9] These two early studies demonstrated the potential of TPN to stimulate tumor cell proliferation and, likely, alter tumor cell cycle kinetics.

Daly et al in 1978 compared an oral protein and carbohydrate diet, TPN, and a restricted diet in hepatoma-bearing Buffalo rats.[10] Compared to the restricted diet, animals given the oral protein and carbohydrate diet or TPN exhibited increases in tumor size but no change in the tumor:host weight ratio. Thus, there was stimulation of tumor growth but not out of proportion to host weight gain. In the Walker 256 carcinosarcoma model in Sprague-Dawley rats, refeeding resulted in a significant increase in tumor volume after only 48 hours.[11] Change from a restricted to oral protein/carbohydrate diet in this model caused a rapid growth spurt of tumors with subsequent return to the normal growth curve.

Growth characteristics of the AC-33 mammary adenocarcinoma are well studied and are clearly shown to be influenced by exogenous nutritional and metabolic factors. Steiger et al in 1975 infused parenteral amino acids with or without hypertonic dextrose for 10 days in Lewis/ Wistar rats with AC-33 mammary tumor implants.[12] Increased tumor protein content was observed in animals receiving parenteral amino acids, independent of hypertonic dextrose, compared to control animals. No significant change in tumor:host weight ratio was found in this study. Oram-Smith et al in 1977 studied the following parenteral solutions in this same animal/tumor model:

a. parenteral amino acids alone,
b. hypertonic dextrose alone,
c. combined parenteral amino acids and hypertonic dextrose and
d. control solution (2.5% dextrose).[16]

N^{15}-glycine infusion techniques measuring tumor protein synthesis rates were increased in all parenteral nutrient groups compared to control animals. Torosian et al demonstrated increased tumor size, increased tumor weight and increased tumor cell proliferation of AC-33 mammary tumor cells in rats receiving total parenteral nutrition.[8,17,18] After only 2 hours of initiating TPN, a significant increase in S-phase tumor cells was observed in this model. No significant cytokinetic alterations were detected beyond 2 hours of TPN with gradual normalization of tumor growth kinetics within 24-48 hours of initiating TPN. Although acceleration of tumor growth kinetics is generally regarded as detrimental, this burst in DNA synthesis can be exploited to improve tumor response to chemotherapy. As predicted from the cytokinetic changes in this tumor model, improved tumor response was observed with two cycle-specific chemotherapeutic agents (methotrexate, adriamycin) following short-term TPN (48 hours).[14] However, no increase in tumor response was found with cycle-nonspecific chemotherapy (cyclophosphamide). Additional studies comparing the effects of glucose, amino acids, or combined glucose/amino acid solutions have determined that amino acids are the critical substrate responsible for these alterations in tumor cell kinetics.[13]

Popp in 1983 reported a dramatic effect of TPN on the growth rate of methylcholanthrene-induced sarcomas in Fischer-344 rats.[15] Tumor growth occurred out of proportion to host lean body mass with increasing levels of TPN. In fact, a direct correlation was observed between tumor weight and the rate of parenteral nutrient infusion in this study.

Isolated animal studies have concluded that no significant stimulation of tumor growth occurs with parenteral nutrition. Kishi et al in 1982 compared TPN to 5% dextrose for 7 days in Wistar rats with Walker-256 carcinosarcoma implants.[16] Although this study concluded that TPN did not accelerate tumor growth, a small number of animals were studied raising the possibility of a type II statistical error. In this report, tumor weight in animals receiving TPN was > 2-fold greater than that seen in control animals. King et al in 1985 compared TPN to an oral diet in hepatoma-bearing ACI-N rats.[17] Although TPN did not increase tumor protein content or H^3-thymidine incorporation, the protein intake of control animals receiving oral diets was 25% greater than that of the experimental group receiving parenteral nutrition. Furthermore, both glucose-based and fat-based TPN increased tumor weight compared to animals receiving oral nutrition.

Nutrient Deficiency Studies

In an attempt to selectively retard tumor growth, several investigators have induced specific dietary deficiencies in tumor-bearing animals. Munro measured the RNA content in hepatoma and host liver in rats fed a 25% protein diet compared with animals fed an isocaloric protein-free diet.[18] No change occurred in RNA content of tumor tissue, but reduced RNA content of host liver was observed in rats ingesting the protein-free diet. In 1977, Ota et al measured protein content and found similar results in Buffalo rats with Morris hepatomas.[19] In animals given protein-free diets, tumor protein content remained unchanged whereas liver protein content decreased. Reilly et al demonstrated decreased growth of methyl-cholanthrene-induced sarcomas in rats during starvation; however, DNA specific activity increased in the tumor and decreased in host liver.[20] Thus, tumors are able to effectively incorporate protein at the expense of host tissues during periods of dietary protein restriction. These studies led to the concept of the tumor as a "nitrogen trap" and documented that tumors are successful competitors with host tissues for nutrient substrates during periods of limited dietary supply.

Imbalances of selective amino acids have similarly resulted in continued tumor growth and host nutritional depletion.[21,22] Studies performed in rats, mice and hamsters with essential amino acid deprivation demonstrate minimal tumor inhibition.[21-23] In fact, accelerated tumor growth has been observed with moderate leucine restriction and concentrated infusion of branched-chain amino acids.[23,24]

Two studies have documented selective tumor inhibition with diets deficient in specific nutrients. Jose and Good fed tumor-bearing mice casein-restricted diets to reduce plasma levels of cysteine, methionine, and tryptophan.[25] A reduction in tumor:host weight ratio occurred with this dietary deficiency. Theurer found selective reduction in the growth rate of BW-20232 adenocarcinoma of C57/BL mice fed diets deficient in phenylalanine, valine, or isoleucine.[23] A proportional reduction in tumor and host weight occurred using diets deficient in tryptophan, threonine, leucine, or methionine in this model. Folate deficiency inhibited growth of the Walker-256 carcinosarcoma, but did not alter growth of the Murphy-Strom lymphosarcoma in rats.[26] However, prolonged folate deprivation causes significant host toxicity and would not be clinically useful. Pantothenic acid deficiency impaired fibrosarcoma and host growth to similar degrees in a rat model.[27] Pyridoxine, zinc, and potassium deficient states have been shown to inhibit tumor growth in experimental models.[28,29] Animals fed diets deficient in vitamin A exhibit increased carcinogenesis of epithelial systems to chemical carcinogens whereas vitamin K deficiency may inhibit spontaneous tumor metastasis in animal models.[30,31] Vitamin E may have a chemopreventative role in the development of cancer by its antioxidant activity while the effect of vitamin C on carcinogenesis remains controversial.[32,33] Although the specific effects of nutrient-deficient diets on tumor cell cycle kinetics has not been precisely defined, selective substrate deprivation clearly can significantly influence the kinetics of tumor growth.

Tumor Metastasis

Spontaneous and induced models of tumor metastasis may be influenced by exogenous nutrient administration. In 1987, a series of studies by Mahaffey and Bryant reported reduced lung metastases in parenterally fed mice with subcutaneous Lewis lung carcinoma implants.[34,35] In these studies, decreased pulmonary metastasis occurred with infusion of either TPN or control electrolyte solutions. These investigators recognized that parenteral fluid load (not nutrient provision) correlated with alterations in pulmonary metastasis in this model. They hypothesized that changes in circulating levels of prostaglandins or other humoral factors may have caused this phenomenon. Torosian et al in 1991 compared three parenteral nutrient solutions to two oral diets in Lobund rats with PA-III prostate adenocarcinoma implants.[36] The TPN solutions were isocaloric and consisted of:

a. carbohydrate alone,
b. carbohydrate and amino acids or
c. carbohydrate, amino acids and lipid.

A standard protein or protein-depleted diet were used as oral diets for comparison; all orally fed animals received isovolumic, electrolyte infusions. Significant acceleration of primary tumor growth and lung metastasis occurred in animals receiving total parenteral nutrition or the standard oral diet compared to protein-depleted controls. Combined TPN with dextrose, amino acids and lipids stimulated tumor metastasis to the greatest extent in this model.

Recently, additional parenteral nutrients were investigated in the MAC-33 mammary adenocarcinoma in rats.[37] Control animals receiving electrolyte solution were compared to animals given isocaloric TPN solutions containing

a. glucose,
b. long-chain triglycerides or
c. a combination of medium- and long-chain triglycerides as the source of nonprotein calories.

Increased primary tumor volume was observed with all TPN solutions compared to the control electrolyte solution. However, lung metastases were greatest in animals receiving the TPN solution containing long-chain triglycerides, intermediate with glucose-based TPN and reduced with combined medium- and long-chain triglycerides. These results indicate that differential effects on primary tumor growth and tumor metastasis can occur with specific parenteral nutrients and that various substrates can have differential effects on the process of tumor metastasis.

Human Studies

Few clinical studies have been reported to objectively define the effect of nutrients on tumor growth in cancer patients. Early retrospective reviews suggested that there was no significant stimulation of tumor growth in cancer patients receiving TPN.[38] Provision of adequate nutrition support can clearly improve host immunologic and nutritional status as indicated by biochemical, arthrometric and body compositional parameters. With the extensive animal research indicating that exogenous nutrients can accelerate tumor growth kinetics, the potential to stimulate human tumor growth with nutrition support remains.

To determine the effect of TPN on tumor protein synthesis, Mullen et al studied 25 patients with upper gastrointestinal cancers who were prospectively randomized to receive TPN or a standard oral diet.[39,40] Tumor fractional protein synthesis rates were determined by N^{15}-glycine infusion techniques in this study. Despite a significant increase in protein/calorie intake in the group receiving TPN, no acceleration of tumor protein synthesis rate was observed in these patients. In patients with metastatic colon cancer, Nixon et al in 1981 noted no direct clinical or biochemical evidence of tumor stimulation with TPN compared to oral nutrition.[41] However, survival was reduced in patients given TPN and could be explained by a greater number of organ metastases (liver, bone, and lung) in the TPN group.

In contrast to the retrospective clinical reviews and previously quoted studies, several reports document significant acceleration of human tumor growth during periods of nutrient provision. One early case report of pediatric neuroblastoma objectively associated increased tumor growth with nutritional repletion and inhibition of tumor growth with nutrient withdrawal.[42] Several studies clearly indicate the ability of parenteral nutrients to alter tumor cell cycle kinetics. Baron et al in 1986 reported a significant change in squamous cancer cell cycle kinetics with TPN.[43] In patients with head and neck squamous cell cancer, tumor aspiration was performed before and after 7-10 days of TPN. Flow cytometric analysis of tumor cells demonstrated a significant increase in hyperdiploid cells after TPN.

Frank et al recently studied a similar clinical population with head and neck malignancies and measured tumor cell proliferation by bromodeoxyuridine (BrdU) incorporation.[44] A significant increase in BrdU incorporation into tumor cells was observed after 7 days of TPN in this study. These studies indicate the potential to accelerate tumor cell proliferation in squamous cell cancers of the head and neck after 1 week of TPN. The ability of these cytokinetic alterations to improve tumor response to cycle-specific chemotherapy has not been studied in clinical populations.

Critical Analysis

Obvious discrepancies exist between the basic animal research and few clinical studies which have been performed to determine if exogenous nutrients can alter tumor cell cycle kinetics. The majority of animal/tumor models indicate that oral or parenteral nutrients can effectively accelerate tumor growth kinetics. Although several cytokinetic studies in humans have documented alterations after TPN administration, tumor growth and tumor protein synthesis rates have not been increased. Significant differences between animal tumors and human malignancies exist and may account for these discordant findings. First, the doubling time of animal tumors is short and ranges from 2-7 days. In contrast, the most rapid solid human tumors double in 30 days—more commonly, doubling time of human cancers approximates several months to years. Second, tumor burden is significantly greater in animals with tumors comprising as much as 60-70% by weight in some animal models. In distinction, human cancers grow to only a fraction of total body weight and do not typically consume sufficient nutrients to starve host tissues by local substrate utilization. Third, tumor immunogenicity is variable in animal models and, in general, is weak or absent in humans. Exogenous nutrients may actually inhibit growth of highly immunogenic tumors by augmenting antitumor immunity. In non-immunogenic tumors, nutrient-induced acceleration of tumor growth can occur without interference by host immune activity. These differences indicate the need to perform further clinical trials before the results of animal studies can be extrapolated to the cancer patient.

Summary

The effect of exogenous nutrients to alter human tumor growth kinetics remains controversial—but the potential clearly exists. Further clinical investigation is needed to determine if specific nutrients can accelerate primary tumor growth, enhance metastasis or alter tumor cell cycle kinetics. If modulation of human tumor cell cycle kinetics can be reliably achieved, nutritional intervention may be used to improve tumor response to chemotherapy or radiotherapy. Basic research and clinical trials are critically needed to determine if tumor cell cycle perturbations by nutrients can result in significant therapeutic impact in the future.

References

1. Tennenbaum A, Silverstene H. Nutrition in relation to cancer. Adv Cancer Res 1953; 1:451-501.
2. Ross MH, Bras G. Lasting influence of early caloric restriction on prevalence of neoplasms in the rat. J Natl Cancer Inst 1971; 51:1095-1113.
3. Green JW, Benditt EO, Humphreys EM. The effect of protein depletion on the host response to transplantable rat tumor Walker 256. Cancer Res 1950; 10:769-774.
4. White FR. The relationship between underfeeding and tumor formation, transplantation, and growth in rats and mice. Cancer Res 1961; 21:281-290.
5. Moore C, Tittle PW. Muscle activity, body fat, and induced rat mammary tumors. Surgery 1973; 73:329-332.

6. Cameron JL, Pavlat WA. Stimulation of growth of a transplantable hepatoma in rats by parenteral nutrition. J Natl Cancer Inst 1976; 56:597-601.
7. Oram-Smith JC, Stein TP, Wallace HW et al. Intravenous nutrition and tumor-host protein metabolism. J Surg Res 1977; 22:499-503.
8. Torosian MH, Tsou KC, Daly JM et al. Alterations of tumor cell kinetics by pulse total parenteral nutrition. Cancer 1984; 53:1409-1415.
9. Cameron JL. Effect of total parenteral nutrition on tumor-host responses in rats. Cancer Treat 1981; Rep 65(Suppl):93-99.
10. Daly JM, Reynolds HM, Rowlands BJ et al. Nutritional manipulation of tumor-bearing animals: Effects of body weight, serum protein levels, and tumor growth. Surg Forum 1978; 29:143-144.
11. Daly JM, Copeland EM, Dudrick SJ et al. Nutritional repletion of malnourished tumor-bearing and nontumor-bearing rats: Effects on body weight, liver, muscle and tumor. J Surg Res 1980; 28:507-518.
12. Steiger E, Oram-Smith J, Miller E et al. Effects of nutrition on tumor growth and tolerance to chemotherapy. J Surg Res 1975; 18:455-61.
13. Torosian MH, Mullen JL, Miller EE et al. Enhanced tumor resopnse to phase-specific chemotherapy by parenteral amino acid administration. JPEN 1983; 7:337-345.
14. Torosian MH, Mullen JL, Miller EE, et al. Adjuvant, pulse TPN, and tumor response to cycle-specific and cycle-nonspecific chemotherapy. Surgery 1983; 94:291-299.
15. Popp MB, Wagner SC, Brito OJ. Host and tumor responses to increasing levels of intravenous nutritional support. Surgery 1983; 94:300-308.
16. Kishi T, Iwasawa Y, Hiroshi I et al. Nutritional responses of tumor-bearing rats to oral or intravenous feeding. JPEN 1982; 6:295-300.
17. King WWK, Boelhouwer RU, Kingsworth AN et al. Total parenteral nutrition with and without fat as substrate for growth of rats and transplanted hepatocarcinoma. JPEN 1985; 9:422-427.
18. Munro HN, Clark CM. The influence of dietary protein on the metabolism of ribonucleic acid in rat hepatoma. Br J Cancer 1959; 13:324-335.
19. Ota DM, Copeland EM, Strobel HW et al. The effects of protein nutrition on host and tumor metabolism. J Surg Res 1977; 22:181-188.
20. Reilly JJ, Goodgame JT, Jones DC et al. DNA synthesis in rat sarcome and liver: The effect of starvation. J Surg Res 1977; 22:281-286.
21. Demopoulous HB, Gerving MA, Bagdoyan H. Selective inhibition of growth and respiration of melanomas by tyrosinase inhibitors. J Natl Cancer Inst 1965; 35:823-827.
22. Jensen OA, Edeberg J, Edmund J. The effect of a phenylalanine-tyrosine low diet on the growth and morphology of transplantable malignant melanomas of the Syrian golden hamster. Acta Pathol Microbiol Scand 1973; 81:559-568.
23. Theurer RC. Effect of essential amino-acid restriction on the growth of female C47BL mice and their implanted BW10232 adenocarcinomas. J Nutr 1971; 101:223.
24. Torosian MH, Stein TP, Presti ME et al Effect of branched-chain amino acids on the host-tumor relationship in parenterally nourished rats. Proceedings of the Federation of American Societies for Experimental Biology 1983; 16:5987.
25. Jose DG, Good RA. Quantitative effects of nutritional essential amino acid deficiency upon immune responses to tumor in mice. J Exp Med 1973; 137:1-9.
26. Nichol CA. The manipulation of metabolism by drugs and nutrients. Cancer Res 1969; 29:2422-2426.
27. Montanez G, Murphy AE, Dunn M. Influence of pantothenic acid deficiency on the viability and growth of rat fibrosarcoma. Cancer Res 1951; 11:834-838.
28. Mihich E, Nichol CA. The effect of pyridoxine deficiency on mouse sarcoma 180. Cancer Res 1950; 19:279-284.
29. Rosen F, Mihich E, Nichol CA. Selective metabolic and chemotherapeutic effects of vitamin B_6 antimetabolites. Vitam Horm 1964; 22:609-641.
30. Moon RC, McCormick DL, Mehta RG. Inhibition of carcinogenesis by retinoids. Cancer Res 1983; 43(Suppl):2469A-2475S.
31. Hilgard P. Experimental vitamin K deficiency and spontaneous metastases. Br J Cancer 1977; 35:891-892.

32. Dunham WB, Zuckenkandl E, Reynolds R et al. Effects of intake of L-ascorbic acid on the incidence of dermal neoplasm induced in mice by ultraviolet light. Proc Natl Acad Sci USA 1982; 79:7532-7536.
33. Black HS, Chan JT. Suppression of ultraviolet light-induced tumor formation by dietary antioxidants. J Invest Dermatol 1975; 65:412-414.
34. Mahaffey SM, Copeland EM III, Economides E et al. Decreased lung metastasis and tumor growth in parenterally fed mice. J Surg Res 1987; 42:159-165.
35. Bryant MS, Copeland EM III, Sinclair KGA et al. Causative factors for decreased pulmonary metastasis in parenterally fed mice. J Surg Res 42:467-474, 1987.
36. Torosian MH, Donoway RB. Total parenteral nutrition and tumor metastasis. Surgery 1991; 109:597-601.
37. Bartlett D, Charland S, Torosian MH. Differential effect of medium- and long-chain triglycerides on tumor growth and metastasis. JPEN 1992; 16(Suppl):30S.
38. Copeland EM, Dudrick SJ. Nutritional aspects of cancer. In: Hickman RC, ed. Current Problems in Cancer. Chicago: Year Book Medical Publishers, 1976:1-51.
39. Mullen JL, Buzby GP, Gertner MH et al. Protein synthesis dynamics in human gastrointestinal malignancies. Surgery 1980; 87:331-338.
40. Stein TP, Buzby GP, Leskiw MJ et al. Parenteral nutrition and human gastrointestinal tumor protein metabolism. Cancer 1980; 49:1476.
41. Nixon DW, Moffitt S, Lawson DH et al Total parenteral nutrition as an adjunct to chemotherapy of metastatic colorectal cancer. Cancer Treat Rep 1981; 65(Suppl):121-128.
42. English WJ, Suskind R, Damrongsak D et al. Can the growth of a neuroblastoma be influenced by a child's nutritional state? Clin Pediatr 1975; 14:868-869.
43. Baron PL, Lawrence W, Jr, Chan WMY et al. Effect of parenteral nutrition on cell cycle kinetics of head and neck cancer. Arch Surg 1986; 121:1831-1286.
44. Frank JL, Lawrence W Jr, Banks WK Jr et al. Modulation of cell cycle kinetics in human cancer with total parenteral nutrition. Cancer 1992; 69:1858-1864.

CHAPTER 8

Plasma Amino Acid Profile in Cancer Patients:

Moving Toward a New Set of Tumor Markers?

Maurizio Muscaritoli, Michael M. Meguid, Carlo Cangiano, Antonia Cascino and Filippo Rossi-Fanelli

Regardless their site of origin, tumors may share the ability to induce a number of metabolic alterations in the host involving aspects of intermediate metabolism, including energy, carbohydrate, fat and protein metabolism.[1,2]

With progressing cancer, the disturbances in protein metabolism are mainly represented by increased whole-body protein turnover, consequent to increased hepatic synthesis and muscle degradation of protein, and increased gluconeogenesis from amino acids.[3,4] In advanced stages, when the cachexia syndrome is present, such metabolic disturbances are probably secondary to the severe protein calorie malnutrition resulting from reduced food intake secondary to anorexia and antineoplastic therapy rather than to the presence of the tumor per se. However, in the early stages when the diagnosis of cancer is made or in the early stages before treatment is started, the changes in protein metabolism are tumor-induced and as such are reflective of and indicators of the effects of the metabolism of the host.

Changes in plasma amino acid patterns reliably reflect the quantitative and qualitative changes in protein metabolism which occur with different pathological conditions. These include chronic liver failure,[5] renal failure,[6] sepsis,[7] diabetes,[8] and pure malnutrition.[9] Similarly, alterations in amino acid metabolism resulting in abnormal concentrations of circulating amino acids also characterize protein metabolism in cancer. The potential role of these changes has only recently become increasingly clear in the diagnostic assessment of patients with malignant diseases.

Studies of Plasma Amino Acid Profiles in Cancer Patients

Changes in plasma amino acids in cancer were initially described in patients with acute and chronic leukemias.[10-12] However, no conclusions regarding the pathophysiologic mechanisms of these alterations or possible diagnostic implications were provided.

More recently, several other clinical studies were reported in which plasma amino acid concentrations were measured in patients with cancers of different origin.[2,13-16] The results of these studies are summarized in Table 8.1. Although these studies consistently show a reduction in gluconeogenic amino acids (GAA) and normal or even increased concentration of the branched-chain amino acids (BCAA), some differences exist relative to some individual amino

Nutritional Support in Cancer and Transplant Patients, edited by Rifat Latifi and Ronald C. Merrell.

Table 8.1. Studies documenting changes in gluconeogenic and branched-chain amino acids

Author	Tumor Type (n) Amino Acids	Gluconeogenic Amino Acids	Branched-Chain
Clarke EF (1978)	Mixed (18)	↓	↑
Levin L (1983)	Mixed (19)	↓	=
Norton JA (1985)	Esophageal (6)	↓↓	=
	Lymphoma (11)	=	=
	Sarcoma (38)	=	=
Heber D (1985)	Lung (29)	↓	=
Naini D (1988)	Esophageal (8)	=	=
	Lung (8)	↓	=
Tayek JA (1992)	Colon (22)	↓	↑

"=" = unchanged; ↑ = decreased; ↑ increased; ↓↓ = greatly increase

acids, which may be due to the nonhomogeneity of the patients studied and the different site of tumor origin. In the most recent study, only sulfur amino acids were measured in 63 patients with mixed types of cancer, and plasma taurine was 50% lower than controls.[17]

Based on current data, it seems likely that in cancer patients, even in the presence of malnutrition, that BCAAs are maintained within normal range despite marked reduction in total amino acid (TAA) concentrations. In contrast, in pure chronic malnutrition, BCAA are known to be avidly consumed for gluconeogenic purposes and are thus characteristically decreased in plasma.[18-20]

The widely reported reduction in GAA concentration in cancer patients is usually attributed to increased hepatic gluconeogenesis induced by the tumor to assure adequate provision of glucose to the tumor itself. But, other factors, including impaired tolerance to carbohydrates, could be in part responsible for altered plasma amino acid concentrations.[8]

Since glucose intolerance is a common feature in cancer patients, we hypothesized that some of the changes in a plasma amino acid profile could be secondary to altered glucose metabolism. This hypothesis was tested by studying the plasma amino acid pattern in a series of untreated cancer patients in whom the degree of glucose tolerance had been carefully evaluated.[21] In this series of miscellaneous cancer patients, (Table 8.2), plasma amino acid pattern was characterized by a significant rise in phenylalanine, tyrosine, free tryptophan, methionine, proline, glutamic acid, and ornithine (Table 8.3). When patients were subdivided according to glucose tolerance (Fig. 8.1), no significant differences were found between normotolerant, glucose-intolerant, and diabetics, indicating that the modifications in carbohydrate utilization do not play a significant role in the observed changes in plasma amino acids.

Another variable that might affect the concentration of plasma amino acids, particularly that of GAA, is the tumor burden. It is conceivable that a bigger tumor mass has a greater need for fuel substrates essential for intrinsic tumor metabolism. However, dividing cancer patients according to their tumor stage, i.e., stages I and II = localized tumor or regional metastasis; and stage III metastatic disease, no significant difference in plasma amino acid profile could be demonstrated (Fig. 8.2). The absence of a direct correlation between tumor burden and changes in plasma amino acid profile is in keeping with a selective influence of the tumor on the host

Table 8.2. Patient characteristics

Cancer	No.	Sex M	F		Age Range (Years)
Lung cancer		13	10	20	34-70
Breast cancer		7	-	7	45-75
NH lymphoma		4	3	1	47-72
Gastrointestinal cancer		3	3	-	39-72
Thyroid cancer		2	1	1	65-68
Miscellaneous cancer		9	3	6	50-77
Unknown origin		5	3	2	58-73
Total		**42**	**23**	**19**	**34-77**

amino acid metabolism. The lack of correlation also suggests that the observed plasma amino acid patterns may not be due to a disturbed amino acid metabolism within the tumor tissue itself but instead are due to the effects of the tumor on host issues.

That the reduction of GAA during tumor growth is an early phenomenon, and thus at least partly unrelated to the degree of tumor burden, was subsequently confirmed by utilizing an animal model. Male Fischer rats were inoculated with methylcholanthrene-induced (MCA) sarcoma cells and plasma amino acids concentrations determined at different periods during tumor growth.[22] The decrease in GAA occurred as early as six days after tumor cell inoculation, when the tumor was not detectable by palpation (i.e., was not "clinically" evident) and the rats were eating normally and had not lost weight. The reduction of GAA became even more evident with progressive tumor growth on days 22 and 26. In tumor-bearing rats such reduction was also statistically significant with respect to pair-fed rats (i.e., rats fed the same amount of food eaten by neoplastic rats and therefore undergoing an acute malnutrition). This reduction would suggest that the decrease observed in tumor bearing rats was tumor-related and not secondary to nutritional depletion. Interestingly, in a similar experiment[23] in which we looked at the temporal changes of large neutral amino acids after tumor cell inoculation in rats, plasma concentrations of BCAA (which are normally preserved in well-nourished neoplastic patients) were also shown to decrease early during tumor growth. This observation underscores the concept that findings in animals cannot *tout-court* be applied to the clinical setting, due to the relevant differences in host-tumor metabolic interactions (tumor weight to host weight ratio, tumor immunogenecity, host immune response, etc.) between humans and a rat model.

Plasma Amino Acid Profiles in Selected Tumor Types

It is known that human tumors originating in different organs may greatly differ in terms of rate of proliferation and their ability to metastasize, thereby influencing the host's metabolism.[24] It is thus conceivable that they may also cause different and specific alterations in host's plasma amino acid profile.

To confirm the hypothesis that patients with different cancers may have different amino acid patterns, we measured the concentrations of 28 plasma-free amino acids in three different

Table 8.3. Plasma amino acid levels in 42 cancer patients (M±SE)

Amino Acids (mmoles/dL)	Controls (42)			Cancer Patients (24)		
Threonine	13.4	±	0.6	12.5	±	0.6
Valine	25.4	±	1.5	25.9	±	0.5
Leucine	13.5	±	0.6	14.8	±	1.2
Isoleucine	7.2	±	0.3	8.9	±	0.4
Methionine	3.3	±	0.3	4.2	±	0.3*
Phenylalanine	6.1	±	0.6	8.5	±	0.4*
Lysine	21.2	±	1.2	21.4	±	1.8
Total tryptophan	4.8	±	0.3	4.2	±	0.1
Free tryptophan	0.42	±	0.02	0.79	±	0.06*
Taurine	8.1	±	0.2	7.8	±	0.8
Serine	13.0	±	0.4	12	±	1
Glutamic acid	3.7	±	0.8	8.9	±	0.7*
Glutamine	52.9	±	0.5	52.6	±	0.4
Proline	20.5	±	1.1	26.6	±	1.4*
Glycine	23.9	±	0.9	22.8	±	1.8
Alanine	35.4	±	1.8	35.2	±	2.1
Citruline	2.9	±	0.4	2.9	±	0.3
Tyrosine	6.3	±	0.4	8.3	±	0.4*
Ornithine	7.8	±	0.6	13.4	±	0.6*
Histidine	7.7	±	0.5	8.4	±	0.4
Arginine	7.5	±	0.3	7.3	±	0.6

*Significantly different ($p<0.05$) versus controls. Reproduced with permission from Cascino A et al. Clin Nutr 1988; 7:213-218.

groups of cancer patients.[25] Twenty-two well-nourished breast cancer patients,24 with gastrointestinal tract cancer and 12 patients with tumors of the head and neck were studied. Patients from the gastrointestinal group and the head and neck group were malnourished, as indicated by recent weight loss of more than 10% and serum albumin concentrations less than 3.5 g/dl. Fasting plasma amino acid concentrations in cancer patients were compared with those obtained in 11, healthy well nourished controls. For data analysis, amino acids were grouped into total amino acids (TAA), essential amino acids (EAA), branched-chain amino acids (BCAA), aromatic amino acids (AAA) and gluconeogenic amino acids (GAA). Patient's data were also tabulated for individual diagnostic group and evaluated by discriminant analysis, as assessed by computing Mahalanobis distances and used to classify the patient into a diagnostic group.[26] The discriminant function identified the amino acids which most closely represented a diagnosis consistent with either control subjects or gastrointestinal, head and neck, or breast cancer.

As seen in Table 8.4, the data showed that in breast cancer, TAA concentrations were significantly higher than in female controls, even though both were in good nutritional state. The TAA concentrations in the gastrointestinal and the head and neck groups did not differ from controls, even though both patient groups were malnourished and controls were well

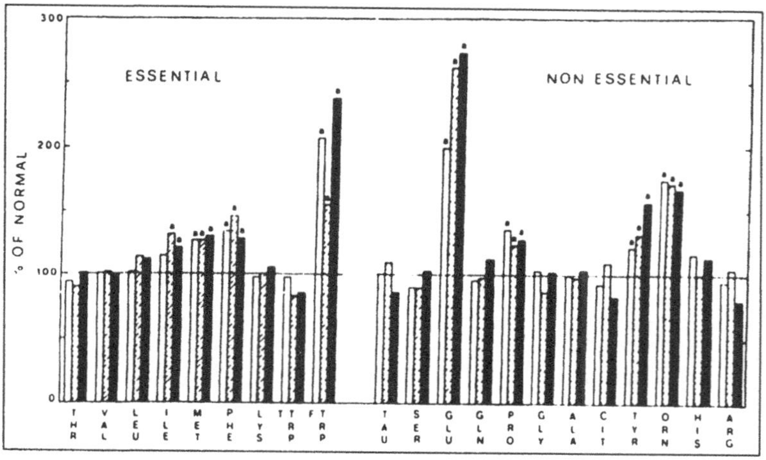

Fig. 8.1. Effect of glucose on AA levels in normal,[16] glucose intolerant,[20] and diabetic[6] cancer patients. Values are expressed as a percentage of normal; normal tolerance; glucose tolerant; diabetic cancer patients. (a) Significantly versus controls. ($p<0.05$)

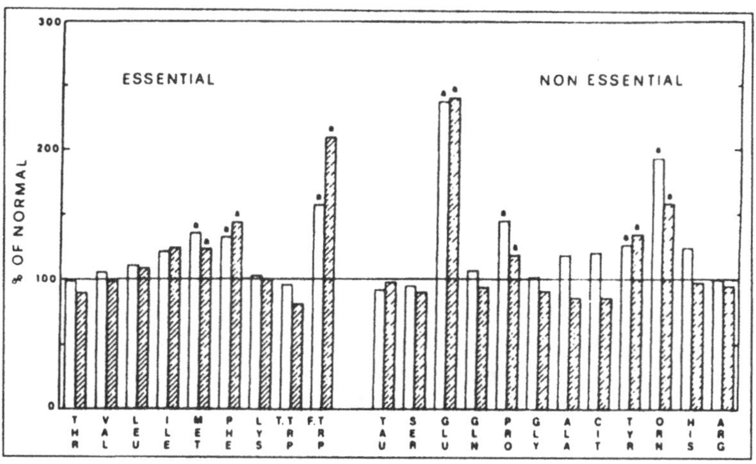

Fig. 8.2. Effect of tumor spread on plasma amino acid levels in the TS I-II group[17] and TS III group[23] of cancer patients. Values are expressed as a percentage of normal; TS I-II; TS III. (a) Significantly versus controls. ($p<0.05$)

nourished. As shown in Figures 8.3 and 8.4, of the 28 amino acids analyzed, there were 7 which were identified as correlating highly with diagnosis. These were glutamine, threonine, histidine, cysteine, alanine, arginine and ornithine. Despite comparable well-nourished status of patients with breast cancer and female controls, the plasma concentrations of TAA, EAA, BCAA, AAA, GAA, and the GAA-TAA ratio in breast cancer patients were all significantly higher than those found in control subjects. Moreover, even if patient, in the gastrointestinal and the head and neck groups were similarly malnourished, plasma GAA-TAA ratios in gastrointestinal patients were significantly higher than those in the head and neck group (Table 8.5).

Table 8.4. Correlation between known diagnosis and predicted diagnosis by discriminant analysis from seven plasma amino acids

	Predicted Diagnosis							
True Diagnosis	Control-Female	Control-Male	GI-Female	GI-Male	Head/Neck-Female	Head/Neck-Male	Breast-Female	Total
Control-Female	5	5	0	0	0	0	0	7
Control-Male	2	2	0	0	0	0	0	4
GI-Female	1	0	5	2	0	1	0	9
GI-Male	1	0	2	12	0	0	0	15
Head/Neck-Female	1	0	0	0	5	0	0	6
Head/Neck-Male	0	0	1	0	1	4	0	6
Breast-Female	0	0	2	0	0	1	19	22

Reproduced with permission. Kubota et al. Cancer 1992; 2343-2348.

In a subsequent study,[27] we examined plasma amino acid profiles in 74 previously untreated lung and breast cancer patients upon diagnosis, when clinical signs of malnutrition were not evident. In untreated lung cancer patients we found a significant reduction of GAA but also a significant increase in glutamic acid, ornithine and free tryptophan. In the group of untreated female breast cancer patients, we failed to observe any alteration in GAA plasma concentrations, while glutamic acid, ornithine, and free tryptophan were significantly higher than in control subjects. In breast cancer patients we also observed a significant rise in taurine levels. In both groups of patients, changes in plasma amino acid were unrelated to age, sex or nutritional state. Considering the profound differences which exist between lung and breast cancer, their tissue organ site, their tendency to metastasize and their influence on host metabolism, one can hypothesize that:

1. the plasma amino acid profile associated with lung cancer is likely related to tumor-induced systemic metabolic alterations, while,
2. the increase of specific amino acids in the presence of breast cancer probably reflects the tumor metabolism per se and would not be secondary to metabolic changes induced by the tumor in the host.

Elevations in plasma aromatic amino acids and methionine occur in the presence of cirrhosis,[5] without hepatocellular carcinoma. These plasma amino acid changes do not occur in cirrhotic patients with hepatocellular carcinoma.[28] Furthermore, plasma methionine, tyrosine and phenylalanine concentrations in patients with hepatocellular carcinoma are much higher than in those with cirrhosis without hepatocellular carcinoma or with normal liver. These findings suggest that there are characteristic plasma amino acid patterns in patients who have cirrhosis with hepatocellular carcinoma which may have clinical importance in the diagnostic assessment of this disease state. This is of particular relevance, considering that currently the improved life expectancy of cirrhotic patients which is brought about by improved quality of supportive care and the prevention of complications has led to an overall increase in the frequency of hepatocellular carcinoma during the course of cirrhosis of the liver.

In summary, these studies confirm that tumors originating in different organs lead to characteristic plasma amino acid profile which differ from normal and from each other, i.e., they are organ specific.

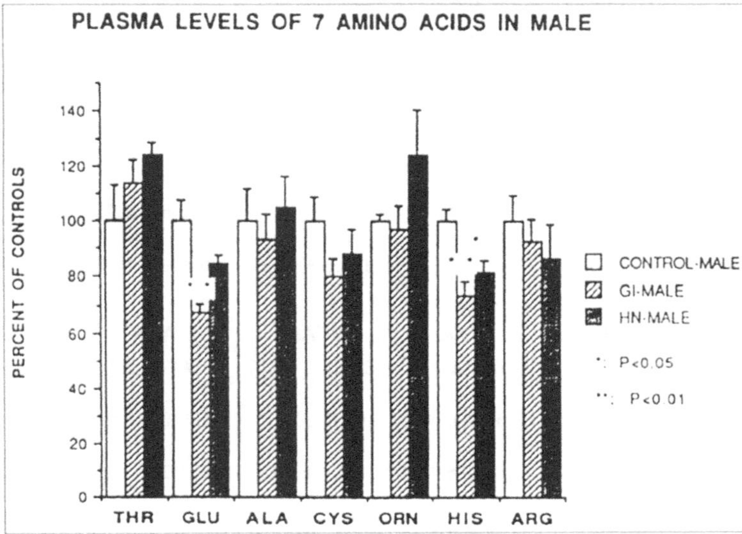

Fig. 8.3. Variation of plasma amino acids in men with gastrointestinal and head and neck cancer expressed as a percentage of control subjects.

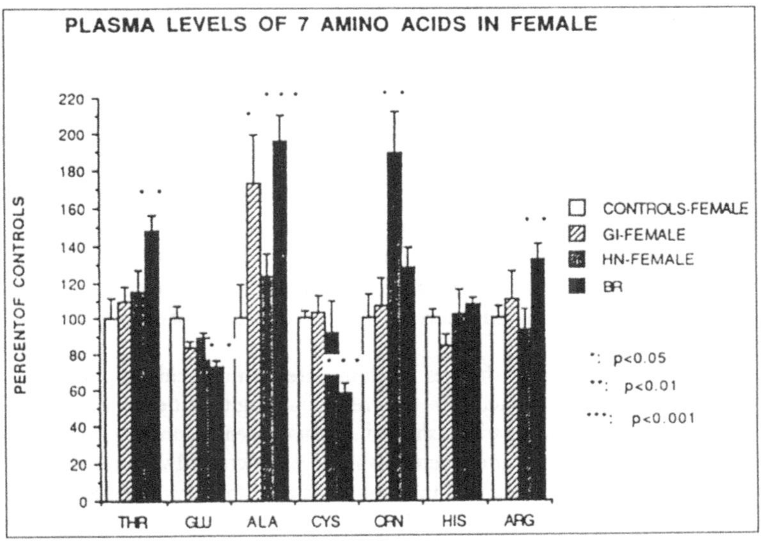

Fig. 8.4. Changes in plasma amino acids in women with breast, gastrointestinal, and head and neck cancer expressed as a percentage of control subjects.

More recently, we revisited the issue of plasma amino acid alterations in the presence of nonsolid tumors by measuring plasma amino acid concentrations in a series of patients with leukemia, at the time the initial diagnosis was made.[29] Forty well-nourished patients with untreated acute myelogenous leukemia (AML) aged 22-78 years, with white blood cells count ranging from 1.08-276.5 x 10^3/cm were studied soon before first remission induction. Patients

Table 8.5. Plasma free amino acid groups

Male Group	Total TAA	EAA	BCAA	AAA	GAA
Control N=5	2677 ± 108	734.5 ± 32.1	363.9 ± 22.3	98.5 ± 51.3	541.9 ± 51.3
GI 2345 (n=15)	± 18645.4	± 34.7	287.8 ± 19.2*	102.6 ± 5.2*	565.8 ± 43.3*
Head/Neck (n=6)	2638 ± 73	683.4 ± 52.0	313.5 ± 33.6	106.9 ± 10.2	574.9 ± 52.2
Female Group	**Total TAA**	**EAA**	**BCAA**	**AAA**	**GAA**
Control (n=6)	2529 ± 115	678.7 ± 48.8	311.4 ± 28.5	98.9 ± 7.1	477.8 ± 28.7
Breast (n=22)	3035 ± 118**	858.2 ± 40.3	394.0 ± 24.2*	120.0 ± 7.6*	792.0 ± 40.1***
GI 2808 (n=9)	±228	737.2 ± 83.7	335.3 ± 48.9	113.2 ± 12.1	730.6 ± 69.7**
Head/Neck (n=6)	2775 ± 127	777.6 ± 53.6	374.5 ± 31.9	110.6 ± 9.4	579.8 ± 45.8

nmol/ml; M±SE; Compared to controls: * $p<0.05$; ** $p<0.01$; *** $p<0.001$; Reproduced with permission from Kubota A et al. Cancer 1992; 69:2343-2348

were then followed-up for 1.8 months after enrollment in the study. Plasma concentrations of glutamic acid, free tryptophan, ornithine and glycine were significantly higher in ANL, while serine, methionine and taurine were significantly reduced when compared to controls. Branched chain amino acids (leucine, isoleucine and valine) concentrations did not differ between leukemic patients, and controls. No gender-related differences in plasma amino acid concentrations were observed among AML, and no correlation was found between plasma-free amino acid concentration and either white blood cell count or LDH activity.

Eighteen months after enrollment in the study, 15 out of the 40 patients were alive (11 were disease-free, 3 had relapsed, 1 resistant to therapy); the remainder 25 patients had died either early during first remission induction (12 patients) or because of relapse or therapy-resistant disease (13 patients).

In order to clarify whether changes in plasma-free amino acids might be predictors of outcome in leukemic patients, data were analyzed according to patients' status at 18 months. Disease-free and resistant/relapse patients showed similar concentrations of amino acids (glutamic acid, free tryptophan, ornithine and glycine) which were increased with respect to controls at diagnosis. However, taurine, serine and methionine tended to be further decreased in patients who had relapsed or with resistant disease (17 pts) with respect to disease-free patients (11 pts).

Taurine reduction might have some relevance in the clinical outcome of AML. This is based on experimental and clinical evidence indicating that taurine is the most abundant intracellular free amino acid in white blood cells.[30] Plasma concentrations of this amino acid have been shown to drop significantly after chemotherapy for hematologic malignancies[31] and its intracellular concentrations were demonstrated to directly correlate to chemosensitivity of leukemia cell line.[32] In addition, taurine supplementation has been shown to accelerate recovery from neutropenia in irradiated mice.[33] This is likely the consequence of its effect as a membrane stabilizer and an antioxidant agent.[34]

The conclusion of this study shows that changes in plasma amino acids which occur in AML are only in part similar to those observed in solid tumors (increased glutamic acid, free tryptophan and ornithine). The reduction of plasma taurine concentration appears to be a typical feature of AML and might be secondary to the deficiency of its precursors serine and methionine.[17] Further in vitro and in vivo studies will clarify whether plasma amino acid profile in leukemic patients is at least in part normalized by induction remission and whether plasma amino acid imbalance reoccurs during relapse. Moreover, it remains to be elucidated if plasma taurine deficiency is common to other hematologic malignancies and if its reduction may have some relevance in the diagnosis and follow up of patients with acute leukemia.

From these data it can be concluded that the differences between control and patients with primary sites of cancer in some plasma amino acid concentrations are independent of nutritional status, and that some cancers induce characteristic changes in the peripheral plasma amino acid profile. These findings suggest that the determination of some peripheral amino acid concentrations in plasma might prove helpful in the diagnosis and follow-up of patients with neoplastic disease.

Plasma-Free Tryptophan Concentrations in Cancer Patients

Our studies of peripheral plasma amino acid profile in cancer patients have consistently shown that circulating plasma-free tryptophan is increased in cancer states. This has led us to postulate that free tryptophan could be a suitable candidate as a tumor marker.

Tryptophan is unique among the plasma amino acids in that it circulates in blood in two forms. Under physiological conditions, about 90% of plasma tryptophan is bound to its natural carrier, albumin; the remainder circulates unbound as free tryptophan.[35] Plasma-free tryptophan concentrations are influenced by the concentrations of albumin and also by the concentration of circulating free fatty acids. Free fatty acids displace tryptophan from albumin by competing for the same binding sites.

High plasma and brain levels of free tryptophan have been reported in cancer-bearing animals.[36,37] More recently, we have shown that the increase in circulating plasma-free tryptophan occurs early during the growth of an experimental tumor.[23,38] Our previous studies have also documented an increase in plasma-free tryptophan concentration in untreated cancer patients[39] and that both plasma[40] and CSF[41] tryptophan concentrations significantly correlate with the presence of cancer anorexia. The presence of anorexia is due to the increased synthesis of serotonin in the brain secondary to increased brain availability of its precursor, tryptophan.[37,41] These data strongly suggest that this amino acid may play a role in the pathogenesis of cancer anorexia. Plasma-free tryptophan concentrations do not correlate with either albumin and/or free fatty acid concentrations in untreated, well-nourished cancer patients.[41] This has led to the postulate that the free tryptophan increase in the tumor-bearing host may be due to a direct effect of the tumor on the binding of tryptophan to albumin.

To verify this hypothesis, free tryptophan, albumin and free fatty acid levels were assayed in 12 lung and 16 breast cancer patients before and 15 days after complete surgical resection of the tumor.[42] Eight subjects undergoing operation for non-neoplastic disease served as a control. Free tryptophan levels significantly decreased after tumor ablation but did not change in controls (Fig. 8.5). Since no correlation was found between free tryptophan and fluctuations in albumin and free fatty acid, it is conceivable that the tumor itself may be responsible for free tryptophan elevation. We have therefore proposed the use of free tryptophan as a marker of neoplastic disease. For this purpose we have recently reviewed the peripheral amino acid profile of more than 200 patients with tumors of the breast, lung, colon, and stomach among other sites of tumor origins, included data from different studies from our group in the last 10 years. We found that plasma-free tryptophan, concentrations were consistently and significantly higher

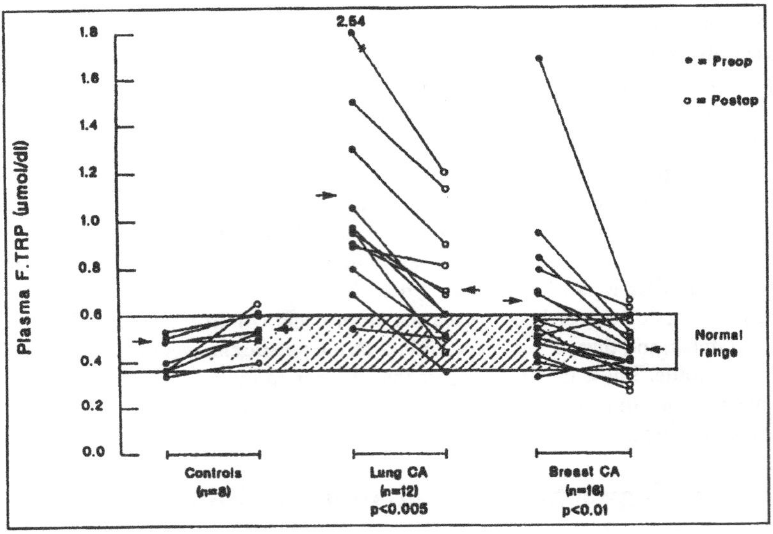

Fig. 8.5. Plasma-free tryptophan in 8 controls, 12 lung and 16 breast cancer patients before and 15 days after surgery. Arrows indicate the group mean. (See Cascino et al.[22])

in the cancer bearing state than in healthy controls. The sensitivity of this marker in predicting the presence of a tumor approximated 80% for gastric and lung cancer and reached 65% for miscellaneous patients with solid tumors.[22] The mechanism(s) whereby this increase occurs remains unknown, but could include cytokines. Similarly to what is observed for other commonly used marker of neoplastic diseases, however, abnormalities of plasma concentrations of free tryptophan may occur in conditions other than neoplastic disease,[5,43-45] thus lowering the specificity of this high-sensitive tumor marker. Studies are currently ongoing in our two respective laboratories (in Syracuse, NY and in Rome, Italy) to confirm the existence of the relationships between free tryptophan and tumor presence in larger series of cancer patients and different types of tumors. Confirmation of such an association would make free tryptophan a novel and useful marker for monitoring the presence of neoplastic disease.

During the last three decades, much evidence has accumulated of a close relationship between the presence of a tumor and the changes in circulating plasma amino acid profile. The sum of these clinical studies clearly indicates that during all stages of cancer growth studied so far, disturbances in the host's protein metabolism are present which are closely reflected by changes in plasma amino acid concentrations. It can be concluded that specific alterations in plasma amino acid patterns characterize cancer states independently of concomitant malnutrition, glucose intolerance and tumor stage. In the cancer patient the nutritional relevance of the altered amino acid profiles continuously challenges those involved in the field of clinical nutrition, but at present the knowledge of single plasma amino acid changes induced by neoplastic disease is rapidly gaining importance as a diagnostic tool. This is suggested by the evidence that tumors originating in different organs may induce different plasma amino acid profiles.

Separation and measurement of plasma amino acids is still a costly and time-consuming procedure which requires both skill and experience. Thus, before routinely employing these promising results as a new diagnostic tool for cancer, further studies are needed in which plasma amino acids are measured in even larger series of cancer patients. Careful measurement of amino acid concentrations in plasma should be made in collating the findings with the diagnosis of the type and stage of the cancer, the patient's nutritional status after therapeutic intervention, and during the course of prolonged follow-up for disease relapse. The availability of a larger data set of plasma-free amino acids in a variety of cancers, stages, gender and nutrition status is rapidly allowing the creation of predictive models to assist the future diagnosis and management of patients. It is likely that based on an understanding of the differences in plasma amino acid profiles more specific and rational biochemical treatment modalities will arise in the nearest future.

References

1. Lundholm K, Edstrom S, Ekman L et al. Metabolism in peripheral tissues in cancer patients. Canc Treat Rep 1981; 65 (Suppl5):79-83.
2. Heber D, Byerly LO, Chlebowski RT. Metabolic abnormalities in the cancer patient. Cancer 1985; 55:225-229.
3. Kurzer M, Meguid MM. Cancer and protein metabolism. Surg Clin North Am 1986; 66:969-1001.
4. Norton JA, Shamberger R, Stern TP, et al. The influence of tumor bearing on protein metabolism in the rat. J surg Res 1981; 30:456-461.
5. Cascino A, Cangiano C, Fiaccadori F et al. Plasma and cerebro-spinal fluid amino acid patterns in hepatic encephalopathy. Dig Dis Sci 1982; 27:828-832.
6. Kopple JD, Jones M, Fukuda S et al. Amino acid and protein metabolism in renal failure. Am J Clin Nutr 1978; 31:1532-1540.
7. Freund HR, Ryan JA, Fischer JE. Amino acid derangements in patients with sepsis: Treatment with branched-chain amino acid rich infusion. Am J Surg 1978; 188:423-430.
8. Wahren J, Felig P, Cerasi E et al. Splanchnic and peripheral glucose and amino acid metabolism in diabetes mellitus. J Clin Invest 1972; 51:1870-1878.
9. Felig P, Owen OK, Wahren J et al. Amino acid metabolism during prolonged starvation. J Clin Invest 1969; 48:584-594.
10. Waisman HA, Pastel RA, Poncher HG. Amino acid metabolism in patients with acute leukemia. Pediatrics 1952; 10:653-659.
11. Kelley JJ, Waisman HA. Quantitative plasma amino acid values in leukemic blood. Blood 1957; 12:635-643.
12. Rudman D, Vogler WR, Howard CH et al. Observations on the plasma amino acids of patients with acute leukemia. Canc Res 1971; 31:1159-1165.
13. Clarke EF, Lewis AM, Waterhouse C. Peripheral amino acid levels in patients with cancer. Cancer 1978; 42:2909-2913.
14. Levin L, Gebers W, Jardine L et al. Serum amino acids in weight losing patients with cancer and tuberculosis. Eur J Canc Clin Oncol 1983; 19:711-715.
15. Naini AB, Dickerson JWT, Brown MM. Preoperative and postoperative levels of plasma protein and amino acid in esophageal and lung cancer patients. Cancer 1988; 62:355-360.
16. Tayek JA, Chlebowski RT. Metabolic response to chemotherapy in colon cancer patients. JPEN 1992; 16 (suppl 6):65S-71S.
17. Gray GE, Landel AM, Meguid MM. Taurine-supplemented TPN and taurine status of malnourished cancer patients. Nutrition 1994; 10:11-15.
18. Arroyave G, Wilson D, Funes CD et al. The free amino acids in blood plasma of children with kwashiorkor and marasmus. Am J clin Nutr 1962; 11:517-524.
19. Holt LE, Synderman SE, Norton PM et al. The plasma aminogram in kwashiorkor. Lancet 1963; 2:1343-1348.
20. Smith SR, Pozefsky T, Chetri MK. Nitrogen and amino acid metabolism in adults with protein-calorie malnutrition. Metabolism 1974; 23:603-618.
21. Cascino A, Cangiano C, Ceci F et al. Plasma amino acids in human cancer: The individual role of tumor, malnutrition, and glucose tolerance. Clin Nutr 1988; 7:213-218.
22. Rossi-Fanelli F, Cangiano C, Muscaritoli M et al. Tumor-induced changes in host metabolism: A possible marker of neoplastic disease. Nutrition 1995; 11(5):595-600.
23. Muscaritoli M, Meguid MM, Beverly JL, et al. Mechanisms of early tumor anorexia. J Surg Res 1996; 60:389-397.
24. Kern KA, Norton JA. Cancer cachexia. JPEN 1988; 12:286-298.
25. Kubota A, Meguid MM, Hitch DC. Amino acid profiles correlate diagnostically with organ site in three kinds of malignant tumors. Cancer 1992; 69:2343-2348.

26. Wilkinson L. SYSTAT: The system for statistics. Evanston, IL; SYSTAT, Inc., 1988; 532-540.
27. Cascino A, Muscaritoli M, Cangiano C et al. Plasma amino acid imbalance in patients with lung and breast cancer. Anticancer Res 1995; 15:507-510.
28. Watanabe A, Higashi T, Sakata T et al. Serum amino acid levels in patients with hepatocellular carcinoma. Cancer 1984; 54:1875-1882.
29. Muscaritoli M, Conversano L, Petti MC et al. Plasma amino acid concentrations in patients with acute myelogenous leukemia. Nutrition 1999; 15(3):195-199.
30. Fukuda K, Hirai Y, Yoshida H et al. Free amino acid content of lymphocytes and granulocytes compared. Clin Chem 1982; 28:1758-1761.
31. Desai TK, Maliakal J, Kinsie JL et al. Taurine deficiency after intensive chemotherapy and/or radiation. Am J Clin Nutr 1992; 55:708-711.
32. Jiang XR, Yang M, Morris CJ et al. High field proton NMR investigations on the metabolic profiles of multidrug-sensitive and -resistant leukaemic cell lines: Evidence for diminished taurine levels in multidrug-resistant cells. Free Rad Res Comms 1993; 19,6:355-369.
33. Abe M, Takahashi M, Takeuchi K et al. Studies on the significance of taurine in radiation injury. Radiation Res 1978; 33:563-573.
34. Hayes KC, Carey RE, Schmidt SY. Retinal degeneration associated with taurine deficiency in the cat. Science 1975; 188:949-951.
35. McMenamy RH, Oncley JL. The specific binding of L-tryptophan to serum albumin. J Biol Chem 1958; 233:1436-1440.
36. Krause R, James JH, Ziparo V et al. Brain tryptophan and the neoplastic anorexia cachexia syndrome. Cancer 1979; 44:1003-1008.
37. Krause R, Humphrey C, vonMeyenfeldt M et al. A central mechanism for anorexia in cancer: A hypothesis. Canc Treat Rep 1981; 65:15-21.
38. Meguid MM, Muscaritoli M, Beverly JL et al. The early cancer anorexia paradigm: Changes in plasma free tryptophan and feeding indexes. JPEN 1992; 16(Suppl 6):56S-59S.
39. Rossi-Fanelli F, Cangiano C, Ceci F et al. Plasma tryptophan and anorexia in human cancer. Eur J Canc Clin Oncol 1986; 22:89-95.
40. Cangiano C, Cascino A, Ceci F et al. Plasma and CSF tryptophan in cancer anorexia. J Neural Trans 1990; 81:225-233.
41. Rossi-Fanelli F, Cangiano C. Increased availability of tryptophan in brain as common pathogenic mechanism for anorexia associated with different diseases. Nutrition 1991; 7:1-4.
42. Cascino A, Cangiano C, Ceci F et al. Increased plasma free tryptophan levels in human cancer: A tumor related effect? Anticanc Res 1991; 11:1313-1316.
43. Sullivan PA, Mumaghan D, Callaghan N et al. Cerebral transmitter precursors and metabolites in advanced renal disease. J Neural Neurosurg Psychiatry 1978; 41:581-585.
44. Askanazy J, Carpentier YA, Michelsen CB et al. Muscale and Plasma amino acids following injury. Ann Surg 1980; 192:78-84.
45. Freund HR, Ryan JA, Fischer JE. Amino acid derangements in patients with sepsis: treatment with branched chain amino acid rich infusions. Ann Surg 1986; 188:423-429.

Chapter 9

Anti-Methionine Cancer Chemotherapy:

L-Methionine and Its Potential Effects for Cancer Therapy

Narihide Goseki and Takeshi Nagahama

Cancer Proliferation and Methionine

Malignant tumors are characterized by a high rate of growth.[1] Since tissue growth requires protein biosynthesis, many studies have attempted to arrest tumor growth by limiting the substrate of protein biosynthesis consumed in the diet.[2-6] However, in these studies it was almost impossible to maintain good nutritional status because the amount of such nutrients taken orally or through the stomach via a feeding tube is inevitably limited. Thus, it is difficult to determine whether the antitumor effects obtained were induced by malnutrition or by unbalanced amino acid nutrients.

Total parenteral nutrition (TPN) has facilitated the provision of high caloric intake with precise nutritional control. Taking advantage of TPN, we have been able to use a so-called amino acid imbalance for cancer treatment as an adjunct to cancer chemotherapy.

L-methionine, a sulfur-containing amino acid, is essential for methylation in the synthesis of RNA, DNA, protein and other biochemical substances. It has been demonstrated that various malignant tumor cells of human origin can not survive in a medium lacking L-methionine in tissue culture studies.[8-10] Kreis and Hession reported that enzymatic deprivation of L-methionine using L-methioninase inhibits the growth of Walker carcinosarcoma 256 in rats.[11]

Recently recombinant methioninase produced by *E. coli* transfected gene from pseudomonas putida was established[12] and preclinical and clinical trial were started.

While, we have prepared a specially formulated amino acid mixture to deplete the methionine level in vivo, creating a TPN mixture devoid of all the sulfur-containing amino acids.[7] Because of its involvement in a methionine-sparing effect described by Finkelstein and Mudd,[13] we also wanted to eliminate L-cysteine in order to create an effective methionine-depleted state.

In this chapter, firstly, the fundamental experimental studies in both nontumor-bearing and tumor-bearing animals of this parenteral treatment with or without anticancer agents are presented and then the results of clinical trials on gastrointestinal tract cancers will be described. And finally, the recent studies of the recombinant methioninase produced by *E. coli* transfected gene from *Pseudomonas putida* are also presented.

Nutritional Support in Cancer and Transplant Patients, edited by Rifat Latifi and Ronald C. Merrell. ©2001 Eurekah.com.

L-Methionine- and L-Cysteine-Free Amino Acid Solution

AO-90, an amino acid solution which does not contain L-methionine or L-cysteine, was prepared by removing L-methionine from commercial Pan-Amin S®(a Vuj-N-type amino acid solution lacking L-cysteine). Accordingly, the only difference between these two solutions is the presence or absence of L-methionine. The composition of both AO-90 and Pan-Amin S® solutions are summarized in Table 9.1.

Methionine-Depleting TPN

By infusing AO-90 amino acid mixture as the sole protein source of TPN, a marked methionine-depleted state can be created safely. Hypertonic glucose was administered as a TPN solution to supply sufficient calories combined with electrolytes, vitamins and minerals. We avoided providing any substances containing the thiol radical, such as glutathione. We named this parenteral treatment methionine-depleting total parenteral nutrition (Met-deplt TPN) or RT-therapy. Because all the sulfur-containing amino acids and the thiol radicals were restricted by this treatment, the term RT-therapy by means of S restriction (...QRSTUV...,→, ...QRTUV...,→RT) was coined by Murakami in 1978.[14]

Influences of RT-Therapy

Experiment in Nontumor-Bearing Dogs[15]

We divided 12 male mongrel dogs into two groups, an AO-90 and a Pan-Amin S group. In the AO-90 group, the animals received an AO-90 amino acid solution, while the Pan-Amin S group received Pan-Amin S® as a protein source. Both groups were kept on TPN without oral intake. The same amount of amino acid nitrogen and glucose were infused for three weeks (Table 9.2). Nutritional status and the blood and urine biochemical analysis including amino acid fraction of the plasma were assessed during the course of the three weeks. Then the animals were sacrificed and liver specimens were taken for a histological study.

Results

No difference was observed in general appearance and activity in dogs in both groups. In the AO-90 group, the body weight decreased by a mean 12.3%, while in the Pan-Amin S group, it increased by 7.2% by the end of the experiment. Serum albumin fell significantly in the AO-90 group from a mean of 2.86-1.39 g/100 ml. Nitrogen balance was negative throughout the experimental period in the AO-90 group with a mean of 6.574 g/day, while it was positive with a mean of +1.512 g in the Pan-Amin S group. The serum level of thiol radical in the AO-90 group steadily decreased from a mean of 0.517-0.08 mmol/l, whereas it remained normal in the Pan-Amin S group. The serum aminogram showed increases in lysine, phenylalanine, threonine and serine and decreases in methionine, aspartic acid , asparagine and cystine ($p<0.05$). Other laboratory parameters, such as blood glucose, blood urea nitrogen, glutamic-oxaloacetic transaminase (GOT) and glutamic-pyruvic transaminase (GPT) were kept in normal ranges during the experiment without a significant difference between both groups. Only serum alkaline phosphatase (Al-P) rose slightly in the AO-90 group; all the other tests of liver function remained normal. Concentrations of protein, glycogen and triglyceride of the liver tissue showed no difference between the groups at the end of the experiment.

Table 9.1. Composition of amino acid solution

Amino Acid	Code Name AO-90 (g/dL)	Pan-Amin S® (g/dL)
L-Arginine	0.66	0.66
L-Histidine	0.30	0.30
L-Isoleucine	0.55	0.55
L-Leucine	1.23	1.23
L-Lysine	1.49	1.49
L-Methionine	—	0.71
L-Phenylalanine	0.87	0.87
L-Threonine	0.54	0.54
L-Tryptophan	0.18	0.18
L-Valine	0.61	0.61
Glycine	1.00	1.00
Total amino acids (g/dL)	7.43	8.14
Total nitrogen (g/L)	11.9	12.6

Otsuka Pharmaceutical Factory, Inc., Tokushima, Japan.

Table 9.2. Daily composition of TPN solutions per kg for two groups of dogs

	Pan-Amin S	AO-90
Glucose (g)	21.0	21.0
Free amino acid (g)	3.60	3.27
Lipid (g)	0.30	0.30
Total volume (ml)	120.96	120.96
Total calories (kcal)	101.7	100.4
Total nitrogen (g)	0.557	0.526
Nonprotein calories/N (kcal/g)	157	166
Calories/ml	0.835	0.823

A histopathological study of the liver tissue revealed a marked difference between the two groups. The liver cells of the Pan-Amin S group showed normal architecture. In contrast, the hepatocytes of the AO-90 group were slightly swollen with marked nucleolar hypertrophy and transparent deposits, resistant to Sudan III and PAS-stainings, were observed in the cytoplasm.

AH-109A Ascites Hepatoma-Bearing Rat Experiment[16]

Seven-week-old male Donryu rats weighing about 180 g were used for this experiment. Each rat was inoculated intraperitoneally (IP) with AH-109A ascites hepatoma cells, which were maintained in rats by weekly IP implantation and harvested seven days after serial transplantation.

Two days after inoculation, the rats were divided into three TPN groups and a freely-fed group. The rats of TPN groups received central venous catheterization at the superior vena cava and were housed individually in metabolic cages. These rats were subdivided into three groups and were kept on TPN for seven days with one of the three nutritional regimens shown in Table 9.3 with no oral intake.

1. AO-90 group (n=13): AO-90 amino acid solution was infused as a protein source.
2. Pan-Amin S group (n=13): Pan-Amin S® amino acid solution was infused as a protein source.
3. Glucose group (n=5): Rats were maintained without any amino acid.
4. Freely-fed group (n=24): Rats were fed on laboratory chow and water.

Nine days after tumor inoculation, the rats were sacrificed to survey the antitumor effect and the adverse side effects including biochemical and pathohistologica studies.

The blood and ascites of rats in each group were analyzed to determine the levels of total protein, albumin, GOT, GPT, Al-P, glucose, albumin/globulin ratio (A/G ratio), and amino acid fraction.

The volume of ascites and the number of tumor cells were examined. Smear samples of the ascites tumors were stained by the methods of Papanicolau and Feulgen, and by azure B staining with DNase pretreatment. These samples were histologically and histometrically examined. The histopathological examination of the liver, bone marrow, spleen, stomach, and small and large intestines were also performed. Histometrically, the sizes of nucleoli of tumor cells and hepatocytes were assessed.

Results

The volume of ascites and the number of tumor cells of the four groups at sacrifice are shown in Figure 9.1. The increases in ascites and tumor cell number of the AO-90 group were less than those of the other three groups with significant differences at $p<0.01$ or 0.05.

Ascites of the AO-90 group was slightly hemorrhagic, while those of the other groups were deeply bloody.

Levels of plasma protein and serum albumin of the AO-90 group were lower than those of the freely-fed group. A slight rise of urea nitrogen was observed in the AO-90 group; however, urea nitrogen was in the normal range in the other three groups. GOT, GPT and AL-P of the AO-90 group were in normal ranges.

An ascites aminogram is summarized in Table 9.4. Methionine and cysteine levels of the AO-90 group were significantly lower than those of the other groups. The levels of serine, valine, leucine and ornithine were higher in the AO-90 group than in the other groups. Similar aminogram changes were observed in plasma of the AO-90 group.

The morphological appearance of the ascites hepatoma cells with Papanicolau staining indicated no apparent differences among the four groups, except for an increase in nucleolar size in the AO-90 group (Table 9.5).

In the liver tissues, an increase in nucleoli of hepatocytes was prominent in the AO-90 group as compared with those in the other groups. By azure B staining with DNase pretreatment, azure B positive substance appeared roughly lucent in cytoplasma of the AO-90 group, which had markedly enlarged nucleoli, while in the other groups, azure B positive substance was not distributed in the cytoplasm but in the nucleoli. And the sizes of nucleoli of both tumor and liver cells of the AO-90 group were significantly larger than those in the other groups (Table 9.5). A slight deposition of lipid was detected in hepatocytes of all three groups, except for the freely-fed group.

Table 9.3. Daily regimens of TPN solutions infused to rats in three experimental groups

	AO-90	Pan-Amin S	Glucose
50% Glucose (ml)	105	105	105
AO-90® (ml)	123	—	—
Pan-Amin S® (ml)	—	123	—
Electrolyte soln. (ml)	22.6	22.6	22.6
Vitamin mixture at (ml)	0.1	0.1	0.1
Sterile water (ml)	—	—	121.5
Total volume (ml)	251	251	251
Glucose (g)	52.5	52.5	52.5
Amino acid (g)	9.14	10.0	—
Na (mEq)	19.0	19.0	19.0
K (mEq)	2.6	2.6	2.6
Ca (mEq)	0.55	0.55	0.55
Cl (mEq)	19.0	19.0	19.0
Mg (mEq)	1.1	1.1	1.1
HPO_4 (mEq)	2.6	2.6	2.6
Total calories (kcal)	247	250	210
Total N (g)	1.46	1.55	0
Nonprotein calories/N	144	135	—

Amounts are listed per kg of body weight.

Experiment in Sato Lung Carcinoma (SLC)-Bearing Rats[17]

Fifty seven-week-old male Donryu rats weighing about 200 g were used in this experiment. Sato lung carcinoma (SLC) (induced in male Donryu rats by 4-nitroquinoline l-oxide and maintained in rats by weekly intraperitoneal implantation) was transplanted into the subcutaneous fatty tissue of the back of the rats (day 0).

These animals were fed a diet of laboratory chow and water ad libitum for 10 days (until day 10) and carefully examined by measuring the tumor size.

On the 11th day after transplantation (day 11), 22 rats with an estimated tumor weight ranging from 0.41 to 2.10 g were subjected to TPN treatment with central venous cannulation. The rats were divided into three different TPN groups, receiving AO-90®, Pan-Amin S®, or no amino acid, respectively. The daily TPN regimens are shown in Table 9.3. The animal groups were as follows:

1. AO-90 group (n=9): AO-90 amino acid solution was infused as a protein source.
2. Pan-Am S group (n=9): Pan-Amin S^R amino acid solution was infused as a protein source.
3. Glucose group (n=4): Rats were maintained without any amino acid.

On day 21 (21 days after SLC transplantation) when the TPN was discontinued, all animals in each group were sacrificed to assess tumor and carcass weights, and then the tissue amino acid fraction was analyzed.

Plasma total protein, albumin, globulin, urea nitrogen, glucose, GOT and GPT were also analyzed.

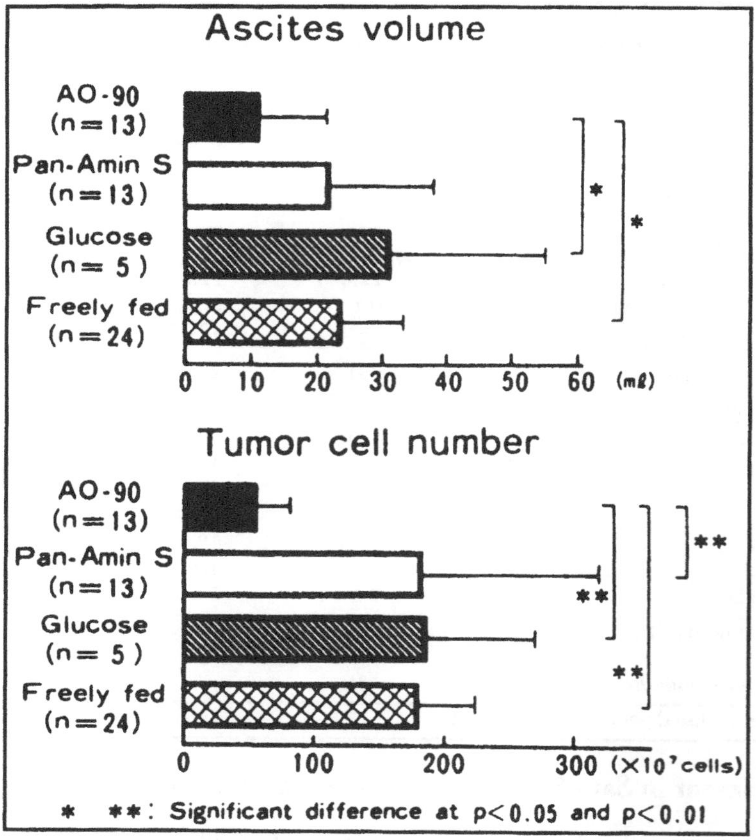

Fig. 9.1. Volume of ascites and number of the tumor cells in AH-109 A ascites hepatoma-bearing rats at the seventh day of TPN with different nutritional regimens.[15]

Results

Four days after the start of TPN, tumor weight in the AO-90 group began to decrease, and at the end of the experiment, tumor weight was significantly less than that of the Pan-Amin S group. However, the tumor weight/carcass weight ratio showed no statistically significant difference among the three groups.

In the plasma, there were decreases in L-cysteine, L-methionine, L-glycine, L-tyrosine, L-tryptophan and L-arginine and increases in L-serine, L-glutamine, L-valine, L-ornithine and L-histidine in the AO-90 group compared with the Pan-Amin S group. In the tumor tissue of the AO-90 group, methionine and tyrosine decreased significantly compared with the Pan-Amin S group, which had a significant increase in serine only.

In plasma biochemical data at the end of the experiment, total protein, albumin, globulin, urea nitrogen, glucose, GOT and GPT values in the AO-90 group showed no significant difference from those of the Pan-Amin S group. Only albumin and the A/G ratio were significantly decreased in comparison with the Pan-Amin S group.

Table 9.4. Ascites aminogram in AH-109A ascites hepatoma-bearing rats

	AO-90 n = 4	Pan-Amin S n = 5	Glucose n = 5
Tau	127.8 ± 50.8	189.7 ± 61.2[e]	118.7 ± 31.2[a]
Asp	29.6 ± 20.7	11.5 ± 4.3[e]	17.4 ± 3.4[a]
Thr	230.4 ± 1266.0[e]	977.5 ± 140.3[f]	148.6 ± 96.0[b,c]
Ser	743.4 ± 348.4[a,f]	107.8 ± 12.7[c]	152.3 ± 91.5[d]
Asn	71.1 ± 17.9	42.2 ± 39.6	48.3 ± 19.4
Glu	182.0 ± 72.2	120.3 ± 12.6	115.7 ± 28.7
Gln	169.5 ± 107.2[a,e]	20.3 ± 7.9[c]	26.9 ± 24.6[c]
Pro	186.1 ± 2.4	146.0 ± 31.8	277.5 ± 186.0
Gly	949.8 ± 391.9	764.2 ± 134.2	699.1 ± 272.4
Ala	917.5 ± 210.9[a]	416.2 ± 52.7[c]	843.1 ± 474.1
Cit	61.4 ± 6.7[b,f]	34.5 ± 4.2[d,f]	13.2 ± 8.9[b,d]
Val	294.9 ± 110.3[a,f]	109.9 ± 15.0[c,e]	53.3 ± 44.9[a,d]
Cys	9.8 ± 5.1[b]	34.6 ± 5.0[d,f]	9.2 ± 7.0[b]
Met	18.4 ± 9.9[b]	87.7 ± 3.6	13.8 ± 12.3[b]
Ile	96.9 ± 56.0[e]	24.0 ± 4.1	19.3 ± 22.4[c]
Leu	227.9 ± 130.1[c]	68.6 ± 11.6	41.4 ± 42.2[c]
Tyr	117.4 ± 111.7	58.5 ± 16.8	34.6 ± 39.3
Phe	140.4 ± 42.8[e]	97.9 ± 5.2	50.4 ± 38.2[c]
Orn	173.3 ± 49.2[b,f]	75.1 ± 14.2[d,f]	30.8 ± 20.9[b,d]
Trp	81.2 ± 19.7[f]	59.9 ± 19.5[f]	21.0 ± 11.4[b,d]
Lys	1003.1 ± 201.2[f]	872.7 ± 66.8[f]	287.7 ± 160.5[b,d]
His	191.6 ± 80.4	108.8 ± 18.8	99.6 ± 65.2
3-MetHis	17.1 ± 6.3	50.2 ± 69.5	11.7 ± 3.8
Arg	150.5 ± 106.4	163.5 ± 20.9[f]	49.0 ± 22.9[b]

Values are means ± S.D. and μmole/liter. [a,b] Significantly different from Pan-Amin S at $p<0.05$ and $p<0.01$, respectively; [c,d] Significantly different from AO-90 at $p<0.05$ and $p<0.01$, respectively.; [e,f] Significantly different from glucose at $p<0.05$ and $p<0.01$, respectively. Courtesy of Goseki et al.[15]

Discussion

In three weeks of Met-deplt TPN (RT-therapy) in mongrel dogs, the dogs exhibited no adverse side effects, neither biochemical nor hematological disorders, except for weight loss and development of hypoproteinemia to some degree. In the SLC-bearing rat experiment, the rats can be kept on Met-deplt TPN for at least 10 days without serious malnutrition. Although, weight loss and the development of hypoproteinemia were inevitable, hepatic disorders, bone marrow suppression and other adverse effects due to this treatment were not observed. Therefore, we do think, this treatment could be safely carried out for a limited term such as two to three weeks in adult patients in general.

In the rats bearing AH-109A ascites hepatoma, increases in the volume of ascites and the number of tumor cells were significantly inhibited in the rats receiving glucose plus AO-90 amino acid solution compared with the control rats given only glucose, glucose plus ordinary amino acid mixture or laboratory rations. These phenomena suggest that the inhibitory effect of methionine deprivation resulted not only from malnutrition in the host animals but also from the development of some deficiency of protein synthesis in the tumor cells. However, tumor growth was not suppressed markedly in the SLC-bearing (solid tumor) rats although the

Table 9.5. Size of nucleoli (μm^2)

	AO-90	Pan-Amin S	Glucose	Freely-fed
Tumor	$3.76 \pm 1.49^{a,c,d}$	$3.01 \pm 1.09^{b,d}$	$3.20 \pm 1.18^{b,d}$	$2.20 \pm 1.16^{a,b,c}$
Liver	$5.85 \pm 2.09^{a,c,d}$	$3.30 \pm 0.93^{b,c,d}$	$4.00 \pm 1.57^{a,b,d}$	$2.76 \pm 1.19^{a,b,c}$

Values are means ± S.D. n = 200. [a] Significantly different from Pan-Amin S group at p<0.01; [b] Significantly different from AO-90 group at p<0.01; [c] Significantly different from glucose group at p<0.01; [d] Significantly different from freely-fed group at p<0.01. Courtesy of Goseki et al.[1]

L-methionine concentrations, both in plasma and tumor tissue, were decreased significantly in comparison with control groups. We conclude that it is impossible to control cancer progression by only Met-deplt TPN since we initially introduced this treatment as an adjunct to cancer chemotherapy.[7]

Analysis of amino acids in plasma and ascites showed marked decreases of L-methionine and L-cysteine levels associated with elevation of some other amino acids such as L-serine, reflecting apparent methionine starvation. Under the circumstances of so-called methionine imbalance, we could not determine the mechanism of inhibiting the proliferation of malignant cells. We think that this treatment affects not only amino acid metabolism but also that of many other substances such as anticancer agents combined with this treatment.

In the liver cell, marked enlargement of the nucleoli was observed. Deposition of a high level of RNA in tumor cells, although not as apparent as that in hepatocytes, resulted in nucleolar enlargement. Bailey et al reported similar findings of inhibiting the transformation process of RNA observed in the hepatocytes of nontumor-bearing rats fed on a nonprotein diet or on a diet lacking in L-methionine and L-cysteine.[18] A phenomenon similar to the inhibition of the transformation of mRNA to rRNA may also appear in the nucleoli of tumor cells, but the degree of enlargement of the nucleoli of tumor cells was smaller than that of the liver cells. Other surveys are necessary to elucidate the mechanism of this phenomenon in tumor cells.

Met-deplt TPN and Thiol Depletion[19]

Experiment

SLC cells were implanted in the subcutaneous fatty tissue of the back of 14 seven-week-old male Donryu rats (day 0). All rats were allowed to ingest food and water freely for five days. By day 6, the tumors had developed in each rat. Nine of these tumor-bearing rats were selected at random and central venous cannulation was performed for TPN treatment. Following the cannulation, the rats were housed individually in the metabolic cage. Then they were divided into two TPN groups; four rats to receive AO-90 and the other five to receive Pan-Amin S® as the sole nitrogen source for TPN. The composition of the two amino acid solutions and the daily TPN regimens are the same as in previous experiments (Tables 9.2, 9.3).

The remaining five rats were fed on a diet of laboratory chow and water, adlibitum, without cannulation.

The rats were grouped as follows:

1. AO-90 group (n=4): The AO-90 amino acid solution was infused as the sole nitrogen source.
2. Pan-Amin S group (n=5): The Pan-Amin S® amino acid solution was infused as the sole nitrogen source.

3. Freely-fed group (n=5): In this group, all the rats were fed on a diet of laboratory chow and water ad libitum.

On day 21, 8 days after the entry of TPN treatment and 12 days after SLC implantation, all rats were sacrificed after weighing, and the tumor and the liver were taken out. The thiol (GSH, GSSG and total GSH) concentrations of the tumor and the liver tissues were also analyzed.

Results

In the AO-90 group, tumor weight, tumor weight/carcass weight, and liver weight were the lower on day 12 with a significant difference at $p<0.01$ in comparison with the freely-fed group.

The thiol concentration with the fractions (GSH,GSSG and total GSH μg/100 mg of tumor and liver tissues) are summarized in Table 9.6. In the AO-90 group, all values of the thiol fractions decreased by more than 40% in the tumor tissue and by more than 80% in the liver tissues as compared with the Pan-Amin S and the freely-fed groups, with a statistically significant difference at $p<0.01$.

Discussion

Hypoxic cell radiosensitizers, e.g., misonidasol (MISO), have been reported to sensitize cells to radiation. It is suggested that one component of the mechanism involved depends on the metabolism and is linked to the depletion of cellular thiols such as GSH.[20,21] Radiosensitizers are also known to have a chemosensitizing effect when combined with radiomimetic alkylating agents or similar drugs.[22-25] It has been confirmed that these phenomena depend on the interaction of the thiol-depleted state and chemotherapeutic agents.[26]

By Met-deplt TPN, the marked depletion of methionine was achieved not only in the plasma but also in the tumor tissue, followed by inhibition of tumor proliferation to a small degree.[17] Although mere amino acid imbalance could not fill the role of an effective treatment for cancer, this study demonstrates that Met-deplt TPN apparently decreases the thiol concentrations both in the tumor and in the liver, resulting in a slight tumor growth inhibition in SLC-bearing rats.

Several thiol-depleting agents other than MISO, such as buthionine-S, R-sulfoximine (BSO) and diethyl maleate (DEM), were examined in vivo. We found that thiol-reactive agents do not exhibit any specification on tumor GSH. In vivo sensitization is complicated by thiol reactions in other organs, especially in the liver, with potential toxicity.[27] To reduce the systemic toxicity and to increase the antitumor effects, combined treatment of different sensitizers at lower dosage levels and at different times during the course of treatment was tried.[28] Some tumor tissue may be more sensitive than normal tissues to melphalan toxicity following GSH depletion by BSO, which suggests that several thiol-modulating agents might enhance the antitumor effects when combined with suitable antineoplastic treatment.[29] We do think that Met-deplt TPN is an effective thiol-modulating treatment and enhances antitumor effects if combined with suitable anticancer agents.

Synergic Effect of Met-deplt TPN on Several Anti-Cancer Agents in Tumor-Bearing Animals

Synergic effects of Met-deplt TPN (RT-therapy) on several anticancer agents such as actinomycin D, nimustine hydrocloride, 5-fluorouracil, mitomycin C, doxolubicin, vincristine and cisplatin were demonstrated in tumor-bearing rats.[19,30-37] This chapter describes the synergic effects on nimustine hydrocloride, 5-fluorouracil, doxolubicin and vincristine.

Table 9.6. Thiol* content in tumor and liver

		Tumor			Liver		
Group	n	Total GSH	GSH	GSSG	Total GSH	GSH	GSSG
AO-90	4	27.5 ± 4.2	26.0 ± 4.0	1.50 ± 0.26	32.0 ± 6.6	30.4 ± 6.2	1.59 ± 0.39
Pan-Amin S	5	45.4 ± 7.2**	43.1 ±7.0**	2.26 ± 0.52**	173.6 ± 14.4**	162.0 ± 11.6**	11.60 ±3.25**
Freely-fed	5	49.5 ± 6.6**	46.7 ±6.2**	2.86 ± 0.52**	193.4 ± 33.1**	181.4 ± 30.1**	11.94 ± 3.11**

Values are means ± SD expressed by μg/100 mg wet tissue; n = number of animals;* Total reduced glutathione (GSH), reduced glutathione (GSH) and oxidized glutathione (GSSG);**p<0.01 vs. AO-90 groups. Courtesy of Goseki et al.[18]

Nimustine Hydrochloride[8]

Experiment

Seven-week-old male Donryu rats weighing approximately 200 g were used for this study. SLC were implanted in subcutaneous fatty tissues of the back of the rats (day 0). These rats were allowed to ingest food and water freely. Subcutaneous solid tumors developed in all rats.

Rats with almost the same tumor size were selected and divided into three experimental groups. Nimustine hydrochloride (ACNU), an alkylating agent, was used as a carcinostatic drug and was administered as follows: (1) 2.5 mg/kg/day x 3(IP), (2) 5.0 mg/kg/day x 3(IP), and (3) 10.0 mg/kg/day x 3(IP).

Central venous cannulation was performed prior to each experiment. Animals were kept individually in a metabolic cage and maintained on TPN for eight days. The animals were subdivided into four groups.

1. AO-90 + ACNU group: The AO-90 amino acid solution was infused as the source of nitrogen and the animals were maintained on TPN. On days 2, 4 and 6 after the start of the TPN treatment, each animal was given a dose of ACNU via the IP route. There were six animals in each dosage group: 2.5 mg/kg x 3(IP), 5.0 mg/kg x 3(IP) and 10.0 mg/kg x 3(IP) of ACNU administration.
2. AO-90 group: AO-90 was infused as the source of nitrogen. TPN therapy was carried out without administration of ACNU. Six animals received a dose of 2.5 mg/kg x 3(IP) dose, six received 5.0 mg/kg x 3(IP), and seven received 10.0 mg/kg x 3(IP) ACNU.
3. Pan-Amin S + ACNU group: TPN was given with Pan-Amin S® and the same doses of ACNU were given to the animals in the same manner as in the AO-90 + ACNU group. There were six animals in each of the three of ACNU dosage studies.
4. Pan-Amin S group: The animals were placed under TPN control with Pan-Amin S® as the source of nitrogen without ACNU administration. There were six animals in both the 2.5-mg/kg x 3 (IP) and 5.0-mg/kg x 3(IP), and seven in the 10.0-mg/kg x 3(IP) ACNU dosage study. The TPN regiments for the rats were the same as in previous experiments (Tables 9.1, 9.3).

During the experiment, the size of each tumor was measured. After the experiment was completed, the blood cell count and blood chemistry such as blood glucose level, total protein, albumin, blood urea nitrogen, creatinine, GOT, GPT and Al-P were analyzed.

The tumors were extracted from the implanted area and the carcass and the tumor were weighed. Then the tumor weight/carcass weight was calculated.

Results

Tumor Response (Tumor Weight)

ACNIU, 2.5 mg/kg x 3(IP), Combined Study

The tumor size in the AO-90 + ACNU group tended to regress on day 14 and thereafter (Fig. 9.2). Upon completion of the experiments, no differences were noted in tumor weight at the dosage level where the carcinostatic effect of ACNU is not clearly defined between the Pan-Amin S + ACNU group and the Pan-Amin S group. In contrast, the AO-90 + ACNU group showed a significant decline in tumor weight as compared with the Pan-Amin S + ACNU group and the Pan-Amin S group at $p<0.05$. The AO-90 group also showed a significant decline in tumor weight compared with both the Pan-Amin S + ACNU and the Pan-Amin S groups. No significant difference was noted in the tumor weight/carcass weight, but a similar tendency was shared by the four groups.

ACNU, 5.0 mg/kg x 3(IP), Combined Study

The tendency for tumor growth inhibition was noted on day 13 and thereafter in the AO-90 + ACNU group, and on day 15 in the Pan-Amin S + ACNU group, judging by the tumor size (Fig. 9.3). Tumor weight upon completion of the experiment, on day 15, suggested that the combined effect in the Pan-Amin S group at this dose level, although different, was statistically significant. In contrast, the decrease in tumor weight was significant in the AO-90 + ACNU group as compared to the Pan-Amin S group ($p<0.05$). A similar tendency was also seen in the tumor weight/carcass weight, without significant differences.

ACNU, 10.0 mg/kg x 3(IP), Combined Study

Tumor growth inhibition was noted on day 12 and thereafter in the AO-90 + ACNU group (Fig. 9.4). Similar effects were noted on day 14 and thereafter in the Pan-Amin S + ACNU group. Tumor weight at the end of the experiment, on day 16, decreased significantly following administration of ACNU in the AO-90 + ACNU group compared to the AO-90 group and the Pan-Amin S group as did tumor weight in the Pan-Amin S + ACNU group compared to the AO-90 group and the Pan-Amin S group with a significant difference at $p<0.01$. These results demonstrate the growth inhibitory effects of ACNU. Similar results were obtained in the tumor weight/carcass weight ratio.

Adverse Effects

Table 9.7 shows the results of blood and plasma analysis in rats which were given 10.0 mg/kg x 3 ACNU with the highest inhibition to the SLC growth among three groups of ACNU combined studies. The results of two other ACNU dosage studies showed almost no difference in comparison with control groups.

Influences on Nutritional Status

As shown in Table 9.7, in the 10.0 mg/kg x 3 ACNU combined study, only A/G showed a minimal decline in AO-90 + ACNU group with a significant difference at $p<0.05$ compared

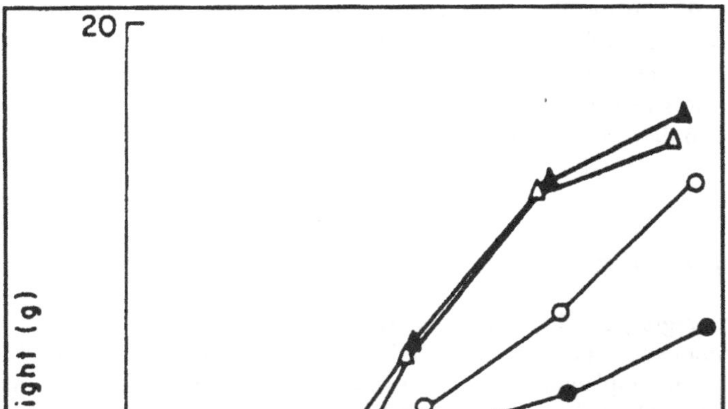

Fig. 9.2. Growth curve of tumor in ACNU 2.5 mg/kg x 3(IP) combined study.[18] *l* = AO-90 + ACNU; *m* = AO-90; *s* = Pan-Amin S + ACNU; *s* = Pan-Amin S; ≠ = ACNU IP

with the Pan-Amin S + ACNU dose group. The same tendencies were also observed in two other ACNU experiments, but no significant difference was observed. There were almost no difference in these values between the AO-90 + ACNU groups and the AO-90 groups in the three ACNU dose experiments.

Influences on Peripheral Blood Cell Counts

There was no significant difference in the WBC count between the AO-90 + ACNU and the Pan-Amin S + ACNU groups in the ACNU at 10.0 mg/kg x 3 combined study, though the values were significantly decreased in both groups ($p<0.01$) compared to those in the AO-90 and the Pan-Amin S groups. In other ACNU dose combined studies, there was no decline in the WBC count across the four groups. There was a slight elevation in the hematocrit, hemoglobin and RBC count in the AO-90 + ACNU group as compared to the control groups in the three ACNU dose combined studies.

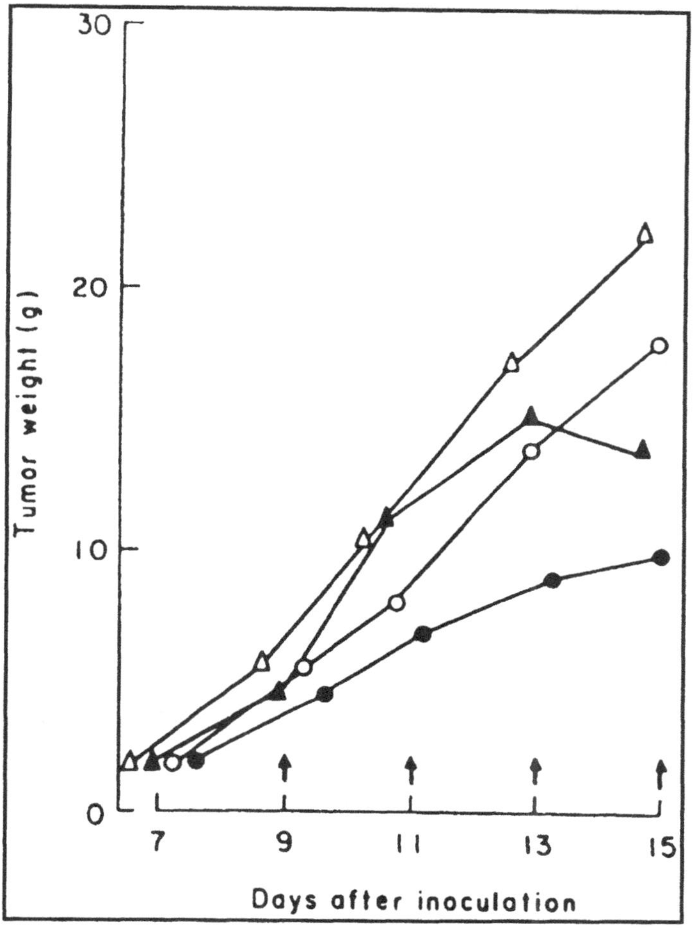

Fig. 9.3. Growth curve of tumor in ACNU 5.0 mg/kg x 3(IP) combined study. Symbols are the same as in Figure 9.2.[18]

Influences on Other Values of Blood and Plasma Analysis

Only Al-P values in the AO-90 + ACNU, the AO-90 and the Pan-Amin S + ACNU groups were elevated compared to those in the Pan-Amin S group with a significant difference at $p<0.05$ or 0.01 in the ACNU at 10.0 mg/kg x 3 combined study. Two other ACNU dose studies did not have this tendency. The GOT and the GPT values did not rise in the AO-90 + ACNU group compared to the control groups in the three ACNU dose combined studies. In contrast, the values of the urea N-group were slightly higher in the AO-90 + ACNU than in the control groups in the three ACNU dose combined studies, without elevation in the creatinine value.

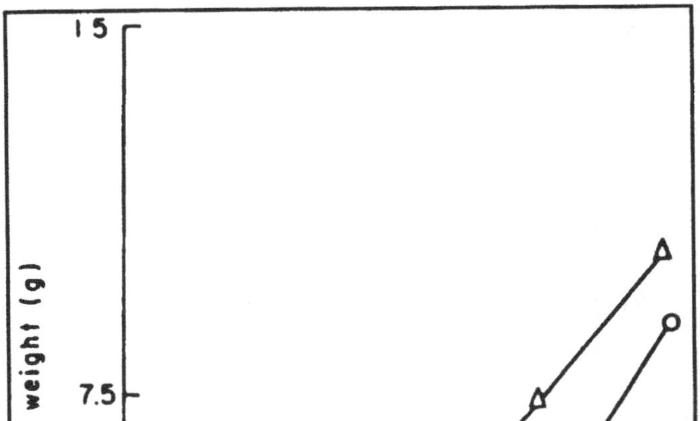

Fig. 9.4. Growth curve of tumor in ACNU 10.0 mg/kg x 3 (IP) combined study. Symbols are the same as in Figure 9.1.

Discussion

Met-deplt TPN inhibited SLC proliferation significantly even in conjunction with a small dose of ACNU, which showed no antitumor effect on the rats treated with the methionine-containing amino acid solution, without increasing the side effects in comparison with control groups. As a result, the antitumor effect of ACNU was enhanced by Met-deplt TPN. Although the mechanisms involved are poorly understood, we believe that it must be related to the decrease in thiol. Thus, we conclude that all the phenomena yielded here are important to the mechanism of Met-deplt TPN even though they are hardly interpreted from a single viewpoint of the thiol-modulating effect.

The results of this study suggest that Met-deplt TPN may enhance the antitumor effect of ACNU without increasing systemic toxicity. According to the results of the current blood analysis and blood chemistry tests, decreases in WBC as a side effect due to ACNU in the AO-90 + ACNU group are similar to those in the Pan-Amin S + ACNU group. In this case, ACNU was administered 10.0 mg/kg x 3 with a combination therapy of Pan-Amin S + ACNU

Table 9.7. Blood count and plasma biochemical analysis in ACNU 10.0 mg/kg x 3 (IP) combined study

Group	AO-90 + ACNU* (n = 5)	AO-90 (n = 7)	Pan-Amin S + ACNU* (n = 6)	Pan-Amin S (n = 7)
Hematocrit (%)	34.4 ± 2.3[a]	33.1 ± 4.1	32.0 ± 2.2	33.3 ± 3.4
Hemoglobin (g/100 ml)	12.1 ± 0.8	11.6 ± 1.1	10.9 ± 1.2	11.6 ± 1.1
RBC ($x10^4/mm^3$)	891 ± 68[a,b]	828 ± 80[c]	763 ± 77	807 ± 61
WBC ($x10^2/mm^3$)	31 ± 5[d,e]	131 ± 43[c]	41 ± 12[e]	119 ± 21
Total protein (g/100 ml)	4.34 ± 0.18[c,d,e]	3.37 ± 0.52[c,e]	4.75 ± 0.21	4.90 ± 0.17
Albumin (g/100 ml)	1.64 ± 0.08[c,d,e]	1.26 ± 0.21[c,e]	2.02 ± 0.13	2.05 ± 0.15
A/G ratio	0.61 ± 0.03[a,c]	0.60 ± 0.06[c,e]	0.74 ± 0.09	0.72 ± 0.07
Glucose (mg/100 ml)	77 ± 17	67 ± 10[c,e]	92 ± 8	93 ± 11
Urea nitrogen (mg/100 ml)	16.4 ± 1.9[f,e]	14.0 ± 1.4[e]	14.1 ± 1.8[e]	10.8 ± 0.6
Creatinine (mg/100 ml)	0.34 ± 0.05	0.30 ± 0	0.33 ± 0.05	0.36 ± 0.05
GOT (U/liter)	130 ± 34[b,c]	370 ± 128[c]	117 ± 14[e]	254 ± 87
GPT (U/liter)	14 ± 2[c,f]	20 ± 5[c,e]	10 ± 1[e]	13 ± 2
Al-P (U/liter)	254 ± 96[b]	182 ± 53[e]	173 ± 28[e]	112 ± 15

Values are means ± S.D.; n = number of animals
* 10mg/kg
[a] $p<0.05$ versus Pan-Amin S + ACNU group.
[b] $p<0.05$ versus Pan-Amin S group.
[c] $p<0.01$ versus Pan-Amin S + ACNU group.
[d] $p<0.01$ versus AO-90 group.
[e] $p<0.01$ versus Pan-Amin S group.
[f] $p<0.05$ versus AO-90 group.

as a control which is known to inhibit tumor growth. However, no decrease in WBC nor adverse effect on liver function were noted on day 15, when ACNU at 2.5 mg/kg was combined; a dose which exhibited tumor growth inhibition exclusively in the AO-90 + ACNU group. Although another side effect of ACNU, namely the elevation in the urea nitrogen level, was observed to a certain extent, no elevation was demonstrated in the creatinine level which supports a hypothesis that the effects on the kidneys are negligible.

The side effects of AO-90 include decreased serum total protein, albumin and A/G ratio, and blood glucose, as well as very slight adverse effects on hepatic function. Nonetheless, AO-90 exhibited a sensitizing effect which acted on tumor growth inhibition through ACNU. The elevation in RBC at 10.0 mg/kg x 3 of ACNU and at 2.5 mg/kg in the AO-90 + ACNU group is hardly related to the inhibitory effect of ACNU on the bone marrow. This is a problem to be solved in the future, though we believe that Met-deplt TPN is a useful thiol-modulating treatment because the antitumor effects of some antineoplastic drugs are enhanced without any apparent elevation in systemic toxicity. Consideration should be given to the selective administration of combined anticancer agents in actual clinical use to the tumor tissue.

5-Fluorouracil[34]

Experiment I

We examined the growth inhibitory effect of a combination therapy of Met-deplt TPN and 5-fluorouracil (5-FU) on Yoshida sarcoma (YS).

Six-week-old male Donryu rats were used for the experiment. Rats were transplanted with YS to the dorsal adipose tissue (day 0), which make the solid-type tumor. Rats were cannulated through the vena cava immediately after tumor transplantation and were maintained on TPN with one of the four regimens combined with or without 5-FU. The rats were not allowed to take any food orally during the TPN period. The animal groups were as follows:

1. AO-90 + 5-FU group(n=6): AO-90 amino acid solution was infused for eight days. The total parenteral nutrition regimens are shown in Table 15.3. 5-FU was mixed with the TPN solution and infused for six days (from day 2-7) at a dose of 15 mg/kg/day (90 mg/kg total).
2. AO-90 group (n=5): Rats received TPN with AO-90 for eight days without 5-FU administration.
3. Pan-Amin S + 5-FU group (n=6): Rats received TPN for eight days with Pan-Amin S®. 5-FU was infused in the same dose and manner as in the AO-90 + 5-FU group.
4. Pan-Amin S group (n=5): Rats received TPN for eight days with Pan-Amin S® without 5-FU.

At the end of the experiment (day 8), rats were killed and the tumor-implanted area was extracted to measure the tumor and carcass weight. Major organs, including the liver and the lung, were histopathologically examined for the presence of tumor cells.

Experiment II

We studied the life span of YS-bearing rats administered with Met-deplt TPN combined with 5-FU. To examine the synergic effect of the Met-deplt TPN and 5-FU, eight-week-old male Donryu rats were transplanted with YS into the dorsal adipose tissue as in experiment I, and then cannulated the vena cava for TPN (day 0).

The experimental groups were AO-90 + 5-FU (n=6), AO-90(n=5),Pan-Amin S + 5-FU (n=6), and Pan-Amin S (n=7). The AO-90 + 5-FU and the Pan-Amin S + 5-FU received 10 mg/kg/day of 5-FU(80 mg/kg total) mixed with TPN solution for eight days and were also given 10 mg/kg/day intraperitoneally, twice on days 10 and 11. These tumor-bearing rats were on TPN for 10 days. After TPN the animals were fed by laboratory chow and water ad libitum.

To examine the tumor extension in rats when they were killed (experiment I) or at death (experiment II), all animals were surveyed pathologically with the following criteria.

(-): No evidence of metastasis with histologic examination.
(±): Histologically, only a small amount of tumor cells were observed.
(+): Histologically, small foci of tumorous formation of cancer cells were demonstrated in the examined organs.
(++): Histologically, the apparent tumor metastasis-forming masses were demonstrated, but no more than one half of the organ tissue.
(+++): Pathologically, marked metastasis in the examined organs was demonstrated resulting in failure of the metastasized organ.

Results: Experiment I

Tumor Weight

Tumor weight and tumor/carcass weight ratio were significantly lower in the AO-90 + 5-FU group than other groups at $p<0.01$ by Student's t test.

Pathohistologic Findings

In the AO-90 + 5-FU group, only one rat showed hematogenous liver metastasis, while in the AO-90 group one rat had metastases apparently in the liver, lung, and kidney (Table 9.8). Other rats in these two groups showed no hematogenous metastasis in histopathology. In contrast, in the Pan-Amin S + 5-FU and the Pan-Amin S groups, all rats developed severe hematogenous metastasis in several organs.

Table 9.8. Pathologic findings of tumor metastasis in each rat of four groups at autopsy

Animal No.	Group	Metastasis Liver	Kidney	Lung	Heart	Spleen
1		—	—	—	—	—
2		++	—	—	—	—
3	AO-90 + 5-FU	—	—	—	—	—
4		—	—	—	—	—
5		—	—	—	—	—
6		—	—	—	—	—
7		—	—	—	—	—
8		±	—	—	—	—
10	AO-90	—	—	—	—	—
11		±	—	±	—	—
12		+	+	+	—	—
13		++	+	+	—	±
15		++	+	—	—	—
16	Pan-Amin S + 5-FU	++	—	+	—	—
17		+	—	++	+	—
18		+	+	+	—	—
19		+++	—	++	—	++
20		+++	—	+++	+	++
21	Pan-Amin S	++	—	++	—	±
22		++	±	++	—	—
23		±	—	++	—	—
24		±	—	++	—	+

See text for legend explanation. Data table courtesy of Goseki et al.[33]

Results: Experiment II

Survival Days

Mean survival days were 45.2±17.6 for the AO-90 + 5-FU group (two rats were killed on day 60), 30.4±18.0 days for the AO-90 group (one rat was killed on day 60), 21.3±8.4 days for the Pan-Amin S + 5-FU group, and 5.9±3.8 days for the Pan-Amin S group. The AO-90 + 5-FU group had a significantly longer survival period than the Pan-Amin S + 5-FU ($p<0.05$) and the Pan-Amin S groups ($p<0.01$)(Wilcoxon's rank sum test).

Tumor Extension at Autopsy

Table 9.9 summarizes the tumor extension of each rat at death and the five criteria used in experiment I, as well as other findings such as tumor weight and survival period. An examination of the tumorous area in death cases revealed that the implanted tumor decreased in severe necrosis and was completely diminished from the implanted site in two rats (33.3%) in the AO-90 + 5-FU group and that only three rats (50%) had slight hematogenous metastasis. The cause of death in this group was organ failure: two rats (33.3%) died of respiratory and hepatic failure attributable to hematogenous metastasis; the other two rats (33.3%) probably died of voluminous bleeding, body fluid loss or infection from the necrotic lesion of the remnant tumor at the implanted site.

Table 9.9. Pathologic findings of tumor metastasis at autopsy in four experimental groups

Group	Animal No.	Metastasis Liver	Kidney	Lung	Heart	Spleen	Tumor Weight (g)	Survival Days	Other Findings	Cause of Death
	1	—	—	—	—	—	None*	<60		None
	2	+++	+++	++	—	—	18.6†	23	Peritoneal dissemination	Meta§
AO-90 + 5-FU	3	—	—	—	—	—	82.4†	46		Tumor[‖]
	4	—	—	+	—	—	79	58		Tumor
	6	±	—	—	—	—	None*	≥60		None
	7	+	+++	+++	+++	—	16.3	24	Pleuritis	Meta
	8	—	—	—	—	—	None†	>60		None
	10	+++	+++	++	—	—	2.0	13		Meta
AO-90	11	—	—	—	—	—	24.1	32	Peritoneal dissemination	Tumor
	12	+++	+++	+++	—	—	25	26	Pleuritis	Meta
	14	++	++	+++	—	—	17.6	21	Pleuritis	Meta
	16	—	++	+++	—	+++	8.2‡	33	Peritoneal dissemination	Meta
	17	+++	+++	+++	++	—	8.1	15		Meta
Pan-Amin S + 5-FU	18	++	+++	+++	—	—	11.9	17		Meta
	19	++	+++	+++	++	—	10.9	18		Meta
	20						48.6	31	Peritoneal dissemination	Meta
	21	+++	+++	+++	+	—	12.8	14		Meta
	22	+	—	—	—	—	24.1	23		Tumor
	23	++	+++	+++	+++	—	7.2	16		Meta
	24	++	++	+++	++	—	5.4	14		Meta
Pan-Amin S	25	—	—	+++	—	—	13.7	17		Meta
	26	—	—	+++	—	—	13.9	17		Meta
	27	+++	++	+++	++	—	3.1	11		Meta
	28	+++	+++	+++	—	—	7.1	13		Meta

-,±, +, ++, +++: cf. text; * Tumor in the transplanted area was diminished from the host animal at autopsy; † Showed complete regression of the tumor; ‡ Most of the tumor demonstrated a decrease in necrosis macroscopically; § Meta, pathologically diagnosed to be dead of the metastasized lesion; ‖ Tumor, pathologically diagnosed to be dead of bleeding, loss of body fluid and/or infection from the transplanted area's tumor. Courtesy of Goseki et al.[33]

The cause of death in the AO-90 group was organ failure resulting from tumor metastasis in three rats (60%) and bleeding from primary tumor lesion in one rat (20%), while the tumor in one rat (20%) was eradicated.

In the Pan-Amin S and the Pan-Amin S + 5-FU groups, one rat (8.3%) died of body fluid loss, bleeding from primary tumor, and infection. The other 11 rats (91.7%) were thought to have died of organ failure due to severe hepatic and pulmonary hematogenous metastasis of the tumor.

Discussion

Met-deplt TPN combined with 5-FU markedly reduced the degree of metastasis and prolonged survival in the studies using YS cells that easily induce metastatic organ failure and death with extensive hematogenous metastasis.

Methionine is a precursor of S-adenosyl methionine, which is a major methyl group donor in transmethylation. Methylation of cell membrane phospholipids and of DNA is related to the metastatic ability of tumor cells. Breillout et al[37] reported that a methionine-deprived diet inhibits metastatic spread of Lewis lung carcinoma and rhabdomyosarcoma tumor cells grafted into animals. They hypothesized that the metastatic process is easily disturbed by a methionine-deprived diet, affecting the cell membrane phospholipid and DNA methylation. Similarly, AO-90 might inhibit metastatic processes. The metastases appear to be more receptive in this therapy than do primary tumors,[37] but the chemosensitizing effect of AO-90 on 5-FU is not fully explainable.

In recent experiments on tumor-bearing rats[38] and human gastric cancer,[39] we analyzed free and total thymidylate synthase (TS-free and TS-total) activities and calculated the thymidylate synthase inhibition rate(TSIR). Met-deplt TPN with 5-FU significantly decreased TS-free and markedly elevated the TSIR in both experiments. Therefore, we conclude that one of the mechanisms of the synergic effect of Met-deplt TPN and 5-FU is biochemical modulation.

Doxorubicin[35]

It was reported that methionine-dependent cancer cells arrest in the late-S/G2 phases of cell cycle when incubated in methionine-lacking medium in tissue culture studies.[9] So, in this experiment, we attempted to evaluate the synergistc effect of Met-deplt TPN with doxorubicin, which acts on the late S/G2 phase cells, in YS-bearing rats. The experiment was designed like the former 5-FU combined study and the results were also similar to those in the study (Fig. 9.5).

Results

Immedeately after receiving Met-deplt TPN with three times ip administration of doxorubicin, YS tumor growth in the inocculated site and the metastasis were markedly inhibited and followed apparent longer survival of the animals in comparison with those of control groups.

Discussion

Hoffman and Stem demonstrated in tissue culture studies that methionine-dependent cancer cells arrest in the late-S/G2 phases of cell cycle when incubated in methionine-lacking medium.[9] We suggested that in this condition, doxorubicin administration may be effective in killing these cancer cells.[10]

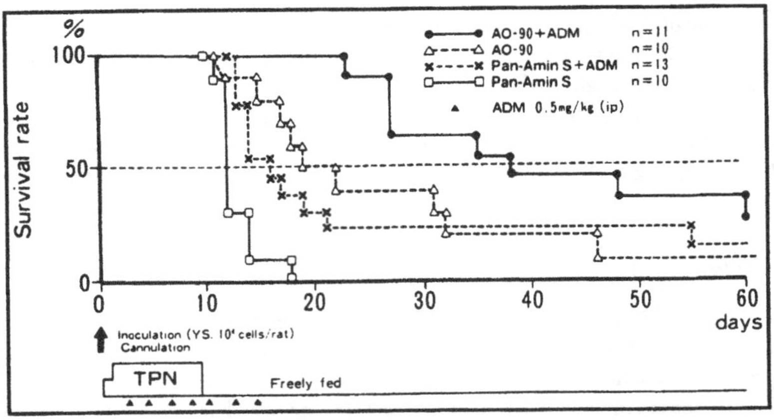

Fig. 9.5. Experimental protocol and results (survival rates and days of the rats in each group) in Doxorubicin combined study. ADM: doxorubicin

We performed two experiments to study the antitumor effect of Met-deplt TPN combined with administration of doxorubicin in YS-bearing rats. As a result, hematogenous YS metastasis was markedly reduced and animal survival was prolonged.

We think that methionine depletion was the main cause of the marked reduction of metastasis in YS in this study and in another study using this parenteral therapy combined with 5-FU in YS-bearing rats.[34] This phenomenon suggests that AO-90 inhibits metastatic processes. It has also been reported that glutathione inhibitors, BSO (buthionine sulfoximine), enhance the effect of doxorubicin.[24] This phenomenon suggests that enhancement of the antitumor effect by a combination of Met-deplt TPN and doxorubicin correlates with the decrease in intracellular glutathione in this study. However, this discussion can not explain the chemosensitizing effect on the doxorubicin of Met-deplt TPN fully. It is necessary to undertake additional fundamental studies to clarify the all mechanism.

Vincristine[37]

The antitumor effect of vincristine (VCR), which acts on mitotic phase cells, was examined with methionine infusion immediately after Met-deplete TPN in YS-bearing rats. Rats were given Met-deplt TPN for 8 days immediately after inoculation with YS cells (days 0-8) which was followed by methionine-infusing regular TPN for 3 days (days 9-11) along with intraperitoneal administration of 0.05 mg/kg/day VCR. All rats were then fed solid food and water ad libitum until their deaths, with 0.1 mg/kg VCR administeration on days 12 and 13. As controls, a Met-deplete TPN only group, Met-infusing TPN groups with and without VCR, and freely-fed group with and without VCR were studied.

The progression of YS was markedly suppressed by Met-depleted TPN with VCR. The median survival day (MSD) was 25 days, significantly longer ($P<0.001$) (generalized Wilcoxon tests) by 11-14 days than any of the other groups.

In conclusion, VCR should have greater efficacy as an anticancer agent when administered together with methionine after Met-deplt TPN.

Discussion

This study, we examined whether such a synergistic anticancer effect of VCR was observed in vivo, when VCR was administered with methionine infusing TPN immediately after Met-deplt TPN in YS-bearing rats.

In the autoradiographic examination using ^{3}H-thymidine on AH-109A-ascites-hepatoma-bearing rats, the tumor cell-cycles in the Met-deplete TPN group (AO-90 group) was markedly delayed and the fraction of labeled mitosis was reduced to less than 70% of the methionine infusing control group (Pan-Amin S group). The cell cycle arrest recovered immediately after initiation of infusion of the methionine-containing Pan-Amin S® amino acid solution. The fraction of labeled mitotic phase cells increased within a short period, and the labeled mitosis wave showed the same pattern as that of methionine infusing Pan-Amin S group (Fig. 9.6).

While, Guo et al demonstrated that YS cells in mice on a methionine-free diet were arrested in the late S-G_2 phases,[42] by Met-deplt TPN, complete tumor cell arrest didn't be demonstrated, but we do believe that the results of this study showed the enhancement of antitumor effects of VCR administration with methionine infusion immediately after Met-deplete TPN in YS-bearing rats.

Clinical Trials in Digestive Organ Cancer[41,43,44]

An early phase II study was conducted in patients with advanced and recurrent gastrointestinal cancer to evaluate the synergic effect of the combined use of AO-90 amino acid solution and anticancer agents (5-FU and MMC).

Thirty-five patients with advanced gastric or colon cancers were subjected to clinical evaluation according to two criteria of the solid cancer chemotherapy,[45] direct efficacy criteria and criteria for improvement of gastrointestinal passage disorder. Of 21 patients appropriately eligible for the direct efficacy criteria, more than partial response was seen in 23.8% (5/21), while 27.8% (5/18) of the patients completely met the evaluation requirements. Of 14 patients with a gastrointestinal passage disorder due to cancer, 57.1% (8/14) effectively responded to the therapy.[44]

In the prospective randomized clinical trial (late phase II trial), 138 patients with inoperable or postoperative recurrent advanced gastric cancer who were receiving a standard therapy of 5-FU and MMC (MF therapy) were treated with AO-90. The patients were randomly allocated to receive either AO-90 (AO/MF group) or a commercially available amino acid solution (C/MF group) by total parenteral nutrition for 14 days as one cycle. Clinical effects were as follows. The responder patient showed complete or partial response, incidence in AO/MF group was 26.3% (15 of 57) and that in C/MF group was 8.1% (5 of 62), and there was a significant difference between the values at $p = 0.013$. As for the effect on patients' survival in stratified four groups, only one group, patients with postogastrectomy recurrent gastric cancer associated with measurable liver and/or lymph node metastases, had a sufficient number of cases and AO/MF group showed a significantly longer 50% survival time than the C/MF group (180.5 days versus 110.0 days, Log-rank test, $p=0.0102$).[46]

Conclusion

In these early and late phase II studies in patients with recurrent or unresectable gastric cancer, Met-deplt TPN was performed with 5-FU and MMC administration for two weeks without severe adverse effect, and was repeated again after a two-week interval in several patients.

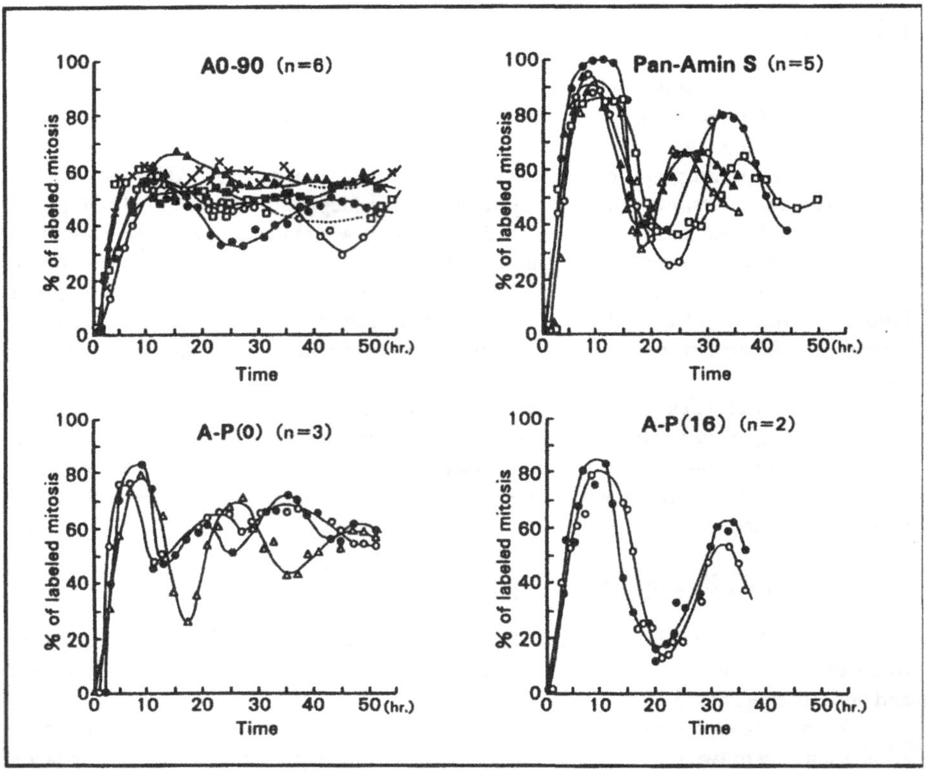

Fig. 9.6. Labeled mitosis waves of the tumor cell in all the rats in AO-90, Pan-Amin S and A-P groups. A-P[0] group showed the wave immediately after switching the amino acids from AO-90 to Pan-Amin S® and AP[16] showed the waves 16 hr. after the exchange. In the AO-90 group, labeled mitotic fraction was less than 70%, and the duration of cell cycle could not be calculated. Immediately after changing the amino acid solution from AO-90 to Pan-Amin S®, the labeled mitotic fraction began to increase and the labeled mitosis wave pattern was similar to that of the Pan-Amin S group with 16 hours (A-P[0] and A-P[16]).

In late phase II study, we had an exciting results as this parenteral treatment showed apparent efficacies to relive the patient's symptoms and resulted in some benefits in patients' survival.

While clinical trials of the combined use with other agents are now being considered, we hope that RT-therapy will be established as an effective anti-methionine cancer chemotherapy in the near future.

Methioninase and Methionine Depletion for Cancer Therapy

Recently enzymatic degradation of methionine achieved by selective methionine-cleaving enzyme, namely methioninase, is on trial.

Methioninase L-methionine-α-deamino-γ-mercaptomethane-lyase, in the presence of 5′-phosphate, catalyzes the alpha, gamma-elimination of methionine to alpha-ketobutyrate, methanethiol, and ammonia. It is composed of four identical subunits with 398 amino acid and has a calculated molecular weight is 42626. This enzyme isolated from *Clostridium sporogens* was reported to inhibit the growth of leukemia cells with no affect on normal fibroblasts capable of using homocysteine in place of methionine for growth. The semipurified enzyme successfully inhibits the growth of Walker carcinosarcoma in the 256 of rats without significant toxicity.[47]

As the production and stability of the enzyme from *Clostridium sporogens* was limiting, recombinant methioninase produced by *E. coli* transfected gene from *Pseudomonas putida* was established for recent preclinical and clinical trial.[48-50]

In vitro experiments using Yoshida sarcoma and, H-460 human nonsmall cell lung cancer r-Methioninase demonstrated selective antitumor efficacy with no apparent affect on a normal fibroblast cell strain. In vitro experiments of this enzyme have demonstrated tumor-specific efficacy for cancer cells and no toxicity for normal fibroblast cell strains (unpublished data).

In vivo experiments in the Yoshida sarcoma-bearing rat, by intravenous continuous administration of r-methioninase, significantly inhibits tumor growth and metastasis in comparison with control rats with no toxicity (unpublished data).

In a clinical early phase I trial on advanced breast cancer patients, it was reported that intravenous administration of r-methioninase successfully decreased the serum level of methionine from 20 μM/ml, the amount before administration, to 0.1 μM/ml without serious acute toxicity.

It is concluded that enzymatic degradation of methionine can be a new therapy for methionine depletion in humans. Evaluation of clinical antitumor efficacy in humans and its synergistic effect should be carried out.

References

1. Dethlefsen LA, Prewitt JMS, Mendelsohn ML. Analysis of tumor growth curve. J Nat Cancer Inst 1968; 40:389.
2. Drummond, JC. A comparative study of tumor and normal tissue growth. Biochem J 1917; 11:325-377.
3. Skipper HE, Johnson JR. A preliminary study of the influence of amino acid deficiencies on experimental cancer chemotherapy. In: Amino Acids and Peptides with Antimetabolic Activity. Chiba Foundation Symposium. Boston: Little, Brown and Co., 1958:38-61.
4. Sugimura T, Brinbaum SM. Winiz M et al. Quantitative nutritional studies with water-soluble, chemically defined diets. VIII. The forced feeding of diets lacking in one essential amino acid. Arch Biochem Biophys 1959; 81:448-455.
5. Lorincz AB, Kuttner RE. Response of malignacy to phenylalanine restriction. Nebr Stata Med J 1965; 50:609-617.
6. Theuer RC. Effect of essential amino acid restriction on the growth of female C57BL mice and their implanted BW10232 adenocarcinoma. J Nutr 1971; 101:223-232.
7. Goseki N, Murakami T, Mori S et al. Effect of intravenous methionine free hyperalimentation combined with anticancer drugs (RT-therapy) on adenocarcinoma of gastrointestinal tract. Jpn J Gastroenterol 1980; 77:112.
8. Ashe H, Clar BR, Hardy DN et al. Ns-methyltetrahydrofolate: Homocysteine methyltransferase activity in extracts from normal malignant and embryonic tissue culture cells. Biochem Biophys Res Commun 1974; T57:417-425.
9. Chello PL, Bertino JR. Effect of methionine deprivation on L5178 murine leukemia cells in culture. Interference with the antineoplastic effect of methotrexate. Biochem Pharmacol 1976; 25:889-892.
10. Hoffman RM, Jacobsen SJ, Erbe RW. Reversion to methionine independence by malignant rat and SV40-transformed human fibroblast. Biochem Biophys Res Commun 1978; 82:228-234.
11. Kreis W, Hession C. Biological effects of enzymatic deprivation of L-methionine in cell culture and an experimental tumor. Cancer Res 1973; 33:1866-1869.
12. Lishko VK, Lishko OK and Hoffman RM. The preparation of endotoxin-free L-methionine-alpha-deamine-gamma-mercaptomethane-lyase (L-methioninase) from *Pseudomonas purida*. Protein Expression and Purification 1993; 4:529-533.
13. Finkelstein JD, Mudd SH. Transsulfation in mammals. The methionine-sparing effect of cysteine. J Biol Chem 1976; 242:873-880.
14. Murakami T. Personal Communication. 1978.

15. Goseki N, Mori S, Menjo M et al. Metabolic derangement by total parenteral nutrition in dog, using an amino acid mixture free from methionine and cysteine—A basic study of cancer therapy by amino acid imbalance. (in Japanese with English abstract) Jap J Surg Metab Nutr 1982; 16:91-100.
16. Goseki N, Onodera T, Koike M et al. Inhibitory effect of L-methionine-deprived amino acid imbalance using total parenteral nutrition on the growth of ascites hepatoma in rats. Tohoku J Exp Med 1987; 151:191-200.
17. Goseki N, Endo M, Onodera T et al. Influence of L-methionine-deprived total parenteral nutrition on the tumor tissue and plasma amino acids fraction and the host metabolism: Experimental study with Sato lung carcinoma-bearing rats. Tohoku J Exp Med 1989; 157:251-260.
18. Bailey RP, Vrooman MJ, Sawai Y et al. Amino acids control of nucleolar size, the activity of RNA polymerase I and DNA synthesis in liver. Proc Nat Acad Sci USA 1976; 73:3201-3205.
19. Goseki N, Endo M. Thiol depletion and chemosensitization on nimustine hydrochloride by methionine-deleting total parenteral nutrition. Tohoku J Exp Med 1990; 161:227-239.
20. Rose CM, Millar JL, Peacock JH et al. Differential enhancement of melphan cytotoxicity in tumor and normal tissue by misonidasole. In: Radiation Sensitizers: Their Use in the Clinical Management of Cancer. Brady LW, ed. New York: Masson, 1980:250-257.
21. Stratford IJ, Adams GE, Horsman WR et al. The interaction of misonidasole with radiation, chemotherapeutic agents, or heat. In: Radiation Sensitizers: Their Use in the Clinical Management of Cancer. Brady LW, ed. New York: Masson, 1980:276-281.
22. Roizin-Towle L, Hall EJ. Enhanced toxicity of antineoplastic agents following prolonged exposure to misonidasole. Br J Cancer 1981; 44:201-207.
23. Clement JJ, Jacobson K. Evaluation of radiosensitizers in combination with chemotherapeutic agents in solid tumor. Int J Radiat Oncol Biol Phys 1982; 8:631-634.
24. Biaglow JE, Varnes ME, Clark EP et al. The role of thiols in cellular response to radiation and drugs. Radiat Res 1983; 95:437-455.
25. Bump EA, Yu NY, Taylor YC et al. Radiosensitization and chemosensitization by diethylmaleate. In: Nygaard OF, Simic MG, eds. Radioprotectors and Anticarcinogens. New York: Academic Press, 1983:297-323.
26. Matsudaira H, Sekiguchi F, Nagasawa H. Sulfhydyl content of rat ascites hepatomas with different sensitivity to x-rays and nitrogen mustard N-oxide. Gann 1967; 58:415-426.
27. Wells PG, Boerth RC, Oates JA et al. Toxicologic enhancement by combination of drugs which deplete hepatic glutathione: Acetaminophen and doxorubicin (Adriamycin). Toxicol Appl Pharmacol 1980; 54:197-209.
28. Tomashefsky P, Aster M, White RD. Relationship between thiol depletion and chemosensitization in transplantable murine tumors. J Natl Cancer Inst 1985; 74:1233-1238.
29. Kramer RA, Greene K, Ahmad S et al. Chemosensitization of L-phenylalanine mustard by the thiol-modulating agent buthionine sulfoximine. Cancer Res 1987; 47:1593-1597.
30. Goseki N, Onodera T, Kosaki G et al. Methionine- and cystine-free amino acid imbalance by total parenteral nutrition as an adjunct to cancer chemotherapy. In: Ogoshi S, Okada A, eds. Parenteral and Enteral Hyperalimentation, Amsterdam: Elsevier Science Pub, 1984:343-355.
31. Goseki N, Onodera T, Tominaga T et al. Inhibitory effect of methionine deprived total parenteral nutrition combined with actinomycin-D on rat experimental tumors. Proceedings of 14th International Congress of Chemotherapy 1985; Anticancer section:438-439.
32. Goseki N, Yamazaki S, Toyoda T et al. Cancer therapy by methionine deprived total parenteral nutrition with mitomycin C and/or S-fluorouracil. Oncologia 1987;20:99-110. (in Japanese with English abstract)
33. Goseki N, Endo M. Thiol depletion and chemosensitization on nimustine hydrochloride by methionine-deprived total parenteral nutrition-experimental studies on Sato lung carcinoma bearing rats. Tohoku J Exp Med 1990; 161:227-239.
34. Goseki N, Endo M, Onodera T et al. Antitumor effect of methionine-deprived total parenteral nutrition with 5-fluorouracil administration on Yoshida sarcoma-bearing rats. Ann Surg 1991; 213:83-88.
35. Goseki N, Yamazaki S, Endo M et al. Antitumor effect of methionine-depleting total parenteral nutrition with doxorubicin administration on yoshida sarcoma-bearing rats. Cancer 1992; 69:1865-1872.

36. Goseki N, Endo M, Yamazaki S. Cancer therapy by methionine-depleting total parenteral nutrition with the administration of antineoplastic agents. Surg Ther 1991; 65:295-302.(in Japanese).
37. Goseki N, Nagahama T, Maruyama M et al. Enhanced anticancer effect of vincristine with methionine infusion after methionine-depleting total parenteral nutrition in tumor-bearing rats. Jpn J Cancer Res 1996; 87:194-199.
38. Breillout F, Hadida F, Echinard-Grain P et al. Decreased rat rhabdomyosarcoma pulmonary metastasis in response to a low methionine diet. Anticancer Res 1987; 7:861-868.
39. Smith K, Berg A, Denekamp J. A companion of the response of pulmonary and subcutaneous tumors to radiation and cyclophosphamide. In: Hellmann K, Eccless SA, eds. Treatment of Metastasis: Problems and Prospects. London: Taylor & Francis Ltd., 1984:169-173.
40. Hibino Y, Kawarabayashi Y, Kohri A et al. The mechanism of potentiation of the antitumor effect of 5-fluorouracil by methionine-free amino acid solution (AO-90) in rats. Jpn J Cancer Chemother, 1994; 21:2021-2028 (in Japanese).
41. Goseki N, Yamazaki S, Shimojyu K et al. Synergistic effect of methionine-depleting total parenteral nutrition with 5-fluorouracil on human gastric cancer: A randomized, prospective clinical trial. Japn J Cancer Res 1995; 86:484-489.
42. Guo H, Lishko K, Herrera H, et al. Therapeutic tumor-specific cell cycle block induced by methionine starvation in vivo. Cancer Res 1993; 53:5676-5679.
43. Goseki N, Onodera T, Mori S et al. Clinical study of amino acid imbalance as an adjunct to cancer therapy. (in Japanese with English abstract) J Jpn Cancer Ther 1982; 17:1908-1916.
44. Sugihara K, Goseki N et al. Early phase II study of the combined use of AO-90 methionine-free amino acid solution and anticancer agents (5-FU and MMC) in patients with advanced recurrent gastrointestinal cancer. (in Japanese with English abstract) Jpn J Cancer Chemotherapy 1990; 17:2405-2413.
45. Japan Society for Cancer Therapy. The criteria for the evaluation of the clinical effects of solid cancer chemotherapy. J Jpn Soc Cancer Ther (Nihon Gan Chiryo Gakkaishi) 1993; 28:101-30.
46. Taguchi T, Goseki N, Kosake G et al. A controlled study of A0-90, a methionine-free intravenous amino acid solution, in combination with 5-fluorouracil and mitomycin C in advanced gastric cancer patients (Surgery group evaluation). (in Japanese with English abstract) Jpn J Cancer Chemother 1995; 22:743-764.
47. Kreis W, Hession C. Isolation and Purification of L-Methionine-α-deamino-γ-mercaptomethane-Lyase (L-Methioninase) from *Clostridium sporogenes*. Cancer Res 1973; 33:1862-1865.
48. Lishko VK, Lishko OV, Hoffman RM. The Preparation of Endotoxin-Free L-Methionine-α-deamino-γ-mercaptomethane-Lyase (L-Methioninase) from *Pseudomonas putida* Protein Exp Purif 1993; 4:529-533.
49. Tan Y, Xu M, Guo H et al. Anticancer efficacy of methioninase in vivo. Anti-Cancer Res 1996; 16:3931-3936.
50. Tan Y, Zavala J Sr, Xu M, Zavala J Jr et al. Serum methionine depletion without side effects by methioninase in metastatic breast cancer patients. Anticancer Res 1996; 16:3937-3942.

CHAPTER 10

Ornithine Alpha-Ketoglutarate Administration in Surgical, Trauma and Cancer-Bearing Patients

Luc A. Cynober and Colette Coudray-Lucas

Introduction

Besides their importance as building blocks for protein synthesis and as energy substrates, both direct and via the synthesis of glucose in gluconeogenesis, some amino acids are also vital metabolic regulators. Supplementation of enteral/parenteral nutrition with these amino acids introduced a new area some years ago, that of pharmacological nutrition.[1] The best known representatives of this category of amino acids are arginine and glutamine, which are discussed in other chapters of this book.

Ornithine α-ketoglutarate (OKG) is a fascinating molecule because it is a precursor not only of glutamine and arginine but also of some other amino acids (such as proline) and keto-acids (α-ketoisocaproate), which are probably important in the control of protein metabolism, and also because it has a potent secretagogic effect on hormones such as insulin and human growth hormone.

Effects and potential mechanisms of action of OKG are discussed here with a special emphasis on surgical and other trauma situations. Readers interested in other action of OKG (e.g., in the field of chronic malnutrition) can refer to a recent general review.[2]

Background

If we consider current concepts in pharmacological nutrition, and especially the role of glutamine and arginine, the usefulness of OKG is obvious. However, it would be bending the truth to say that OKG was designed deliberately for this application; the story of OKG is actually quite complicated.

OKG was conceived by Robert Molimard who synthesized it during the sixties with the aim of improving the neurological status of patients with hepatic encephalopathy at a time when it was thought that ammonia played a major role in the pathogenesis of the coma. The rationale of α-ketoglutarate was to trap ammonia, forming glutamate degraded in turn in ureagenesis, this pathway being activated by the presence of ornithine. As expected, OKG is a potent antihyperammonemic agent.[3-5] Unfortunately, clinical results were not impressive and

Nutritional Support in Cancer and Transplant Patients, edited by Rifat Latifi and Ronald C Merrell.

no correlation was found between normalization of the ammonia levels and the coma status. However, in the course of utilization of OKG in patients with liver failure, an improvement in the nutritional status of these patients was observed.[6,7] Accordingly in the early eighties the benefit of OKG therapy was evaluated in other situations characterized by malnutrition. We published the first results on OKG administered by the enteral route to burn patients in 1984 followed the next year by P. Furst's group using OKG by the parenteral route in surgical patients.[9,10]

Physical and Chemical Properties of OKG

OKG is a salt formed from one molecule of α-ketoglutarate (αKG) and two molecules of ornithine (ORN). Its molecular mass is 410. The pKs of αKG (1.9 and 1.4) and ORN (10.8) give a pH of 7 in aqueous solution over a wide range of OKG concentrations (up to 5 mM).

The stability of OKG in solution (containing glucose, electrolytes and trace elements) was studied at 4° and 24°C over 21 days, a period far exceeding the usual storage times of parenteral OKG-containing solutions. ORN was found to be stable at both temperatures for 21 days. αKG was also stable at 4°C but less so at 24°C (98.9% of initial value on day 15, ns; 93.7% on day 21, $p < 0.05$ vs initial values).[11]

Action of OKG in Trauma, Surgical and Cancer-Bearing Patients

Burn Injury

Burn injury is the trauma situation in which the action of OKG has been best defined since several prospective studies and one retrospective study have been published, all giving concordant results.

The first controlled study concerned enterally-fed burn patients with a burn surface area (BSA) ranging from 21-60%.[8] Six patients served as controls, and eight patients received 10 g OKG twice a day in boluses. A time-dependent increase in basal Orn plasma concentrations was observed in the OKG group, indicating, as confirmed in rats,[12] a decrease in plasma clearance of Orn after burn injury. In addition, a drop in venous concentrations of some gluconeogenic amino acids more marked in the OKG group, and probably related to a decrease in muscle amino acid efflux (see below), was observed. Plasma phenylalanine, a marker of protein catabolism in injury,[13] returned more rapidly to normal in the treated patients. In the second prospective study, enteral nutrition began on day 2, and the patients were less seriously burned (BSA 16-31%).[14-16] Seven patients received 10 g OKG/day as a bolus, and seven served as controls. Arterial and venous blood samples were obtained from days 2-13 for the determination of arteriovenous amino acid differences. An additional venous blood sample was taken 1 h after nutrition restart for glucose, insulin, and glucagon determinations. Nitrogen balance returned to equilibrium more rapidly in OKG-treated patients (on day 13) and hyperphenylalaninemia was lower in this group. Muscle arteriovenous differences were less negative in the OKG group, significantly so for alanine, glycine, lysine, and hydroxyproline. Retinol-binding protein levels were significantly less markedly decreased in the OKG group; this was not related to any difference in the acute phase response (serum α1-acid glycoprotein levels were similar in the 2 groups). Finally, glucose intolerance induced by nutrition was only observed in the control group.

More recently, Donati et al[17] reported data on 40 burn patients (BSA = 40-60%), 21 of them being treated by OKG. Although the placebo was not isonitrogenous (maltodextrine was used), the patients had roughly the same caloric and nitrogen intake, confirming previously reported results. OKG improved nitrogen balance and visceral protein concentrations. In addition, the authors demonstrated that OKG administration had a positive effect on would healing, on the quality of the graft and on the re-epithelization of donor sites. In two large prospective, randomized double-blind studies where the two groups of patients (OKG vs casein-supplemented) had isonitrogenous and isocaloric intake, our group[18,19] confirmed data from Donati's group, i.e., that OKG improves wound healing, and, of interest, indicated that OKG decreases urinary hydroxyproline output in urine.[18]

A retrospective study (2 yr) was also performed.[20] One hundred thirty-six patients met the inclusion criteria (admission to the burn unit within 48 h of injury, survival at least 5 days, and submitted to enteral nutrition). Forty-three percent of the patients received OKG, and the two groups were comparable for BSA. The results indicated a mean reduction of 16 days in the length of stay in hospital in the OKG-treated subgroup of severely burned patients (BSA > 20%) (statistically nonsignificant due to wide dispersion of the data). Mortality in the OKG-treated group showed a trend ($p > 0.09$) toward a significant reduction.

To address questions that cannot, for obvious ethical reasons, be resolved in human studies, a burned rat model was developed. Hypercatabolism was obtained with a double injury, i.e., immersion of the dorsum for 12 sec in water at 95°C, followed by 24 h fasting.[20] OKG (5 $g/kg^{-1}/day^{-1}$) was tested, mixed in a low-calorie hyponitrogenous diet (Osmolite; 1.2 $gN/kg^{-1}/day^{-1}$, 210 $kcal/kg^{-1}/day^{-1}$ in 3 boluses) against three different groups (healthy fed ad libitum, healthy + food restriction + glycine as isonitrogenous control, burn + food restriction + glycine). The results indicate that OKG administration significantly reduces the increase in muscle catabolism (measured ex vivo), the drop in muscle weight, and counteracts the drop in the muscle Gln pool induced by burn injury.[21] In addition, OKG treatment increased the Gln pool in the proximal intestine. A ^{14}C-valine administered loading dose technique indicated that OKG counteracted the decrease in fractional synthesis rate (FSR) of proteins in the jejunum and increased the FSR in the liver of burn rats.[22] Another study indicated that OKG was as efficient as an isonitrogenous diet containing Gln in replenishing the muscle Gln pool. Finally, compared to isonitrogenous control (in form of glycine), OKG was shown to increase plasma, muscle and liver Gln concentrations and to limit hyperphenylalaninemia.[23]

In summary, in burn injury, a highly catabolic state, the action of OKG involves both muscle and splanchnic areas, with a decrease in muscle protein catabolism and an increase in liver and intestine anabolism.

Trauma and Sepsis

Four randomized clinical studies are available,[24-27] one of them being a cross-over study. OKG was administered at 20-25 g/day by the parenteral[24,25] or the enteral route.[24,26,27] In these four trials nitrogen balance was significantly better in the OKG group[24,25,27] or during the OKG-administration phase in the cross-over study.[26] Transthyretin was found to be improved in the OKG group by Demarcq et al[24] but not by Mertes et al.[25] OKG also limited the trauma-induced drop in plasma glutamine.[27]

In a rat model with bilateral femur fracture, it has been shown[28] that OKG administration for 7 days (mixed in a liquid diet; Bioserv F1259) increases food intake and nitrogen balance and abrogates the decrease in muscle glutamine content induced by trauma.

Surgery

Vinnars and associates conducted a series of impressive clinical studies. In the first study, patients were studied for 5 days after colorectal surgery.[9,10] Seven patients received total parenteral nutrition (TPN) (165 $kg/kg^{1}/day^{-1}$; 0.15 g $N/kg^{-1}/day^{-1}$), whereas the other six received an isocaloric, isonitrogenous diet including 25 g OKG to replace 2.7 g of the nitrogen in the

TPN. Nitrogen balance, urinary 3-methylhistidine, creatinine, and ammonia levels were improved in the OKG-treated patients. Furthermore, in the OKG group, plasma Orn and Pro increased relative to preoperative values; Arg was increased and branched-chain amino acids were decreased in muscle. Muscle Gln levels were slightly (but not significantly) improved in the OKG group.

In a second study with a similar protocol involving cholecystectomized patients receiving more nitrogen (0.2 g $N/kg^{-1}/day^{-1}$) but fewer calories (135 $kJ/kg^{-1}/day^{-1}$), the efficacy of OKG (0.35 $g/kg^{-1}/day^{-1}$) on nitrogen balance was confirmed.[29,30] This effect was probably better than reported because the investigators underestimated the amount of nitrogen in OKG. Furthermore, the usual postoperative decreases in muscle polyribosome percentage and Gln levels were abolished in the OKG-treated group. Finally, OKG was as efficient as Gln perfusion itself with regard to repletion of muscle Gln.[31]

More recently, the same group showed that the intravenous administration of α-KG alone has a similar action on ribosome profiles and muscle Gln content to that of OKG.[32,33] However, patients treated with α-KG received 194 mg $\alpha KG/kg^{-1}/day^{-1}$, whereas those treated with OKG received only 126 mg $\alpha KG.kg^{-1}.day^{-1}$ (i.e., 350 mg OKG, given that OKG contains 2 Orn per αKG).[31] Because αKG uptake by muscle after intravenous administration is dose-dependent,[34] we cannot conclude from these data that the effects of OKG are solely mediated by αKG.

Although there are very good reasons to think[35] that OKG may have a protective effect on the intestine, up to now there have been no studies in fields such as intestine adaptation after major resection or after small bowel transplantation. Two papers recently addressed these issues.[36,37]

Adult Wistar rats underwent a resection of the proximal 50% of the small intestine. Then the rats were fed intragastrically with a nutritive mixture supplemented with OKG (1g/kg) or an isonitrogenous amount of casein hydrolysate. OKG supplementation led to accelerated adaptative hypertrophy as judged by variations of villus height, crypt depth, and ornithine decarboxylase mRNA.[36]

Isogenic male Lewis rats were submitted to an orthotopic bowel transplantation and then were fed for 14 days with a diet supplemented or not with OKG (1.4 g/kg/day). Addition of OKG significantly reduced bacterial translocation in mesenteric lymph nodes and in the liver and improved the protein/DNA index in the intestine.[37]

Although not directly related to the topic, the study by Moukarzel at al[38] should be cited. Children under total parenteral nutrition were submitted to different periods of feeding with or without OKG (15 g/day) supplementation. Periods with OKG supplementation were characterized by higher height velocity and increased IGF-Sm-C plasma levels.

Cancer

The effects of OKG in cancer situations have been addressed only recently perhaps because an important question was unsolved: whether such supplementation profits the host or the tumor. Two studies have been performed recently[39,40] which provide a clear answer.

In a first study,[39] enteral supplementation with OKG (4 g/d) (or isonitrogenous amount of glycine) was investigated in Sprague-Dawley rats bearing Yoshida ascites hepatoma 130. Tumor-bearing rats were compared with ad libitum and pair fed controls. OKG improved muscle protein balance by reducing breakdown by 33% and overall amino acid release of incubated muscle (epitrochlearis) by 46%.

In a second study,[40] rats bearing Morris hepatoma 7777 were given OKG both pre- and postoperatively. Compared to glycine-supplemented rats, OKG-treated rats showed a more

Table 10.1. Effects of OKG on amino acid and protein metabolism in stress situations

Effect	Situation	Reference
- Increases muscle protein synthesis	surgery	30
- Decreases muscle protein catabolism	burn	18,20,21
	surgery	9
- Increases liver protein synthesis	burn	21
- Increases intestine protein synthesis	burn	21
- Improves nitrogen balance	burn	15,17,18,21
	trauma/sepsis	24,27
	surgery	9,30
- Increases nutritional markers (visceral proteins: ALB, TTR, RBP)	burn	15,17
	trauma/sepsis	24
- Decreases muscle amino acid output	burn	14
- Improves muscle glutamine content	burn	20,22,23
	trauma	28
	surgery	29

positive nitrogen balance, higher concentrations of glutamine in muscles and accelerated protein deposition in the intestine.

In the two studies OKG did not affect tumor growth. Hence it appears, at least in the two models studied, that OKG, as glutamine, supports the host without deleterious effect on tumor growth.

In summary, all studies performed in acute situations indicate a beneficial effect of OKG on nitrogen balance (Table 10.1). Whether this effect is due to an anabolic[30] or anticatabolic action[9,21] is controversial and probably depends on the predominant underlying mechanism of malnutrition (increased muscle catabolism in severe trauma and sepsis versus decreased protein synthesis in less severe insult).

OKG, Wound Healing and Immunity

Impairment of wound healing and immunity is a classical feature of injury and is involved in trauma-associated morbidity and mortality. Key nutrients (such as glutamine) or mediators (such as nitric oxide or polyamines) may be involved in this pathological process. Several studies provide evidence that OKG improves both wound healing and immunity.

Beside the studies performed in burn patients described above, some studies specifically addressed the issue of wound healing. In a double-blind trial, 25 of 52 patients who had undergone reconstructive surgery received 10-15 g OKG/day orally with their oral nutrition. The treated patients showed more rapid wound healing (18 ± 1 vs 26 ± 3 days, $p < 0.01$) and a significantly lower rate of complication.[41] Similar results were obtained in patients undergoing head and neck surgery; OKG orally administered reduced the healing period and the number of complications.[42] Further to this action on connective tissues, it has been established that OKG stimulates the growth of cultured human fibroblasts.[43]

Earlier studies in humans[44] indicated an action of OKG on immunity, with an increase in the lymphocyte count, response to mitogens, and synthesis of immunoglobulins. Later, it was shown that OKG administration to endotoxemic rats[45] or to burn rats[48] suppresses the injury-induced thymic involution. Noteworthy in both studies, variations of Arg and Gln levels on one hand and of thymus weight on the other correlate. This gives credit to Newsholme's hypothesis describing the muscle as a reserve of key amino acids (i.e., Gin and Arg) to support immune cell function.[47] Also in burn rats, we shown[48] that OKG administration maintains neutrophil polynuclear functionality. Finally in a rat model of bacterial translocation we showed a reduced bacterial dissemination in the spleen of OKG-treated rats.[49]

Mechanism of Action

The mechanism of action of OKG is probably multifactorial, linked to the stimulation of anabolic hormone secretion and the production of Orn and/or α-KG metabolites.

Stimulation of Insulin and Human Growth Hormone (hGH) Secretion

Insulin and hGH are potent anabolic hormones, and because their secretion is increased by OKG, they are probably involved in the mechanism of action of this compound.

In patients recovering from gastrointestinal surgery, a glucose perfusion containing OKG leads to plasma peaks of insulin and hGH higher than after glucose alone.[2] In the same way, parenteral infusion of OKG increases hGH secretion in healthy adults[50] and children[51] and this action is stronger than that induced by Arg perfusion.[52] Increased hGH has also been observed after enteral administration of OKG (10 or 20 g)[53] and an increase in IGF-I/Sm-C has been reported in infants under total parenteral nutrition.[38]

The effect of OKG on insulin secretion has been observed in cirrhotic patients[54] and in healthy subjects.[55-57]

The action of OKG on insulinemia observed after oral administration to healthy subjects who have first received a standardized breakfast[56] is much higher than in 12 h fasted healthy subjects.[57] The causes of these discrepancies are unclear. However, because pancreatic intracellular αKG levels appear to be crucial for amino acid stimulated insulin release,[58] it may be that, in fasted subjects, excessive consumption of aKG as an energy substrate prevented full expression of Orn-mediated insulin secretion.

In burn patients, who classically present with hyperglycemia together with hyperinsulinemia and peripheral insulin resistance,[59] another type of OKG action is observed; high insulin levels are not modified, but OKG improves glucose tolerance.[16]

In summary, OKG undoubtedly stimulates insulin and hGH secretion, but this effect appears to be highly dependent on nutritional and metabolic status.

Involvement of OKG Metabolites

Several in vivo studies concerning humans and animals have indicated that the administration of OKG enhances the synthesis of several molecules (see above for details) known to inhibit protein catabolism or stimulate protein synthesis. A recent study[60] on burn patients indicates clearly that metabolite production is both dependent upon the dose (proline) and the rate of administration (glutamine, arginine).

Glutamine (Gln)

Gln is produced from both αKG and Orn via Glu, which is converted into Gln via Gln synthetase in muscle, brain, and perivenous hepatocytes. In these latter cells, the rapid metabolism of αKG into Glu and Gln has been emphasized by two studies on isolated perfused rat liver.[63,64]

Recent studies[23,65-68] provide strong evidence that αKG and Orn interact strongly to modify Gln fluxes. Although, it was clearly shown[23] that replacing ornithine by arginine (either on an isomolar or on an isonitrogenous basis), it was impossible to mimic the effects of OKG on glutamine pools (muscle, liver, plasma) in burn rats.

Two recent studies compared the effects of Orn, αKG and OKG in rats with bilateral femur fracture[67] or burn injury.[68] In both cases the OKG combination was more efficient to modulate Gln pools that Orn or αKG alone. The key sites for this αKG/Orn interaction may be the intestine[65,66] and the kidney.[69]

The key role of Gln in the control of protein metabolism, especially in muscle as an energy substrate in rapidly dividing cells and as a precursor of purines and pyrimidines, is well established and is the subject of specific chapters in this book and therefore does not require further discussion here.

Polyamines

The aliphatic polyamines (putrescine, spermine, and permidine) produced from Orn[70,71] are potent inducers of protein synthesis.[72] OKG stimulates the synthesis and secretion of albumin by isolated rat hepatocytes.[73,74] This effect is inhibited by difluoromethylornithine, an inhibitor of Orn decarboxylase and thus of polyamine formation. Similar results have been obtained in isolated rabbit liver perfused with Orn.[2] In the same way, difluoromethylornithine inhibits the OKG-induced proliferation of fibroblasts in culture.[43] The action of OKG on wound healing could thus be dependent, at least in part on endogenous polyamine synthesis.

Finally, the full activation of T lymphocytes depends on polyamine synthesis, and we cannot exclude the possibility that OKG action on immune function is dependent, as least in part, on the polyamine pathway.

Proline (Pro)

Pro has two important anabolic functions. It is the precursor of hydroxyproline, a qualitatively and quantitatively important amino acid in collagen which has a fundamental role in tissue repair and wound healing.[75]

Proline has a stimulating action on hepatocyte DNA and protein synthesis.[76] It has been shown that Pro is necessary to support DNA synthesis in isolated hepatocytes; a medium containing small amounts of Pro only permits low DNA synthesis. The latter is increased when amino acids are added but not if Pro is lacking.[76]

Another study confirms that Pro plays an important role in hepatic protein synthesis. Perez-Sala et al[77] evaluated the hepatic polyribosomal profile in isolated hepatocytes from fed and starved rats and the effects of individual amino acids administered in vivo to starved rats. Starvation led to a large decrease in the number of polyribosomes, and only three amino acids—alanine, Pro and Orn—were able to limit this decrease.

Arginine (Arg)

Arg is the end product of Orn metabolism in the urea cycle and gives rise to Orn again via the reaction catalyzed by arginase.[35] According to Rose's classification, Arg is a semiessential amino acid, i.e., it is required for optimal growth in young animals but not in adults. By extension, it has been postulated that in pathological situations, such as trauma, requiring increased intake for protein synthesis and wound healing, Arg becomes an essential amino acid.[78]

Arg has many metabolic effects in common with Orn. Both have hormonal secretagogic effects, stimulating the secretion of growth hormone, insulin, and glucagon.[2,78] In trauma situations Arg prevents weight loss, maintains normal growth, and accelerates wound healing (see ref. 67 for review).

Furthermore, Arg displays immunological actions, some of which are also found with OKG, such as increased thymocyte blastogenesis in response to concanavalin A and phytohemagglutinin, increased delayed hypersensitivity, and decreased thymus involution in stressed rats.[78] Interestingly, the immunological action of arginine and ornithine is not shared by citrulline.[79]

However, arginase is a widely distributed enzyme, and the question is whether administered Arg acts via Orn formation and the resulting Orn metabolites. There is as yet no answer to this question, but data from Daly et al[80] are suggestive; postoperative patients supplemented enterally with Arg (25 g/day) exhibited equivalent increases in plasma Orn and Arg (259μm and 213 μm, respectively.) Increase in ornithinemia during Arg-supplemented nutrition have also been observed in burned rats.[81] Furthermore, after perfusing isolated rat livers with 2 mM Arg, we observed a marked release of Orn into the perfusion medium.[55] However, it has been clearly shown that arginine induced tumor growth (in a Ward color tumor model) is not mediated by ornithine and polyamine production.[82]

Alternatively, OKG action could partly be mediated by the formation of Afg that is subsequently metabolized into citrulline; the nitrite oxide formed appears to play a key role in macrophage activation and in various other immune cells.[83]

Other Molecules Possibly Involved

αKG stimulates branched-chain keto acid (BCKA) synthesis. This is an indirect process since it results from transamination reactions, the direction of which depends on the relative quantities of the substrates and reaction products. It has been shown that the administration of OKG to healthy subjects results in an increase in the plasma concentration of Glu and a reduction in that of leucine.[57] It has also been found that OKG increases plasma BCKA in burn patients.[84] Because branched-chain amino acid transaminase is located in the muscle, increases in BCKA in the plasma after OKG administration indicates a high production rate of these substances, above local requirements for energy production.

The action of BCKA on protein metabolism is well known and has been the subject of several studies (for review, see ref. 85). In postoperative patients, OKG and BCKA have the same effects on nitrogen metabolism, including a decrease in muscle protein catabolism and an increase in hepatic protein synthesis.[88]

At the intestinal level, OKG administration induces the synthesis of δ-aminobutyric acid.[69] The significance of this process is not known.

Finally, it has recently been shown that calcium-α-ketoglutarate acts as a potent phosphate binder and hence corrects secondary hyperparathyroidism in hemodialyzed patients.[87] Evaluation of the relevance of this action deserves further studies.

References

1. Cerra FB. Role of nutrition in the management of malnutrition and immune dysfunction of trauma. J Am Coll Nutr 1992; 11;512-518.
2. Cynober L. Ornithine a-ketoglutarate. In: Cynober LA, ed. Amino Acid Metabolism and Therapy in Health and Nutritional Disease. 1995:385-398.
3. Molimard R, Charpentier C, Lemonnier F. Modifications de l'amino acidémie des cirrhotiques sous l'influence de sels d'ornithine. Ann Nutr Metab 1982; 26:25-36.

4. James IM, Dorf G, Hall S et al. Effect of ornithine α-ketoglutarate on disturbances of brain metabolism caused by high blood ammonia. Gut 1972; 13:551-555.
5. Michel H, Oge P, Bertrand L. Action de l'α-cétoglutarate d'ornithine sur l'hyperammoniémie de cirrhotique. Presse Méd 1988; 79:867-871.
6. Tremolieres J, Scheggia E, Flament C. Effets de l'α-cétoglutarate d'ornithine sur le bilan azoté et sur la vitessoe d'oxydation de l'éthanol. Cah Nutr Diet 1972; 7:2-7.
7. Gay G, Villaume C, Beaufrand MJ et al. Effects of ornithine α-ketoglutarate on blood insulin, glucagon and amino acids in alcoholic cirrhosis. Biomedicine 1979; 30:173-177.
8. Cynober L, Saizy R, N'guyen Dinh F et al. Effect of enterally administered ornithine α-ketoglutarate on plasma and urinary amino acid levels after burn injury. J Trauma 1984; 24:590-596.
9. Leander UI, Fürst P, Vesterberg K et al. Nitrogen sparing effect of Ornicetil® in the immediate postoperative state. II:Plasma and muscle amino acids. Clin Nutr 1985; 4:45-51.
10. Vesterberg K, Vinnars I, Leander UI et al. Nitrogen sparing effect of Ornicetil® in the immediate postoperative state. II:Plasma and muscle amino acids. Clin Nutr 1987; 6:213-219.
11. Anglade P, Arnaud PH, Gviond F et al. Stabillité de l'a-cétoglutarate d'ornithine dans les solutions pour nutrition parentérale. Gastroentérol Clin Biol 1990; 14:370 (abstract).
12. Vaubourdolle M, Jardel A, Coudray-Lucas C et al. Metabolism and kinetics of parenterally administered ornithine and α-ketoglutarate in healthy and burned animals. Clin Nutr 1988; 7:105-111.
13. Coudray-Lucas C, Cynober L, Lioret N et al. Origins of hyperphenylalaninemia in burn patients. Clin Nutr 1985; 4:179-183.
14. Cynober L, Blonde F, Lioret N et al. Arterio-venous differences in amino acids, glucose, lactate and fatty acids in burn patients: Effect of ornithine α-ketoglutarate. Clin Nutr 1986; 5:221-226.
15. Cynober L, Lioret N, Coudray-Lucas C et al. Action of ornithine α-ketoglutarate on protein metabolism in burn patients. Nutrition 1987; 3:187-191.
16. Vaubourdolle M, Cynober L, Lioret N et al. Influence of enterally administered ornithine α-ketoglutarate on hormonal patterns in burn patients. Burns 1987; 13:349-356.
17. Donati L, Ziegler F, Pongellini G et al. Nutritional and clinical efficacy of ornithine a-cétoglutarate in severe burn patients. Clin Nutr 1999; 18:307-311.
18. De Bandt JP, Coudray-Lucas C, Lioret N et al. A randomized controlled trial of the influence of the mode of enteral ornithine α-ketoglutarate administration in burn patients. J Nutr 1998; 128:563-569.
19. Coudray-Lucas C, Le Bever H, Cynober L et al. Ornithine α-ketoglutarate improves wound healing in severe burn patients: A prospective randomized double-blind trial versus isonitrogenous control. Crit Care Med 2000: In Press.
20. Cynober L. Amino acid metabolism in thermal burns. J Pen 1989; 13:193-205.
21. Vaubourdolle M, Coudray-Lucas C, Jardel A et al. Action of enterally-administered ornithine alpha-ketoglutarate on protein breakdown in skeletal muscle and liver of the burned rat. JPEN 1991; 15:517-520.
22. Le Boucher J, Obled C, Farges MC et al. Ornithine α-ketoglutarate modulates tissue protein metabolism in burn-injured rats. Am J Physiol 1997; 273:E557-563.
23. Le Boucher J, Coudray-Lucas C, Lasnier E et al. Enteral administration of ornithine α-ketoglutarate or arginine α-ketoglutarate: A comparative study of their effects on glutamine pools in burn-injured rats. Crit Care Med 1997; 25:293-298.
24. Demarco JM, Delbar M, Trochu G et al. Effets de l'α-cétoglutarate d'ornithine sur l'état nutritionnel des malades de réanimation. Cah Anesthésiol 1984; 32:3-6.
25. Mertes N, Mollmann M, Pfisterer M et al. Nitrogen sparing effect of ornicetil supplemented TPN in hypercatabolic septic or polytraumatized patients. In: Soeters PB, Wilson JHP, Holm E, eds. Advances in Ammonia Metabolism and Hepatic Encephalopathy. Amsterdam: Elsevier 1988:141-153.
26. Nicolas F, Rodineau P. Essai controle croisé de l'α-cétoglutarate d'ornithine en alimentation entérale. Ouest Med 1982; 35:711-713.
27. Jeevanandam M. Ornithine α-ketoglutarate in trauma. Clin Nutr 1993; 12:70-71.
28. Jeevanandam M, Ali MR. Altered tissue amino acid levels in traumatized, growing rats due to ornithine α-ketoglutarate supplemented oral diet. J Clin Nutr Gastroenterol 1991; 6:23-32.
29. Wernerman J, Hammarqvist F, Ali MR, et al. Glutamine and ornithine α-ketoglutarate but not branched-chain amino acids reduce the loss of muscle glutamine after surgical trauma. Metabolism 1989; 38 Suppl:63-66.

30. Wernerman J, Hammarqvist F, Von Der Decken AA et al. Ornithine α-ketoglutarate improves skeletal muscle protein synthesis as assessed by ribosome analysis and nitrogen balance postoperatively. Ann Surg 1987; 206:674-678.
31. Vinnars E, Hammarqvist FF, Von Der Decken A et al. Role of glutamine and its analogs in posttraumatic muscle protein and amino acid metabolism JPEN 1990; 14:1258-1298.
32. Wernerman J, Hammarqvist F, Vinnars E. α-ketoglutarate and postoperative muscle catabolism. Lancet 1990; 335:701-703.
33. Hammarqvist F, Wernermann J, Von Der Decken AA et al. α-ketoglutarate preserves protein synthesis and free glutamine in skeletal muscle after surgery. Surgery 1991; 109:28-36.
34. Roth E, Karner J, Roth-Merten A et al. Effect of α-ketoglutarate infusions or organ balances of glutamine and glutamate on anaesthetized dogs in the catabolic state. Clin Sci 1991; 80:625-631.
35. Cynober L. The role of arginine and related compounds in intestinal function. In: Kinney JM, Tucker HN, eds. Organ Metabolism and Nutrition: Ideas for Future Critical Care. New York: Raven Press 1994; 245-255.
36. Czernichow B, Nsi-Emvo E, Galluser M et al. Enteral supplementation with ornithine α-ketoglutarate improves the early adaptative response to resection. Gut 1997; 40:67-72.
37. De Oca J, Bettonica C, Cuadrado S et al. Effect of oral supplementation of ornithine α-ketoglutarate on the intestinal barrier after orthotopic small bowel transplantation. Transplantation 1997; 63:636-639.
38. Moukarzel AA, Goulet O, Salas JS et al. Growth retardation in children receiving long-term total parenteral nutrition: Effects of ornithine α-ketoglutarate. Am J Clin Nutr 1994; 60:408-413.
39. Le Bricon T, Cynober L, Baracos VE. Ornithine α-ketoglutarate limits muscle protein breakdown without stimulating tumor growth in rats bearing Yoshida Ascites Hepatoma. Metabolism 1994; 43:899-905.
40. Le Bricon T, Cynober L, Field CJ et al. Supplemental nutrition with ornithine α-ketoglutarate in rats with cancer-associated cachexia: Surgical treatment of the tumor improves efficacy of nutritional support. J Nutr 1995; 125:2999-3010.
41. Bouchon Y, Merle M. L' α-cétoglutarate d'ornithine per os dans la prévention des complications locales de la chirurgie plastique. Ann Chir Plast Esthet 1984; 29;385-366.
42. Pradoura JP, Carcassonne Y, Spitalier JM. Incidence de l'oxoglutarate d'ornithine (Cétornan) sur la réparation cutané des malades de carcinologie cervico-faciale opérés. Cah. ORL 1990; 25:61.
43. Vaubourdolle M, Salvucci M, Coudray-Lucas C et al. Action of ornithine α-ketoglutarate on DNA synthesis by human fibroblasts. In vitro Cell Cev Biol 1990; 26:187-192.
44. Pasquali JL. La stimulation lymphocytaire in vitro par le pokeweed mitogéne chez les sujets normaux et les sujets dénutris: Influence de sels d'ornithine. Pathol Biol 1983; 31:191-194.
45. Lasnier E, Coudray-Lucas C, Le Boucher J et al. Ornithine α-ketoglutarate counteracts thymus involution and glutamine depletion in endotoxemic rats. Clin Nutr 1996; 15:197-200.
46. Le Boucher J, Farges MC, Minet R et al. Modulation of immune response with ornithine α-ketoglutarate in burn injury: An arginine or glutamine dependency? Nutrition 1999; 15:773.
47. Newsholme EA, Newsholme D, Cure R et al. A role for muscle in the immune system and its importance in surgery, trauma, sepsis and burns. Nutrition 1998; 4:261-268.
48. Roch-Arweiler M, Tissot M, Coudray-Lucas C et al. Immunomodulatory effects of ornithine α-ketoglutarate in rats with burn injuries. Arch Surg 1996; 131:718-723.
49. Schlegel L, Coudray-Lucas C, Barbut F et al. Intestinal impact of ornithine α-ketoglutarate in an experimental model of bacterial translocation. Clin Nutr 1996; 15 suppl, 15 (abstract).
50. Elalouf M, Mizrahi R, Zannetti A et al. Test d'exploration de la fonction somatotrope par l'ornithine chez l'adulte. Rev Fr Endocrinol Clin 1980; 21:357-360.
51. Lecointre CL, Dailly R. Stimulation de l'hormone de croissance par l'ornithine en pathologie pédiatrique: Étude de la résponse insulinique. Quest Med 1981; 34;1197-1204.
52. Donnadieu M, Combourieu M, Schimpff RM. Comparaison de différentes épreuves de stimulation pour l'étude de la fonction somatotrope chez l'enfant. Pathol Biol 1971; 19:293-301.
53. Payne-James J, Grimble G, Cahill E et al. Enteral administration of ornithine-oxoglutarate (OKGA) in man: Effects on hormone profiles and nitrogen (N) metabolism. JPEN 1989; 13(suppl1):836 (abstract).
54. Lambert P. Effets de l'α-détoglutarate d'ornithine sur les sécrétions d'insuline, d'hormone somatotrope, de glucagon et de cortisol dans la cirrhose. Thesis, Université de Montpellierp Montpellier, France 1982.

55. Krassowski J, Rousselle J, Maeder E et al. The effect of ornithine α-ketoglutarate on insulin and glucagon in normal subjects. Acta Endocrinol 1981; 98:252-256.
56. Cynober L, Vaubourdolle M, Dore A et al. Kinetics and metabolic effects of orally administered ornithine α-ketoglutarate in healthy subjects fed with a standardized regimen. Am J Clin Nutr 1984; 39:514-519.
57. Cynober L, Coudray-Lucas C, De Bandt JP et al. Action of ornithine α-ketoglutarate, ornithine hydrochloride and calcium α-ketoglutarate on plasma amino acid and hormonal patterns in healthy subjects. J Am Coll Nutr 1990; 9:2-12.
58. Lenzen S, Schmidt W, Rustenbeck I et al. 2-ketoglutarate generation in pancreatic β-cell mitochondria regulates insulin secretory action of amino acids and 2-ketoacids. Biosci Rep 1986:163-169.
59. Shuck JM, Eaton RP, Schuck LW et al. Dynamics of insulin and glucagon secretion in severely burned patients. J Trauma 1977; 17:706-712.
60. Le Bricon T, Coudray-Lucas C, Lioret N et al. Ornithine α-ketoglutarate metabolism after enteral administration in burn patients: Bolus compared with continuous infusion. Am J Clin Nutr 1997; 65:512-518.
61. Leverve X, Caro LHP, Plomp PJAM et al. Control of proteolysis in perifused rat hepatocytes. FEBS Lett 1987; 219:455-458.
62. Hunter A, Downs CE. The inhibition of arginase by amino acids. J Biol Chem 1945; 157:427-446.
63. De Bandt JP, Cynober L, Lim SK et al. Metabolism of ornithine α-ketoglutarate and arginine in isolated perfused rat liver. Brit J Nutr 1995; 73:227-239.
64. Stoll B, Haussinger D. Functional hepatocyte heterogeneity: Vascular 2-oxogluarate is almost exclusively taken up be perivenous, glutamine synthetase-containing hepatocytes. Eur J Biochem 1989; 181:709-716.
65. Winkler S, Hotzenbein Th., Karner J et al. Kinetics of organ specific metabolism of a bolus injection into the jejunum of glutamine, α-ketoglutarate, ornithine and ornithine a-ketoglutarate. Clin Nutr 1993; 12:57-58.
66. Cynober L. Metabolic interaction between ornithine and α-ketoglutarate as a basis for the action of ornithine α-ketoglutarate. Clin Nutr 1993; 12:54-56.
67. Jeevanandam M, Holaday NJ, Petersen SR. Ornithine α-ketoglutarate supplementation is more effective than its component salts in traumatized rats. J Nutr 1996; 126:2141-2150.
68. Coudray-Lucas C, Lasnier E, Le Boucher J et al. Ornithine α-ketoglutarate effect in glutamine generation. Evidence of component interaction. JPEN 1996, 20 suppl: 22 (abstract).
69. Welbourne TC. α-ketoglutarate, ornithine and growth hormone displace glutamine dependent for ammoniagenesis and enhance renal base regeneration and function. Clin Nutr 1993; 12;49-50.
70. Raul F, Gosse F, Galluser M et al. Functional and metabolic changes in intestinal mucosa of rats after enteral administration of ornithine α-ketoglutarate salt. JPEN 1995; 19:145-150.
71. Jeevanandam M, Holaday N, Ali MR. Altered tissue polyamine levels due to ornithine α-ketoglutarate in traumatized growing rats. Metabolism 1992; 41:1204-1209.
72. Grillo MA. Metabolism and function of polyamines. Int J Biochem 1985; 17:943-948.
73. Lescoat G, Theze N, Fraslin JM et al. Influence of ornithine on albumin synthesis by fetal and neonatal hepatocytes maintained in culture. Cell Differ 1987; 21:21-29.
74. Lescoat G, Desvergne B, Pasdeloup N et al. Effects of ornithine α-ketoglutarate on albumin secretion by adult rat hepatocyte co-culture. In: Guillouzo A, ed. Liver Cells and Drugs. Paris: INSERM/ John Libbey Eurotext 1988; 164:431-436.
75. Adams E. Metabolism of proline and of hydroxyproline. Int Rev Connect Tissue Res 1970; 5:1-91.
76. Nakamura T, Termaoto H, Tomita Y et al. L-proline is an essential amino acid for hepatocyte growth in culture. Biochem Biophys Res Comm 1984; 122:884-891.
77. Perez-Sala D, Parilla R, Ayuso MS. Key role of L-alanine in the control of hepatic protein synthesis. Biochem J 1987; 241:491-498.
78. BarbulA. Arginine: Biochemistry, physiology and therapeutic implications. JPEN 1986; 10:227-238.
79. Rettura G, Barbul A, Levenson SM et al. Citrulline does not share the thymotropic properties of arginine and ornithine. Fed Proc 1979; 38:289 (abstract).
80. Daly JM, Reynolds J, Thom A et al. Immune and metabolic effects of dietary arginine supplementation after burn. Arch Surg 1987; 122:784-789.
81. Saito H, Trocki O, Wang SL et al. Metabolic and immune effects of dietary arginine supplementation after burn. Arch Surg 1987; 122:784-789.

82. Grossie VB, Nishioka K. A parenteral nutrition regimen with ornithine substituted for arginine alters the amino acid, but not polyamine content of the Ward colon tumor. Nutr Cancer 1997; 27:102-106.
83. Moncada S, Palmer RMJ, Higgs A. Biosynthesis of nitric oxide from L-arginine: A pathway for the regulation of cell function and communication. Biochem Pharmacol 1989; 38:1709-1715.
84. Aussel C, Cynober L, Lioret N et al. Branched-chain keto acid plasma concentrations following burn injury: Effects of enteral feeding supplemental with ornithine α-ketoglutarate. Proc 4th World Congress on Intensive Critical Care Medicine Jerusalem 1985; (abstract).
85. May RC, Mitch WE. The metabolism and metabolic effects of ketoacids. Diabetes Metab Rev 1989; 5:71-82.
86. Walser M. Rationale and indications for the use of α-keto analogues. JPEN 1983; 8:37-41.
87. Zimmermann E, Wassmer S, Steudle V. Long-term treatment with calcium alpha-ketoglutarate corrects secondary hyperparathyroidism. Mineral Electrolyte Metab 1996; 22:196-199.

Chapter 11

Nutritional Support after Small Bowel Transplantation

S. Janes and S. V. Beath

Small bowel transplantation has been attempted since the 1950s but with survival times of less than two weeks in dogs and human subjects. It did not become established until the development of more effective immunosuppression regimes in the 1980s and 1990s.[1-6] In the meantime, patients with chronic intestinal failure have been managed with parenteral nutrition (PN) which has developed to the point where 75% of patients can expect to survive 10 years.[7] There are 33 centers carrying out small bowel transplantation world wide (communication Dr. D Grant, 5th International Symposium on Intestinal Transplantation, Cambridge 1997), but only patients who are experiencing complications with parenteral nutrition or feel that their quality of life is intolerable are selected as candidates for intestinal transplantation.[8-10] In the UK, around 250 patients are identified as being on home PN which equates to 4 per million and 50% of these could be considered potential recipients for intestinal transplantation.[11-13]

The reason successful intestinal transplantation has been hard to achieve is because of several unique characteristics: the great mass of lymphoid tissue in the gut renders it highly immunogenic; accurate identification of rejection episodes in the gut is difficult because of the presence of large numbers of lymphocytes under normal circumstances and the patchy nature of rejection; and the gut is continually exposed to bacteria, fungi and food antigens resulting in high rates of sepsis when gut integrity is damaged (as during rejection).[14] Furthermore, the newly engrafted bowel seems to take longer than other organs to recover from the effects of ischemia and hypoxia incurred during harvesting and preservation, and intestinal function may take many months to stabilize (compared with an average of 2-3 weeks after liver transplantation). The current 5-year survival after intestinal transplantation is 50%, although there is a "center effect" with the large North American centers achieving better figures than this and smaller centers with less than 10 patients achieving less (personal communication Dr. David Grant, 5th International Symposium on Intestinal Transplantation, Cambridge 1997).

The postoperative care of intestinal transplant recipients centers on fluid balance, meticulous attention to immunosuppression and a transfer from parenteral to enteral feeding.[15] The fact that only patients experiencing significant morbidity are selected for transplantation means that nutritional support after transplantation must ensure good patient rehabilitation and stimulate graft adaptation simultaneously. The goal of small bowel transplantation is to achieve independence of parenteral nutrition and enhance quality of life for the patient and normal growth velocity for pediatric patients.[16] This chapter will concentrate on nutritional support for small bowel transplantation which can be split into three phases:

Nutritional Support in Cancer and Transplant Patients, edited by Rifat Latifi and Ronald C. Merrell. ©2001 Eurekah.com.

1. Recovery from the effects of ischemia, preservation and implantation.
2. Weaning from parenteral nutrition and establishment of enteral nutrition.
3. Establishment of oral intake of a wide range of foods including whole protein, lactose and long chain triglyceride.

Recovery from Ischemia and Preservation

4 (Post-Op Day 0-14)

The technique used for harvesting small bowel is similar to other abdominal organs in that the graft is mobilized ready for excision and then perfused with a preservation solution at 4°C which flushes out all blood.[17] The graft is kept on ice until re-implantation 6-12 hours later. Even with rapid cooling and the use of a preservation solution high in glucose and lactate, changes secondary to ischemia occur. The production of free radicals from hypoxic tissue induces polymorph margination from capillaries into the lamina propria which becomes congested with a mixed inflammatory infiltrate which produces yet more biochemical changes including elevated phospholipase A2.[18] This process occurs slowly at 4°C, but after 12 hours there is a risk that excessive inflammation may be associated with severe graft dysfunction, sloughing of the mucosa and hyperacute rejection. This is in contrast to liver and kidney tissue which are stable in preservation solution at 4°C for 24 and 48 hours, respectively. Therefore, transplant teams arrange to harvest the intestinal graft and reimplant as soon as possible, even so some loss of villi occurs especially at the apices where the effects of hypoxia are initially manifested. Preservation solutions are an area of very active research with solutions being developed which neutralize chemotactic molecules produced in hypoxic tissue. However, the solution currently used in human intestinal transplantation is generally University of Wisconsin solution which is able to preserve small bowel allografts isolated from the circulation for up to 12 hours maximum.

Early intestinal dysfunction is, therefore, inevitable and takes the form of a paralytic ileus as a result of the surgical handling, lasting 2-5 days. From day 5-15 a secretory diarrhea related to mucosal damage regeneration, develops. The secretory diarrhea is usually mild (10-30 ml/kg/day) and self-limiting.[5,6] Occasionally, especially with end ileostomies a severe secretory diarrhea develops (greater than 100 ml/kg/day) which requires treatment with octreotide (e.g., Sandostatin Sandoz, 1-3 µg per kg per hour intravenously or 50 µg subcutaneously 2-4 times a day), although long-term treatment should be avoided because of reports of hepatic dysfunction.

As soon as bowel sounds are heard or the ileostomy is seen to peristalse, dioralyte solution 10-20 mls/hour is introduced. It is not usually possible to introduce feed until there is evidence of absorption of dioralyte (i.e., ileostomy output is less than enteral intake) which may occur anytime from day 5 postoperatively depending on the speed of recovery from graft preservation and implantation. The principles used in initiating enteral feeding after small bowel transplantation are similar to those in rehabilitation of patients after major intestinal resection.[19] Nutrients are introduced separately and in small increments so that tolerance and adequate absorption are confirmed before adding more nutrients to the feed. This approach, which is called modular feeding, has been useful in enhancing rapid adaptation of the newly engrafted bowel and is also important in enabling quicker identification of rejection episodes which are manifested by malabsorption of the feed, which might otherwise be attributed to changes in the feed. Table 11.1 shows details of the introduction of enteral feeding for pediatric patients at our unit. Our initial modular feed provided 0.2 kcal/ml and 233 mosmol/kg, containing protein, carbohydrate and electrolytes, with lipid omitted. The concentration is increased by adjusting

Table 11.1. Nutrient content of enteral feeds in early post-operative period

			post-op day			
gram per 100 ml of feed	**9**	**12**	**15**	**18**	**21**	**24**
Carbohydrate (Total)	4.0	5.0	6.0	7.0	7.0	10.0
As glucose polymer	4.0	5.0	6.0	6.0	6.0	8.0
As lactose	0	0	0	0	0	0
As sucrose	0	0	0	1.0	1.0	2.0
Protein (Total)	1.1	1.1	1.1	2.2	2.2	2.2
As hydrolysed whey (75%)	0.83	0.83	0.83	1.65	1.65	1.65
As glutamine (25%)	0.28	0.28	0.28	0.55	0.55	0.55
Fat	0	0	0	1.0	3.0	3.0
As medium chain triglyceride	0	0	0	1.0	3.0	3.0
As long chain triglyceride	0	0	0	0	0	0
Energy (kilocalories per ml)	0.21	0.25	0.29	0.46	0.64	0.77

one ingredient at a time, adding fat during the third week of feeding and gradually progressing towards an energy density of 1 kcal/ml and osmolality of 440 mosmol/kg (Fig. 11.1).

Weaning Off Parenteral Nutrition

(Post-Op Day 15-40)

From day 15, the calorie density of the enteral feed is steadily increased to approximately 0.8 kcal/ml. In order to achieve this, medium chain triglyceride[20] and disacharides are introduced individually with increments every 24-48 hours.[19] The volume of the feed is usually kept low (i.e., 10-20 ml/kg per day) whilst the feed is gradually built up. However, once the energy density of the feed is comparable to the energy density of the PN, the latter is reduced while the feed is increased by the same volume. This allows a smooth change over from PN and maintains the patient's nutritional parameters.[16] Fluid balance usually remains a problem especially with end ileostomies and intravenous dextrose saline (i.e., 25-50 ml/kg/day) may still be required at night for some months.[21] This system allows total flexibility and enables the addition of chosen nutrients at the optimal time with respect to evolving recovery and function of the graft. For details of the types of nutrients used please see Table 11.2.

Components of Postoperative Enteral Feed

Protein

The initial protein source used in our unit is hydrolysed whey (hydrolysed whey protein maltodextrin mixture—HWPMM, SHS), which is well absorbed even in the presence of exocrine pancreatic dysfunction (common in enteral understimulation[22]). The HWPMM provides 55 g protein per 100 g and some carbohydrate. Increased intestinal permeability is also common in the first few weeks after transplantation, so the hydrolysed protein is useful, with 28% of the peptides having a molecular weight of greater than 1000 Daltons.[23] Hydrolysed protein is as trophic to gut mucosa as whole protein,[24] which is an advantage after intestinal transplant. In pediatric cases, we aim to provide greater than the reference nutrient intake for

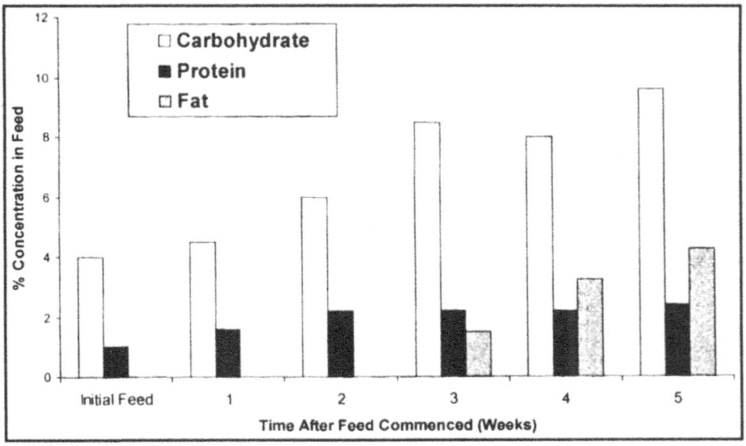

Fig. 11.1. Sequential additions of nutrients to enteral feed for child under 5 years of age weeks 1-6 post small bowel transplant.

protein, based on estimated requirements for sick children,[25] i.e., 3 g/kg for infants and 2 g/kg for children.

Carbohydrate

Glucose polymer (Super Soluble Maxijul, SHS) is the principal source of carbohydrate, as it is well tolerated due to its ease of absorption and low osmolality. When the limit of tolerance of glucose polymer is reached (usually between 8-12 g per 100 ml feed), disaccharides (i.e., sucrose and lactose) may be incorporated to utilize alternative channels of absorption which increases energy density of the feed.[19]

Lipid

During retrieval and re-implantation, the lacteals are interrupted so that long chain triglyceride (LCT) malabsorption is inevitable postoperatively. In the rat model reconnection of lacteals can be demonstrated after 4 weeks,[26] but this appears to be delayed for at least 3 months in humans,[27] where the coefficient of fat absorption on a mixed diet is rarely greater than 80% before six months postoperatively. Thus, medium chain triglyceride (MCT) emulsion (Liquigen, SHS) is the main lipid used in initial feeds because it is absorbed directly into the portal vein and provides an excellent alternative energy source.[28] Three to five grams of fat per 100 mls of feed was introduced from the third week. However, restricting long chain triglyceride (LCT) is likely to induce essential fatty acid deficiency which are important bioactive molecules especially in the central nervous system and intercellular signalling in the immune system.[29] Intralipid contains high concentrations of linoleic and linolenic acid which can be given intermittently (i.e., 500 mg/kg intralipid once per week) intravenously, until lacteal drainage is reestablished at about six months postoperatively.

Substances Promoting Adaptation

Two novel substances were included in the feed used in our Unit: glutamine and pectin. Glutamine is a preferred fuel source for enterocytes and is important in maintaining mucosal integrity and preventing bacterial translocation.[22,30,31] Animal models with short bowel have shown increased mucosal weight and length of villi when fed glutamine. Glutamine provided 25% of the protein within our modular feed.

Table 11.2. Types of nutrients used in nutritional support after small bowel transplantation

Component	Name	Supplier
Carbohydrate		
glucose polymer	Maxijul super soluble	SHS*
lactose	Lactose BP	Thornton & Ross
sucrose	sugar	supermarket
Protein		
hydrolysed whey	Hydrolysed whey protein maltodextrin mixture	SHS*
glutamine	L-Glutamine	SHS*
Fat		
medium chain triglyceride	Liquigen	SHS*
long chain triglyceride	Calogen	SHS*
Vitamins & Minerals	Pediatric seravit	SHS*
Pectin	Citrus pectin powder	Citrus Colloid Limited

*Scientific Hospital Supplies

Ornithine alpha-ketoglutarate is a precursor of glutamine and may also have a useful role in promoting adaptation of the intestinal graft. Rats fed enterally for 7 days with feed enriched with ornithine alpha-ketoglutarate demonstrated an increase villus height and protein turnover in the tibialis muscle after extensive small bowel resection (poster presentation Dr. F. Dumas et al, 5th International Symposium on Intestinal Transplantation, Cambridge 1997). A human study evaluating the effect of 15 g ornithine alpha-ketoglutarate added to parenteral nutrition and administered to children with growth failure, showed an increase in glutamine, glutamate, IGF-1 and height velocity (3.8 cm/yr to 6.5 cm/yr).[32]

Pectin is a source of fermentable dietary fiber, which is thought to increase mucosal hyperplasia and increase height of villi through the trophic action of the short chain fatty acids derived from bacterial metabolism of pectin.[33,34] An additional role of pectin is to prolong intestinal transit time, which enhances nutrient absorption in a similar way to loperamide.[35] One gram of powdered pectin was added per 100 ml of feed.

Electrolytes

To optimize feed absorption, electrolytes are included at similar concentrations as in oral rehydration solutions[36] and are then adjusted to meet individual requirements. Sodium requirements may be as high as 10 mmol/kg/day at first when the ileostomy output is high and sodium enriched. A comprehensive vitamin and mineral supplement (Pediatric Seravit, SHS) is added to the feed to meet the dietary reference values for each patient.[37]

Proprietary Feeds

There is no suitable single feed which meets all these requirements; however examples of proprietary feeds that are used in children post small bowel transplant are Pregestimil (Mead Johnson) and Neocate (Scientific Hospital Supplies UK Limited). Pregestimil is a semi-elemental infant formula, comprised of hydrolysed casein, glucose polymer and both medium- and long-chain fats (55%MCT: 45%LCT). Neocate is an elemental formula comprised of amino acids glucose polymer and long chain fat. Whichever feed is chosen, it is commenced at low concentration and osmolality, e.g., half strength (0.33 kcal/ml) and the density gradually increased, according to tolerance, before advancing the volume of feed given. It is our practice to change from a modular feed to MCT Pepdite between two and six months as most children remain dependent on tube feeding for at least six months, and a proprietary feed is easier to manage in the home environment. Adults are usually able to progress onto a normal diet within 6 months if graft function is good, because they do not usually exhibit food aversion. However, adults who require calorie supplementation may receive Nutrison (Nutricia Clinical Foods Limited) or some other complete feed overnight.

Motility

The gut smooth muscle normally functions as an electrical syncytium with specialized cells in the stomach and proximal duodenum acting as a pacemaker for the migrating myoelectric complexes (MMC) which sweep through the length of the gastrointestinal tract. Since extrinsic denervation of the transplanted bowel is inevitable, motility depends on the intrinsic nervous system of the gut[38] and the transplanted gut develops its own contractile pattern which is independent of the native gut. There have been few studies in man, but MMC activity appears to be absent, and the transit of intestinal content seems to depend on local intestinal contractions which are peristaltic for 10 cm or less.[39] There is some evidence in the canine model, that reconnection of the native enteric nervous system with the donor enteric nervous system occurs after 12-20 months and that MMC's reappear, but it is not known if this occurs in man.[40] Thus motility of the transplanted intestine is sensitive to local factors such as luminal distension and nutrient content, rather than vagally mediated postprandial patterns of inhibition. Clinically, we have found that patients have satisfactory gastric emptying followed by rapid transit (1-2 hours) as the chyme passes along the transplanted intestine to the distal stoma. In our center, this pattern of motility appears to be established two weeks postoperatively. The transit time can be slowed using loperamide (50 mg/kg per dose[35]), but care must be taken in the dose regimen to avoid inducing a paralytic ileus (unpublished observations). Transit time is conveniently measured using carmine red dye taken orally, and transit times of less than 4 hours are treated with increasing doses of loperamide to a maximum of 200 mg/kg/day.

Establishment of Normal Diet

4 (Post-Op 2-12 Months)

A low fat, cow's milk protein free diet is allowed as soon as the patient requests it (in practice usually at least one week postoperatively), but the majority of calories will be derived from the modular feed. A cow's milk protein diet is adopted as there is increased intestinal permeability within the first few weeks posttransplantation. A feed using protein hydrolysates is somewhat unpalatable and older children and adults usually require administration nasogastrically, but babies will often drink it. Many small bowel transplant recipients have poorly developed feeding skills and may be afraid to swallow food.[14] These children may take 12-18 months to learn to be confident about eating and the input of a multi-disciplinary including dietitian, speech therapist and clinical psychologist is crucial,[41] beginning even before transplantation.[12,20] Of 30 pediatric patients from Pittsburgh, 22% required tube feeds at 1-3 years post-transplant, 20% at 3-5 years and two of three patients at 5-7 years posttransplant (poster presentation B Kosmach et al, 5th International Symposium on Intestinal Transplantation, Cambridge 1997). The speed at which a normal diet is adopted is dependent on several factors: patient preference, graft function including motility, fat absorption and type of ileostomy. Patients with a history of pseudo-obstruction and infants tend to remain on specialized feeds longer. However, older children and adults may be able to take a normal diet from three months, although fat malabsorption may cause high stomal output, which has to be compensated for by increased oral intake or overnight intravenous dextrose and saline. By six months the coefficient fat absorption is much improved with approximately 85-90% of conventional long chain fat being absorbed in dogs,[42] but there are only limited studies in man.[27] Some patients, particularly those with an end ileostomy, have a continuing requirement for intravenous fluids, after PN has been stopped, owing to large volumes of fluid and electrolytes lost from the ileostomy. Of 22 patients from Pittsburgh, one third required intravenous fluid overnight at one year posttransplant.[21] However, even end ileostomies achieve better sodium and water reabsorption after 1-2 years.[43]

Monitoring

Even before the patient is fully weaned from PN, close monitoring of graft function and nutritional support is essential. The daily volume of ileostomy output and presence or not of reducing substances is extremely useful in early detection of rejection and other complications such as cytomegalovirus (CMV) enteritis and other systemic infections, as well as determining changes to the feed[21] (see Fig 11.2). Provided good graft function is present (defined as when the ileostomy output does not exceed input and there are no more than 1% reducing substances[16]), then the energy density and volume of the feed can be increased.

In addition to anthropometric indices, biochemical markers of intestinal function such as albumin, essential fatty acids and trace elements should be assessed regularly (see Table 11.3). At one year post transplant, 22 pediatric patients from Pittsburgh, who were off PN, had increased their height centiles, were appropriate weight for height and had maintained both muscle and fat stores (poster presentation Dr. GM Rovera et al, 5th International Symposium on Intestinal Transplantation, Cambridge 1997). Specific markers of intestinal function such as permeability may be assessed by the differential absorption of lactulose and mannitol[44] or chromium labelled EDTA[45] and differential fat absorption by extraction of lipids from ileal output.[46] Disaccharidase activity in the transplanted small bowel is normal with two weeks of operation (unpublished observations) although during severe rejection disacharidase activity may be reduced.[47] Endoscopy and mucosal biopsies are taken frequently (once or twice a week) to monitor for rejection which occurs at some stage in over 75% patients.[3,5,14] Eosinophilic infiltration of the small bowel lamina propria has also been reported to occur in over 50% of patients in the first four months after transplant and in the absence of rejection may represent food sensitization.[48]

Complications after Intestinal Transplant and Implications for Nutritional Support

The main complications of small bowel transplantation are rejection and the consequences of heavy immune suppression required to prevent it (see Table 11.4). Rejection is most likely to occur in the first month and is manifested by malabsorption of enteral feeding. Patients are usually still receiving some PN which can be increased to prevent weight loss, while the rejection episode is treated with methylprednisolone.[15] Severe secretory diarrhea is uncommon fortunately, because it is life threatening and may require octreotide to control it. However, a low grade secretory state may be present at first manifested by sodium rich ileal output (sodium >100 mmol/L), which is important to recognize as the patient will require sodium supplements. Cytomegalovirus (CMV) enteritis has been a major problem especially in adult small bowel transplant recipients that some transplant centers will not use CMV positive donors.

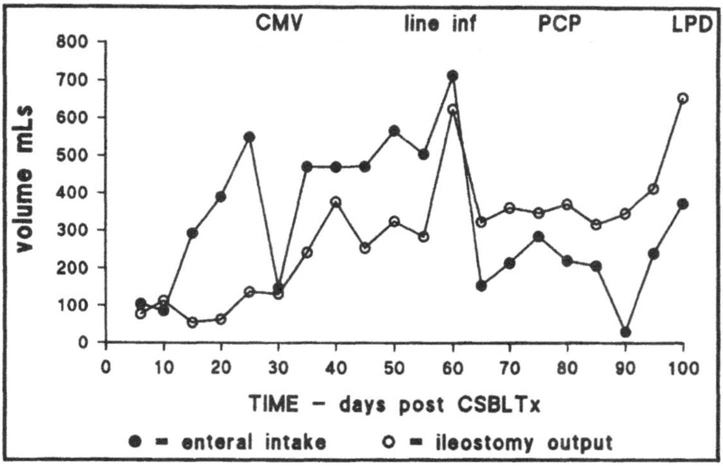

Fig 11.2. Diagram illustrating major complications and effect on absorption of enteral feed in a pediatric patient.

Table 11.3. Nutritional monitoring after small bowel transplantation

Daily	Weekly	Monthly
plasma electrolytes	zinc	copper, selenium, manganese
ileostomy output (mls)	magnesium	essential fatty acids
reducing substances in ileal fluid	liver function	viral serology, polymerase chain reaction e.g. EBV [44]
enteral intake (mls)	coagulation intestinal permeability e.g. lactulose:mannitol coefficient of fat absorption	
weight	triceps skin-fold mid-arm circumference height ileoscopy*	

* gastroscopy and ileoscopy are done every three months to screen for lymphoproliferative disease.

CMV produces multiple ulcers in the graft which often bleed and may perforate. Although, CMV responds to hyperimmune globulin and ganciclovir, it is liable to recur and a temporary return to PN may be needed for a few weeks during treatment. The frequency of infections after intestinal transplantation is directly related to the intensity of immune suppression and high levels of tacrolimus (trough level greater than 30 ng/mL) are associated with opportunist infections such as pneumocystis, and ultimately lymphoma triggered by Epstein-Barr virus (EBV).[49,50] With increasing experience it has been realized that satisfactory gut function can be achieved using lower exposure to tacrolimus (trough levels of approximately 15 ng/mL after 3 months). Paradoxically, insufficient immune suppression is also associated with infection in the form of septicaemia related to translocation of enteric organisms through the leaky mucosa of rejecting bowel.

Table 11.4. Complications after intestinal transplantation

Complication	typical time of onset post-op day
Rejection	7-10
Secretory diarrhea	5-20
CMV enteritis	40
GI infections* e.g.rotavirus	any time, often adenovirus after discharge
Lymphoproliferative disease	180 days

* may be severe leading to loss of graft

Conclusion

Small bowel transplantation is an alternative to patients on parenteral nutrition with irreversible intestinal failure, but the necessity to use intense immune suppression and the susceptibility of the transplanted intestine to functional instability causing malabsorption of drugs, fluids and food means that this is far from being a routine operation and requires sophisticated nutritional support. However, in well prepared patients supported by an experienced multidisciplinary team of surgeons, physicians, nurses, dietitians, play specialists, feeding psychologist and liason staff this operation achieves around 65% four-year survival[51] and is now a viable option for patients with chronic intestinal failure.

Acknowledgment

We are grateful to colleagues in the Liver Unit and Gastroenterology Department especially Dr. D.A. Kelly and Prof. I.W. Booth for their support, to Mrs. Rosie Jones for much of the original dietetic protocol, and to Mrs. Anita MacDonald for reviewing the manuscript.

References

1. Grant D, Wall W, Mimerault R et al. Successful small bowel-liver transplantation. Lancet 1990; 335:181-4.
2. Kelly DA, Buckels JAC. The future of small bowel transplantation. Arch Dis Child 1995; 72:447-451.
3. Grant D. Current results of intestinal transplantation. Lancet 1996; 347:1801-3.
4. Beath SV and Mayer AD. Small bowel transplantation. Hospital Update 1996:27-31.
5. Kocoshis SA. Small bowel transplantation in infants and children. Gastroenterology Clinics of North America 1994; 23:727-42.
6. Goulet O, Jan D, Brousse N et al. Intestinal transplantation. J Ped Gastroenterol Nutr 1997; 25:1-11.
7. Messing B, Crenn P, Beau P et al. Long-term survival and parenteral nutrition-dependency of adult patients with nonmalignant short bowel. Transplant Proc 1998 In Press.
8. Colomb V, Goulet O, De Potter S et al. Liver disease associated with long-term parenteral nutrition in children. Transplant Proc1994; 26:1467.
9. Dollery CM, Sullivan ID, Bauraind O et al. Thrombosis and embolism in long term central venous access for parenteral nutrition. Lancet 1994; 344: 1043-45.
10. Beath SV, Needham SJ, Kelly DA et al. Clinical features and prognosis of children assessed for isolated small bowel (ISBTx) or combined small bowel and liver transplantation (CSBLTx). J Pediatr Surg 1997; 32:459-61.

11. Ingham Clark CL, Lear PA, Wood S et al. Potential candidates for small bowel transplantation. Br J Surg 1992; 79:676-9.
12. Beath SV, Booth IW, Murphy MS et al. Nutritional care and candidates for small-bowel transplantation. Arch Dis in Child 1995; 73:348-50.
13. Beath SV, Brook GA, Buckels JAC et al. Demand for paediatric small bowel transplantation in the United Kingdom. Transplant Proc 1998 in press.
14. Tzakis AG, Todo S, Reyes J et al. Clinical intestinal transplantation: focus on complications. Transplant Proc 1992; 24:1238-40.
15. Beath SV, Kelly DA, Booth IWB et al. Post operative care of children undergoing combined small bowel and liver transplantation. Brit J Int Care 1994; 4:302-8.
16. Janes S, Beath SV, Jones R et al. Enteral feeding after intestinal transplantation: the Birmingham experience. Transplant Proc 1997;29:1855-6.
17. Casavilla A, Selby R, Abu-Elmagd K et al. Logistics and technique for combined hepatic-intestinal retrieval. Ann Surg 1992; 216:605-9.
18. Sonnino RE, Wong L, Franson RC. Early secretory events during intestinal graft preservation. Transplant Proc 1998; in press.
19. Warner BW,Ziegler MM. Management of the Short Bowel Syndrome in the Pediatric Population. Pediatric Clinics of North America 1993; 40:6,1335-1350.
20. Gracey M, Burke V, Anderson CM. Medium chain triglycerides in pediatric practice. Arch Dis Child 1970; 45:445-52.
21. Rovera GM, Graham TO, Hutson WR et al. Nutritional management of intestinal allograft recipients. Transplant Proc 1998; in press.
22. Vanderhoof JA, Langnas AN, Pinch LW et al. Short Bowel Syndrome—A Review. J Pediatr Gastroenterol Nutr 1992; 14:359-370.
23. MacDonald A. Which formula in cow's milk protein intolerance? The dietitian's dilemma. Euro J Clin Nutr 1995; 49,Suppl 1,S56-S63.
24. Vanderhoof JA, Grandjean CJ, Burkley KT et al. Effect of casein versus casein hydrolysate on mucosal adaptation following massive bowel resection in growing rats. J Pediatr Gastroenterol Nutr 1983; 2:617-21.
25. Shaw V and Lawson M. Principles of paediatric dietetics. In: Shaw V and Lawson M, eds. Clinical Paediatric Dietetics. 1st ed.Oxford:Blackwell Science Limited,1994:3-12.
26. Liu H, Teraoka S, Ota K et al. Successful lymphangiographic investigation of mesenteric lymphatic regeneration after orthoptopic intestinal transplantation in the rat. Transplant Proc 1992; 24:1113-14.
27. Mousa H, Bueno J, Griffith J et al. Intestinal motility in children after small bowel transplantation. Transplant Proc 1998; in press.
28. Senior JR. Medium Chain Triglycerides. Philadelphia: University Pennsylvania Press, 1968.
29. Sanders TAB. Essential and Trans-fatty acids in nutrition. Nutrition Research Reviews. 1988; 1:57-78.
30. Hambridge KM, Krebs NF, Sokol RJ. Energy and Nutrient Requirements. In: Roy CR, Silverman A and Alagille D, eds. Pediatric Clinical Gastroenterology. 4th ed. Missouri: Mosby—Year Book,Inc., 1995; 1005-1019.
31. McAnena OJ, Moore FA, Moore EE et al. Selective Uptake of Glutamine in the Gastrointestinal Tract: Confirmation in a Human Study. Brit J Surgery 1991; 78:480-2.
32. Moukarzel AA, Goulet O, Salas JS et al. Growth retardation in children receiving long-term parenteral nutrition: Effects of ornithine alpha-ketoglutrate. Am J Clin Nutrition 1994; 60:408-13.
33. Allard JP and Jeejeebhoy KN. Nutritional Support and Therapy in the Short Bowel Syndrome. Gastroenterology Clinics of North America 1989; 18:3,589-601.
34. Thompson JS. Management of the Short Bowel Syndrome. Gastroenterology Clinics of North America 1994; 23:2,403-420.
35. Sandhu BK, Tripp JH, Milla PJ et al. Loperamide in severe protracted diarrhea. Arch Dis Child 1983; 58:39-43.
36. Greenough III WB. Oral rehydration therapy: An epithelial transport success story. Arch Dis Child 1989; 64:419-22.
37. Department of Health Report on Health and Social Subjects No 41. Dietary reference values for food energy and nutrients for the United Kingdom. London:HMSO, 1991.

38. Sarr MG, Kelly KA. Myoelectric activity of the autotransplanted canine jejuno-ileum. Gastroenterol 1981; 81:303-10.
39. Sarr M, Hakim N. Motility of the transplanted gut. In Enteric Physiology of the transplanted intestine 1994; 5:28-54. RG Landes Austin, TX U.S.A.
40. Quigley EMM, Spanta AD, Rose SG et al. Long-term effects of jejunoileal autotransplantation on myoelectric activity in canine small intestine. Dig Dis Sci 1990; 35:1505-17.
41. Harris G, Booth IW. Feeding problems and eating disorders in children and adolescents. In: Cooper PJ and Stein A, eds. Monographs in clinical pediatrics. 1st ed. Harwood Academic Publishers, Reading 1992; 5:61-84.
42. Thompson JS, Rose SG, Spanta Ad et al. The long-term effect of jejunoileal autotransplantation on intestinal function. Surgery1991; 111:62-8.
43. Ein SH. The pediatric ostomy. In: Walker WA, Durie PR, Hamilton JR et al, (eds). Pediatric Gastrointestinal Disease. Philadelphia,Toronto: BC Decker Inc, 1991:1767-78.
44. Lim et al. HIV and permeability. Scand J Gastroenterology 1993; 28:573-80.
45. Grant D, Hurlbut D, Zhong R et al. Intestinal permeability and bacterial translocation following small bowel transplantation in the rat. Transplantation 1992; 52:221-224.
46. Beath SV, Willis KD, Hooley I et al. New method for determining faecal fat excretion in infancy. Arch Dis Child 1993; 69:138-40.
47. Akhtar K, Deardon D, Pemberton PW et al. Study of mucosal brush border enzyme activity in porcine small bowel transplantation. Transplant Proc 1996; 28:2556-7.
48. Putnam PE, Bueno J, Kocoshis SA et al. Tissue eosinophilia after small bowel transplantation in children. Transplant Proc 1998; in press.
49. Hann I. UKCCSG lymphoproliferative disease study. Available from Dept. Epidemiology and Public Health, University of Leicester.
50. Green M, Reyes J, Jabbour N et al. Use of quantitative PCR to predict onset of Epstein-Barr viral infection and post-transplant lymphoproliferative disease after intestinal transplantation in children. Transplant Proc 1996; 28:2759-60.
51. Langnas AN, Antonson DL, Kaufman SS et al. Preliminary experience with intestinal transplantation in infants and children. Transplant Proc 1996; 28:2752.

CHAPTER 12

Nutritional Support of Patients with Liver Transplant

Rifat Latifi, Giacomo Basadonna and Amadeo Marcos

Liver transplantation (LT) has evolved into the most successful form of treatment for end-stage liver disease (ESLD), with operative survival exceeding 90% for the first graft, retransplant free survival greater than 85% at one year,[1,2] and predicted actuarial survival over 75%.[3] Although the indications for LT have become more standardized (in adults as well as in children), 89% of adult patients undergoing LT have advanced cirrhosis (secondary to primary cholestatic liver disease, alcoholism, hepatitic C/non-A-non-B, autoimmune hepatitis, 0 or hepatitis B), followed by fulminant hepatic failure (5.5% -7.0%), metabolic disease (4%), malignant and benign neoplasms (3%-6% and 0.5% respectively), biliary atresia (0.5%), and other miscellaneous indications (1.8%-2%).[3] On the other hand, the most common indications for LT in children are extrahepatic atresia (>50%) and alpha-1-antitrypsin deficiency (9%-14%), followed by metabolic diseases (12%-13%), cirrhosis (7%) and other indications.[4]

Patients who have undergone LT, or are awaiting LT, represent the most complex medical challenges. These patients are often plagued with multiple comorbid diseases and require meticulous, well-planned and executed, highly individualized medical care. Since the liver plays a major function in metabolic homeostasis, nutritional and metabolic support becomes one of the most important therapeutic interventions. Malnutrition and its associated sequella are common features in patients with chronic liver disease in general, and most patients awaiting LT are no exception. Malnutrition in this group of patients is a consequence of increased nutrient requirements in the presence of hypermetabolism, anorexia and inadequate oral intake, associated with concomitant impaired digestion, absorption, and assimilation of nutrients.

Maintaining adequate nutrition represents a significant therapeutic challenge, considering the profound metabolic alterations that impair the incorporation of nutrient substrate into tissue. A vicious cycle then occurs, in which the patient is profoundly hypoproteinemic, reflecting the depleted stores and synthetic activity of the liver, and hypercatabolic, while at the same time provision of protein may result in hyperammonemia and aggravation of existing encephalopathy. Although poor nutritional status increases morbidity and mortality perioperatively, it is difficult to correctly identify and quantify the impact of individual nutrient therapy on patients with ESLD requiring LT.

Nutritional Support in Cancer and Transplant Patients, edited by Rifat Latifi and Ronald C. Merrell.

Malnutrition in Patients with Chronic Liver Disease

The complex metabolic derangements that accompany liver failure reflect the magnitude of the problems associated with liver failure (Table 12.1).[5] The frequency and the degree of malnutrition varies among patients studied.[6] In one study, malnutrition was present in all 74 patients undergoing LT.[7] Patients with primary biliary cirrhosis retained their hepatic synthetic function better than others despite extreme wasting of muscle and fat.[8] Others[9] found malnutrition present less frequently (79%) and no differences in nutritional status with regard to the etiology of chronic liver disease. The etiologic factors of malnutrition in chronic liver disease are multiple and synergistic in nature.[10] Among these factors, anorexia, nausea, vomiting, poor dietary habits, malabsorption, frequent large volumes therapeutic paracenteses, and inadequate protein and caloric intake are the most common. The mechanisms of malnutrition varies and may include catabolism and altered hepatic metabolism of key nutrients: carbohydrates, fat and protein.[10-12]

The hepatocellular dysfunction that causes unique changes of protein, carbohydrate and fat metabolism becomes more prominent during the fasting state,[10,11] when a starvation type metabolism develops. Hepatic glycogen depletion occurs rapidly (10-12 h instead of 36-48 h) resulting in accelerated and premature protein catabolism.

Hepatic glucose production and peripheral glucose oxidation are significantly decreased. This prevents fasting hypoglycemia. Serum free fatty acids are increased secondary to increased peripheral lipolysis, which in turn stimulates ketogenesis.[12] Ketogenesis, however, ultimately fails with worsening of liver failure. In addition to protein, fat, and carbohydrate metabolic abnormalities, impaired hepatic function results in vitamin and mineral deficiencies. When patient condition deteriorates and nutrient metabolism becomes insufficient to maintain body homeostasis, as in the late stages of cirrhosis, or infulminant hepatic failure, liver transplantation becomes the only current, long-term live saving intervention.

Hepatic Encephalopathy

When the liver fails or blood is shunted past the liver, hyperammonemia results and brain function deteriorates. This frequent complex syndrome in patients with ESLD, known as hepatic encephalopathy, is manifested by varies signs and symptoms that range from rapidly developing delirium, convulsions and coma in acute failure to more gradually impairment of intellect that eventually leads to stupor and coma in chronic liver failure[13] (Table 12.2). Progression and severity depend on whether the disease is in an acute or chronic stage and on the magnitude of the insult to the liver. Although, the exact biochemical mechanisms and neurochemical pathways of this complex condition are not known, one or more of several mechanisms may adversely affect the brain function of patients with liver failure[14] (Table 12.3).

The blood-brain barrier (BBB) consists of tight junctions of endothelial cells in the cerebral capillary bed. If substances are to affect the brain and tissues, they must penetrate this barrier. Transport of substrates through the BBB is strictly regulated and depends on their lipid solubility and on specific transport systems. For example, amino acids and glucose have their own unique and distinct carrier systems for transportation into the brain. The effectiveness of this physiologic barrier protects the organisms from the potentially adverse effects of serious metabolic changes. In HE, however, the BBB commonly breaks down, as demonstrated by the uptake of substances that otherwise are not taken up by the brain.[15] In addition, hepatic encephalopathy significantly alters the BBB which affects specific transport systems of neutral amino acids, glucose, ketone bodies, and basic amino acids. Transport of neutral amino acids into the brain significantly increases, whereas the transport of glucose, ketone bodies, and basic amino acids decreases.[16,17] On the other hand, BBB alteration may be consequence of a generally

Table 12.1. Metabolic alterations in chronic liver disease*

Alteration	Mechanism
Increased plasma glucagon	Portosystemic shunting Impaired hepatic degradation Hyperammonemia
Increased plasma aromatic amino acids	Hyperinsulinemia Increased peripheral insulin resistance Decreased effective insulin to glucagon ratio Impaired hepatic function
Increased plasma epinephrine and cortisol	Impaired hepatic function
Decreased liver and muscle carbohydrate	Accelarated glycogenolysis Impaired glycogenesis
Accelarated gluconeogenesis	Hyperglucagonemia
Hyperglycemia	Portosystemic shunting Increased glucose production Decreased insulin-dependent glucose uptake Decreased insulin-dependent hepatic glycolysis
Hyperammonemia	Deamination and accelerated bacterial degradation of protein in the colon
Increased plasma aromatic amino acids	Decreased hepatic clearance Increased release into circulation Hypoalbuminemia, hyperbilirubinemia Decreased incorporation of aromatic amino acids into proteins
Increased plasma methionine, glutamine, asparagine, histidine	Decreased hepatic clearance
Decreased plasma branched-chain amino acids	Hyperinsulinemia Excessive uptake Increased use of BCAA as energy source

*Used with permission from Latifi R, Killam J, Dudrick SJ: Nutritional support in liver failure. Surg Cli North Am 1991; 71:567-578.

nonspecific increased permeability, which can expose the brain to a variety of neurotoxic substances circulating in the blood and possibly cause cerebral edema.

Portal-systemic shunting of blood is associated with hyperammonemia, increased glutamine concentration in the brain, an altered plasma neutral amino acid pattern, and high levels of several large neutral amino acids in the brain. An altered amino acid pattern is caused by liver disease as much as by portal-systemic shunting. The amino acids of great importance in hepatic encephalopathy are tyrosine, phenylalanine, and tryptophan because they are involved in the synthesis of the catecholamines dopamine, noradrenaline, and serotonin and are potential precursors of "false" neurotransmitters, such as octopamine or tryptamine.[18] These amino acids together with leucine, isoleucine, valine, methionine, threonine, and histidine compromise the group of large neutral amino acids that have a common blood-brain transport system. Other amino acids, such as glutamate, aspartate, taurine, and glycine, can also act as neurotransmitters. Because some of these amino acids are precursors for neurotransmitters and other potentially neuroactive substances, high central nervous sytem levels of these amino acids may contribute to the development of encephalopathy. Changes in the blood-to-brain transport of

Table 12.2. Clinical grading of hepatic encephalopathy

Grade	Mental Status and Tremors	EEG Findings
I. Prodome (frequently diagnosed retrospectively)	Euphoria, occasional depression; fluctuant, mild confusion; slowness of mentation and affect; slurred speech; disordered sleep rhythm; tremor slight	Usually no EEG changes
II. Impending coma	Accentuation of state I: drowsiness; inappropriate behavior; inability to maintain sphincter control; tremor present (easily elicited)	Generalized slowing pattern, abnormal
III. Stupor	Asleep most of the time but can be roused; incoherent speech; marked confusion; tremor usually present (patient may not be able to cooperate)	Abnormal
IV. Deep cerebral coma	May or may not respond to painful stimuli, and tremor usually absent	Abnormal

neutral and basic amino acids, in part, may be caused by an increase in the V-max of the respective carrier system, which is possible a consequence of hyperammonemia.[19]

Amino Acids in Hepatic Encephalopathy

Hepatic encephalopathy is associated with increased plasma and brain concentrations of the aromatic amino acids (AAA): phenylalanine, tyrosine and tryptophan (free form) and decreased concentrations of the branched-chain amino acids (BCAA): valine, leucine and isoleucine.[18-23] Because large neutral amino acids share a common transport carrier in crossing the BBB, decreased BCAA concentrations in the blood may in fact facilitate increased transport AAA into the brain.[23] High concentrations of the aromatic amino acids and methionine could limit the cerebral uptake of BCAA because the higher AAA concentrations result in greater competition at the BBB for the carrier-mediated transport system used by BCAA. When brains of normal dogs were perfused with AAA a hepatic-like coma and an abnormal neurotransmitter pattern were induced. These findings are, in part, responsible for the "false" neurotransmitter-amino acid hypothesis of hepatic encephalopathy.[18] The hepatic-like coma induced by AAA may be prevented when AAA and BCAA were infused simultaneously.[24]

During liver failure or diversion of the blood around the liver, amines and their amino acid precursors accumulate in the circulation and enter the brain and the peripheral autonomic nervous sytem, replacing the true neurotransmitter norepinephrine with weak or "false neurotransmitters", such as octopamine, tyrosine, phenylethylamine, of phenylethanolamine. On the other hand, patients with liver failure have increased levels of norepinephrine and the other catecholamines.[25] Nonetheless, it is thought that this imbalance of neurotransmitters has serious consequences which may be manifested as brain dysfuction and high cardiac output, low peripheral vascular resistance states, as well as hepatorenal syndrome. Whether amino acid imbalance precipitates encephalopathy, possibly, by promoting the synthesis of toxic aromatic amines, is not clear, but attempts to correct this imbalance have been the rationale for metabolic and nutritional therapy of patients with liver failure in which the protein ration is enriched with BCAA and low in AAA is administered (Tables 12.4-12.5).

Table 12.3. Mechanism that may affect brain function in hepatic encephalopathy

Mechanisms	Consequences
Disruption of integrity of the brain-blood barrier	Exposure of the brain to undesireable neuroactive compounds
Alterations of specific transport systems	Increased transport of neutral amino acids Decreased transport of glucose, ketone bodies, and basic amino acids
Accumulation of neurotoxic substance in the blood	Functional alterations of the brain
Changes in the substrate supply	Alterations and catabolism of normal neurotransmitters in the brain
Lack of nutrients, e.g., glucose	Impairment of brain energy metabolism

The administration of BCAA to rats with portocaval shunts reduces the concentration of tryptophan, serotonin, and 5-HIAA in brain indoles.[20] Formation of "false" or "weak" adrenergic neurotransmitters may contribute to hepatic coma, possibly by displacing norepinephrine and dopamine from their storage granules at nerve terminals.

Nutritional Assessment

Accurate assessment of nutrition status of patients with chronic liver disease is very important, albeit characterized by many encumbrances to performing and interpreting clinical and biochemical studies. Identifying and differentiating the impact of malnutrition from the effects of liver disease and the therapy represent the main predicament. Common methods of determining the nutritional status of these patients have obvious practical limitations. Total body weight and measurements of tricepts skinfold thickness and mid-arm circumference are influenced by ascites and edema. Hypoalbuminemia is an important indicator of nutritional status and a good index of hepatic functional reserve. The absolute rate of albumin synthesis is significantly lower in patients with cirrhosis and correlates well with the Child-Turcotte score and with its Pugh modification. However, when analyzed separately the rate of albumin synthesis does not correlate with serum albumin concentration, intravascular albumin mass, or with other clinical indexes of liver function or integrity. Hypoalbuminemia decreases the colloid oncotic pressure, which contributes to the accumulation of ascites and edema, a persistent characteristic of decompensated liver function. Close metabolic monitoring of these patients is of utmost importance. In addition to routine biochemical indices which include transferring, prealbumin and retinol-binding protein, plasma levels of vitamins, trace elements and apolipoprotein A-IV may be useful in a comprehensive nutritional assessment.

The new metabolic tool that is coming in the clinical practice is the measurements of the hepatic mitochondrial redox potential,[26] but its role in CLD has not been studied to date. The hepatic mitochondrial redox potential represents the ratio of acetoacetate to β-hydroxybutyrate and can be expressed and measured as the arterial blood ketone body ratio (AKBR). The AKBR is not a specific measurement of liver insufficiency, but does allow the severity of liver damage to be graded and reflect the state of perfusion of splanchnic bed. Different clinical and biologic phenomena are associated with a decrease in the AKBR, of which most prominent are enhanced catabolism, impaired oxygen utilization, hypoperfusion and deterioration of the immune response.

Protein malnutrition is a characteristic feature of patient with cirrhosis. Moreover, functional alterations and histologic abnormalities of the liver are known consequences of malnutrition.

Table 12.4. Use of branched-chain amino acids in hepatic encephalopathy: the rationale

When gluconeogenesis and ketogenesis are depressed, BCAAs* may furnish as much as 30% of energy requirements for skeletal muscle, heart and brain.
The BCAAs may regulate the movement of other amino acids across the myocyte membrane.
The BCAAs increase hepatic protein synthesis when given with glucose and decrease aromatic amino acid concentrations.
Theoretically, BCAAs may improve peripheral catecholamine synthesis.
The BCAAs complete with aromatic amino acids for transport across the blood-brain barrier.

*BCAA=Branched-chain amino acids

Table 12.5. General treatment of hepatic encephalopathy

Identify and treat other medical problems
Balance protein intake
Avoid antidepressants and hypnotics unless severe mania is present
Avoid extensive paracentesis or abrupt removal of fluids by dialysis
Avoid vigorous diuresis
Avoid use of acetazolamide
Slowly correct hyponatremia
Do not overhydrate
Monitor hemodynamic status and blood gases closely
Monitor arterial blood ketone body ratio
Avoid lumbar puncture if possible
Supplement vitamins
Avoid use of strong cathartics

Protein deprivation profoundly depletes liver protein stores and adversely affects the breakdown and conversion of polysomes to free ribosomes. On the other hand, in chronic liver disease, alterations in visceral protein synthesis, cellular immunity, and total lymphocyte count may be present independently of protein malnutrition. Plasma protein levels correlate inversely with the degree of liver damage and are its best indicators. Serum concentrations of AAA can indicate the severity of chronic and acute liver disease as well. Furthermore, because patients with hepatic encephalopathy have the lowest BCAA:AAA ratio, this determination may serve as an index of liver function impairment. Other markers such as lean body mass and fat stores are not reliable indicators of structural liver damage. Measurement of nitrogen balance has its limitation, because it is difficult to differentiate impaired hepatic protein synthesis from accelerated breakdown of endogenous protein.

Standard tests of hepatic metabolic capacity and blood flow (aminopyxine breath test, galactose elimination capacity and clearance of idiocyanine green) are less sensitive indicators of prognosis than the Child-Pugh classification of liver disease.[7] The recognized, but less applied clinically prognostic nutritional index, which relies mostly on plasma protein lacks the predictive value in patients with ESLD.[8]

Peritransplant Nutrition: Clinical Studies

A retrospective study of 160 adults with liver transplant malnutrition score was one of the six variables that highly correlated with patient survival. High malnutrition score was associated with increased postoperative morbidity and mortality.[27] The degree of postoperative malnutrition has been shown to predict the postoperative morbidity and mortality in liver transplant patients.[9] This study demonstrated a relationship between moderate to severe malnutrition and increased morbidity (increased requirements for ventilatory support, and ICU and hospital stay) and mortality.[9]

Identifying the correct caloric and protein requirements in patients before and after undergoing LT is not an easy task. In a prospective study of 16 adults patients scheduled to undergo LT, nitrogen balance, 24-hour urinary cratinine, 3-methyl histidine and resting energy expenditure (REE) were determined before transplantation and on days 1,2,5,14 and 28 posttransplant.[28] Only 15% of patients in this study achieved a positive balance, although parental nutrition (containing 1.29 gm/protein/kg, 30% and 70% of calories were given as fat and dextrose respectively) was started within 24 hours. Calculation of malnutritional needs on this study was based on the Harris-Benedict equation, and parenteral feeds were continued until solid oral intake reached 1200 calories in 24 hours in females and 1500 in males. These investigators have suggested to use the Harris-Benedict formula for calculation of nutritional requirements and to add 20% more calories in order to provide nutritional support in LT patients.

In another, most recent study,[29] 150 patients with ESLD undergoing LT prospectively were assessed and followed for an average of 46 months after LT. Body composition analysis (24 hour urinary creatinine excretion, anthropometric measurements and bioelectrical impedance analysis), changes of REE and other variables were analyzed. Patients with severe hypermetabolism (changes in REE >20%) and reduced body cell mass (<35% of body weight) had reduced survival after LT. Based on the degree of hypermetabolism and malnutrition a prognostic risk profile predicted the survival. Patients with high risk profile had a 5-year survival of 88% (P<.01). Others, have also found that nutritional status indices predict survival after LT in children.[30]

Although nutritional assessment in adult patients with ESLD is difficult and complex, and the current indices cannot be applied optimally to all patients, it is clear that hypermetabolism and malnutrition are present in a significant number of patients undergoing LT and that perioperative nutritional and metabolic status of these patients has great prognostic value.

Metabolic Changes Following Liver Transplant

Hypermetabolism as a systemic manifestation of cirrhosis is closely related to splanchnic hemodynamics and malnutrition.[31] Following LT patients are in a hypercatabolic state, however their postoperative metabolic state (as measured by REE) is a reflection of their preoperative state.[32,33] The degree of hemodynamic abnormality, when studied intraoperatively, correlates with the stage of the disease.[34] LT does not reverse the existing splanchnic and systemic hemodynamic abnormalities of ESLD completely.[31] The splanchnic hyperemia persists for at least two weeks, and the closure of porto-collateral circulation occurs at a much slower rate, even though the portal pressure returns to normal.[35] Total liver blood flow markedly increases after LT, although the effects of persistent high liver blood flow on metabolism are not known.[31]

Immediately following LT (in the first 6 hours) glucose utilization by the graft is impaired until mitochondrial redox potential improves.[36] During this period, the liver preferentially uses fatty acid oxidation for ATP generation. After 6 hours, if transplanted livers have normal function, substrate utilization shifts from fat to glucose. On the other hand, failing grafts will continue to utilize fat. These significant metabolic changes may be followed by the serial measurements of arterial ketone body ratio (acetoacetate/3-beta-hydroxybuturate) or AKBR. The AKBR reflects the hepatic mitochondrial redox potential (oxidized nicotinamide adenine dinucleotide (NAD)/reduced nicotinamide adenine dinucleotide (NADH). The ketone body ratio reflects one of the most fundamental regulatory factors of energy production in the liver. The AKBR changes may be used to predict the function of transplanted liver. Low values of AKBR associated with low levels of ketone bodies should be regarded as a strong indicator of graft failure.[37,38] Furthermore, fatty acid oxidation and ketogenic pathways are accelerated to compensate for energy deficits immediately after LT. Administration of small quantities of glucose in post operative period has been suggested.[38] Glucose infusion may be increased as mitochondrial redox potential recovers (AKBR >0.07). Low AKBR values (<0.07) have been associated with increased mortality and morbidity.[39] Following LT low levels of some amino acids such as glutamine, asparagine, citrulline and taurine have been reported.[40,41]

All of the existing studies have involved patients undergoing standard cadaveric donor LT. The use of partial liver grafts is becoming increasingly common, however. These grafts are typically less than 50% of normal adult liver mass. The ability of these organs to utilize nutrient substrates in the early postoperative period has not been systematically studied and the optimal nutritional support for these patients in unknown.

Nutrition Status of Donors: Does it Matter?

The nutritional status of donors has not received adequate attention, although it has been suggested that the nutritional state of the donor may affect the outcome of LT.[42] Often times organ donors receive little, if any, nutritional support during their hospitalization, and are subjected to acute starvation that may last for days in an ICU. This period could substantially deplete the liver energy stores. The role of acute nutritional repletion on the outcome of LT was examined experimentally.[42] Donor pigs were divided into three groups and pretreated for seven days before harvesting their livers. Group I was given intravenous saline; Group II was fed orally regular animal diet; and Group III was fasted, but given 20% glucose intravenously. Harvested livers were stored for four hours in cold Euro-Collins solution before transplantation. The glycogen content of the liver at harvesting was consumed completely in Group I, but was well preserved in Groups II and III. The ATP content of the liver in all three groups were similar at harvesting and were markedly reduced four hours after cold preservation. The amount of ATP recovered one hour after reperfusion was 26% of that before preservation for Group I, and 48% and 73% for Goups II and III, respectively. The mean survival for Group III was 37.2 days vs 5.8 ± 0.7 days and 9.8 ± 2.0 days in Group I and II ($P<0.01$). Other investigators,[43,44] have also shown that livers from nutritionally repleted animals have improved function when compared with livers from fasted animals after cold preservation. It has been demonstrated that glycogen stores could be rapidly repleted in pigs by intraportal infusion of glucose over 3 hours.[45] Studies in humans using this technique[46] have shown that:

1. glycogen stores in human allograft are good,
2. glycogen is needed throughout the transplantation procedure,
3. newly synthesized glycogen is superior to preformed glycogen, and
4. glycogen repletion improves outcome of liver transplantation in humans.

It appears that glucose infusions lower peak transaminase levels, and protects against the effects of prolonged warm ischemia.[46] Rapid hepatic glycogenation was shown to be beneficial in maintaining adenine nucleotide levels in a large animal model.[47] Moreover, livers harvested from hyperglycemic donors receiving insulin infusion during the procurement operation have significantly lower reperfusion hepatocellular injury.[48] On the other hand, others have reported that livers from fasted rats were more tolerant to long-term hypothermia than livers from fed donors.[49,50] More contrasting data were reported recently, were both extensive donor fasting and glucose feeding enhanced outcome in LT.[51]

Nonetheless, prevention of the nutritional status of the donor liver prior to harvesting may be a method to improve function of the livers for transplantation, as the evidence suggests that nutritional status of organ donors matters. Furthermore, nutritional modulation of hepatic reperfusion may have greater influence on liver preservation of LT, and nutritional supplementation may precondition the liver.[52] Until a large human randomized clinical trial answers this question definitively, organ doors should be treated principally the same as other critically ill patients: once fully resuscitated, start the nutritional support, and when possible use the gastrointestinal tract to deliver nutrition.

How to Feed Liver Transplant Patients

Attempts to reverse malnutrition in patients with advanced cirrhosis are very complex and often are compounded by nutrient substrate intolerance and encephalopathy requiring some protein restriction. The principle question to be answered is not do these patients need to be fed, but rather how to do it and when to start? Until recently TPN was the preferred form of postoperative nutritional support in LT as it provides all the nutrients even in a face of significant gut dysfunction. In a randomized prospective, partially blinded, study, 28 patients were studied for seven days following liver transplant.[53] Patients were randomized into three groups: Group I (N=10) was given isotonic intravenous fluids with glucose; Group II (N=8) was given standard TPN providing 1.5 g/protein/kg/day; and Group III (N=10) isocaloric isonitrogenous TPN fortified with 3.5% branched-chain amino acids. All patients were similar with regard to their disease stage, muscle wasting, albumin levels and jaundice. Nitrogen balance was significantly improved in both TPN groups (Group I vs Group II, $P<0.0001$; Group I vs Group III, $P<0.0011$). Furthermore, both TPN groups were extubated earlier than the control and had a shorter ICU stay (an average 2.4 days), although these differences did not achieve statistical significance. This study[53] demonstrates that patients post LT can tolerate aggressive nutritional support and protein load of 1.5 gm/kg/d without evidence of precipitating encephalopathy. Urea nitrogen and creatinine were also not effected by protein intake. Achievement of nitrogen balance in patients with liver transplant has been reported previously.[54,55]

Since TPN has its inherent potential problems (metabolic, infectious, mechanical) the provision of nutrients enterally when possible is an attractive concept. The efficacy and the tolerance of early enteral feeding via a double-lumen nasojejunal tube, placed at the time of transplant, was studied in 14 LT patients.[56] These patients were compared with 10 patients who received TPN as a main of provision of nutritional support. Tube feeding was started within 18 hours after transplantation, while TPN within 24 hours for most patients. These investigators concluded that with early tube feeding after LT, the nutritional status can be maintained as efficiently as with TPN, although nitrogen balance studies were not performed. All patients in this study underwent studies of intestinal absorptive capacity and intestinal permeability before and after transplant. These studies showed no significant differences in intestinal permeability between the enterally and parenterally fed patients within the first seven days of provision of nutritional support.[56] This technique of provision of early enteral feeding may be difficult to achieve in cases when choledocho-enteric biliary drainage is performed. In another study,[57] early enteral feeding immediately after LT was shown to be safe but early postoperative feeding did not effect the ICU or hospital stay, or ventilatory dependence. Thirty one (out of 51) patients completed the study. Tube feeding was tolerated well by 14 patients—and these patients had a better nitrogen balance on post transplant day 4 than control ($P<.03$) and greater cumulative 12-day nutrient intake ($P<0.002$) and lower viral infection rate ($P<.05$).

The major deficiency of this study is that the control group received no nutritional support until they were able to eat.

The means and technique of delivering enteral feedings has been a matter of debate.[58] Because of associated discomfort of the nasojejunal tubes to the patient and other technical difficulties, routine placement of feeding jejunostomies at the same time of transplant has been recommended.[59] Many surgeons avoid this procedure due to the high incidence of complications and infrequent need for prolonged tube feeding. Although 15% of the patients had some type of complication related to placement of J-tubes, placement of these tubes at the time of operation can be done safely, with very low risk of serious complications and without adding significant operative time. J-tubes may, however, be best reserved for those patients who are at risk for prolonged postoperative malnutrition (i.e., prolonged ventilation). Naoenteric tubes are generally tolerated in the short term and do not carry the risk of J-tubes.

Based on these and other studies, enteral nutrition should be the primary mode of nutrition support whenever possible. When to start the feeding is another question. In general if the patients is fully resuscitated as documented by normal lactic acid, base excess and gastric pHi (when measured) enteral nutrition support may be started and advanced as tolerated. On the other hand, TPN should be reserved for patients that cannot be fed enterally or their resuscitation is prolonged and/or complicated by multiple returns to the operating room, for those with prolonged ileus, abdominal distention, abdominal sepsis, or chylous ascites.[60]

Conclusion

Malnutrition is very common in patients undergoing liver transplantation and poor nutritional status is associated with increased morbidity and mortality. Correction and prevention of malnutrition and metabolic support of these patients in their perioperative period is of utmost importance. Enteral feeding should be initiated as soon as possible after LT. Enteral feeding should be used, even when all nutritional needs cannot be met with this technique, and requires supplementation with TPN. Use of specialized nutritional formulas need to be evaluated further, while individualized dietary consult with the patient and the family is of great importance and should be part of the therapeutic interventions.[61] The use of type and the amount of lipids in nutrition support of patients with liver failure and transplant have not been studied. It is clear, however that too much lipids, or when most of lipids are given in a form of omega-6-fatty acids have deleterious effects on the immune status of the critically ill patients. The nutritional therapeutic interventions should be oriented toward reducing patient's protein catabolism and to provide sufficient nutrient substrates in a form of amino acids and dextrose to improve nitrogen balance without aggravating the hepatic encephalopathy.[7]

References

1. Lake JR.ed: Advances in liver transplantation. Gastroenterolo Clin North Am 1993: 22;2.
2. Health Resources and Services Administration, bureau of health resources report of center-specific graft and patient survival rates. Rockville, MD: US Dept of Health and Human Services, 1992:8-10.
3. Clavien PA, Krk AD;Liver transplantation. In Sabiston, DA, Jr, ed, Textbook of Surgery, 15th ed, W.B. Saunders, 1997:461-473.
4. Belle SH, Beringer KC, Detre KM. Trends in liver transplantation in the United States. In Tersaki PI and Cecka JM, eds: Clinical Transplants. Los Angeles, UCLA Tissue Typing Laboratory, 1993.
5. Latifi R, Dudrick SJ. Hepatic encephalopathy: Metabolic and nutritional implications. In: Latifi R, ed. Amino Acids in Critical Care and Cancer. R.G. Landes Company, Austin, 1994.
6. Porayko MK, DiCecco SR, O'Keefe SJD. Impact of malnutrition and its therapy on liver transplantation. Seminars in Liver Disease 1991; 11:4:305-314.
7. DiCecco SR, Wieners EJ, Wiesner RH et al. Assissment of nutritional status of patients with end-stage liver disease undergoing liver transplantation. Mayo Clin Proc 1989; 64:92-102.

8. Pikul J, Sharpe MD, Lowndes R et al. Degree of preoperative malnutrition is predictive of post-operative morbidity and mortality in liver transplant recipients. Transplantation 1994; 57:469-472.
9. Munoz SJ. Nutritional therapies in liver disease. Semin Liv Dis 1991; 11:278-290.
10. Romign JA, Endert E, Sauerwein HP. Glucose and fat metabolism during short-term starvation in cirrhosis. Gastroenterology 1981; 100:1017-1024.
11. Petrides AS, DeFonzo RA: Glucose and insulin metabolism in cirrhosis. J Hepatol 1989; 8:107-114.
12. Riggio O, Merli M, Cantafora A et al. Total and individual free fatty acids concentrations in liver cirrhosis. Metabolism 1984; 33:646-651.
13. Latifi R, Killam RW, Dudrick SJ. Nutritional support in liver failure. Surg Clin North Am 1991; 71:567-578.
14. Morgan MY: The treatment of chronic hepatic encephalopathy. Hepato-gastroenterol 1991; 38:377-387.
15. Horowitz ME, Schafer DF, Molnar P et al. Increased blood-brain transfer in a rabbit model of acute liver failure. Gastroenterology 1983; 84:1003-1011.
16. Mans AM, Biebuyck JF, Shelly K et al. Regional blood-brain barrier permeability to amino acids after portocaval anastomosis. J Neurochem 1982; 38:705-717.
17. James JH, Escourrou J, Fischer JE. Blood-brain neutral amino acids transport activity is increased after portocaval anastomosis. Science 1978; 200:1395-1397.
18. Fischer JE, Baldessarini R. False neurotransmitters and hepatic failure. Lancet 1971; 2:75-80.
19. Cardelli-Cangiano P, Cangiano C, James JH et al. Uptake of amino acids by brain micro vessels isolated from rats after portocaval anastomosis. J Neurochem 1981; 36;627-632.
20. Soeters P, Wilson JHP, Meijer AF et al. Advances in ammonia metabolism and hepatic encephalopathy. Amsterdam: Elsvier, 1988.
21. Howkins RA, Jessy J, Mans AM et al. Effect of reducing brain glutamine synthesis on matabolic symptoms of hepatic encephalopathy. J Neurochem 1993; 60:1000-1006.
22. Munro HN. Interaction of liver and muscle in the regulation of metabolism in response to nutritional and other factors. In: Arias I et al, eds. The liver: biology and pathobiology. New York: Raven Press, 1982.
23. Bergeron M, Layrargues Pomier G, Butterworth RF. Aromatic and branched-chain amino acids in autopsied brain tissue from cirrhotic patients with hepatic encephalopathy. Met Brain Dis 1989; 4:169-176.
24. Rossi-Fanelli F, Freud H, Krause R et al. Induction of coma in normal dogs and its prevention by the additional of branched-chain amino acids. Gastroenterology 1982; 83:664-670.
25. Mizock BA, Sabelli HC, Dubin A et al. Septic encephalopathy. Evidence for altered phenylalanine metabolism and comparison with hepatic encephalopathy. Arch Inter Med 1990; 150:443-499.
26. Yamamoto Y, Ozawa K, Okamoto R et al. Prognostic implications of postoperative suppression of arterial ketone body ratio. Time factor involved in the suppression of hepatic mitochondrial redox potential. Surgery 1990; 107:289-294.
27. Shaw BW Jr, Wood RP, Stratta RJ et al. Stratifying the causes of death in liver transplant recipients. Arch Surg 1989; 124:895-900.
28. Plevak DJ, DiCecco SR, Wiesner RH et al. Nutritional support for liver transplantation: Identifying caloric and protein requirements. Mayo Clin Proc 1994; 69:225-230.
29. Seleberg O, Bottcher J, Tusch G et al. Identification of low-risk patietns before liver transplantation: A prospective cohort study of nutritional and metabolic parameters in 150 patients. Hepatology 1997; 25:652-657.
30. Redox B, Multer M, Kardorff R et al. Liver transplantation in children with chronic end-stage liver disease. Transplantation 1996; 62:1071-1076.
31. Henderson JM: Abnormal splanchnic and systemic hemodynamics of end-stage liver disease: What happens after liver transplantation? Editorial. Hepatology 1993; 17:514-516.
32. Shanbhogue RLK, Bistrain BR, Jenkins RL et al. Increased protein catabolism without heper metabolism after human orthotoic liver transplantation. Surgery 1987; 101:146-149.
33. Muller MJ, Loyal M, Schwarce M et al. Resting energy expenditure and nutritional state in patients with liver cirrhosis before and after liver transplantation. Clin Nutr 1994; 13:145-152.
34. Paulsen AW, Lintmalm GBG. Direct measurement of hepatic blood flow in native and transplanted organs, with accompanying systemic hemodyanamics. Hepatology 1992; 16:100-111.
35. Navasa M, Feu F, Garcia-Pagan JC et al. Hemodynamic and Humoral changes after liver transplantation in patients with cirrhosis. Hepatology 1993; 17:355-360.

36. Ozaki N, Ringe B, Bunzendahl N et al. Ketone body ratio as an indicator of early graft survival in clinical liver transplantation. Clin Transplant 1991; 5:48-54.
37. Takada Y, Ozawa K, Yamaoka Y et al. Arterial ketone body ratio and glucose administration as an energy substrate in relation to changes in ketone body concentration after liver related liver transplantation in children. Transplantation 1993; 55: 1314-1319.
38. Ozaki N, Ringe B, Gubernatis G et al. Changes in energy substrates in relation to arterial ketone body ratio after liver transplantation. Surgery 1993;113:403-409.
39. Shimahara Y, Yamamoto N, Kobayashi N and Ozawa K: Hepatic mitochondrial redox potential and its application in metabolic care and nutrition support in liver failure. In Latifi R, Dudrick SJ, eds: Surgical Nutrition: Strategies in critically ill patietns. Springer-Verlag/R.G. Landes, New York/ Austin, 1995:171-194.
40. Iapichino G, Rodrizzani D, Bonetti G et al. Early metabolic treatment after liver transplant: Amino acid tolerance. Inten Care Med 1995; 21:802-807.
41. Iapichino G, Ronzoni G, Bonetti G et al. Determination of the best amino acid input after orthotopic liver transplantation. Minerva Anestiol 1992; 9:593-508.
42. Sadamori H, Tanaka N, Yagi T et al. The effects of nutritional repletion on donor for liver transplantation in pigs. Transplantation 1995; 60:317-321.
43. Morgan GR, Sanabria JR, Clavien PA et al. Correlation of donor nutritional status with sinusoidal lining cell viability and liver function in the rat. Transplantation 1991; 51:1176-1183.
44. Boudjema K, Lindell SL, Belzer FO et al. Effects of methods of preservation of livers from fed and fasted rabbits. Cryobiology 1990; 28:227-236.
45. Cywes R, Clavien PA, Sanabria JR et al. Glycogen repletion and metabolism during the porcine hepatic allograft retrival and preservation (abstract). Hepatology 1991; 14:574.
46. Cywes R, Greig PD, Sanabria JR et al. Effects of intraportal glucose infusion on hepatic glycogen content and degradation, and outcome of liver transplantation. Ann Surg 1992; 216:235-247.
47. Cywes R, Greig PD, Morgan G et al. Rapid donor nutritional enhancement in a large animal model. Hepatology 1992; 16:1271-1279.
48. Shulman G, Rossetti. Influence of the route of glucose administration on hepatic glycogen repletion. Am J Physiolo 1989; 257:E681-685.
49. Sumimoto R, Southard JH, Belzer Fo et al. Livers from fasted rats acquire resistance to warm and cold ischemia. Transplantation 1993; 55:728.
50. Sankary H, Foster P, Brown E et al. Relevance of the nutritional status of donors in viability of transplanted hepatic allograft. Transplantation 1992; 54:170.
51. Lindell SL. Hansen T, Rankion M et al. Donor nutritional status—a determinant of live preservation injury. Transplantation 1996; 61:239-247.
52. Helton WS: Nutritional issues in hepatobiliary surgery. Sem Liv Dis 1994; 14:140-157.
53. Reilly J, Mehta R, Teperman L et al. Nutritional support after liver transplantation: Prospective study. JPEN 1990; 14:386-391.
54. O'Keefe S, Williams R, Calne R. Catabolic loss of body protein after human liver transplantation. BMJ 1980; 280:1107-1108.
55. Johnson P, O'Grady J, Calvey H et al. Nutrition in liver transplantation. In: Calne R, ed: Liver transplantation. Gune and Straton. New York 1987:113-117.
56. Wicks S, Somasundaram S, Bjarnason I et al. Comparison of enteral feeding and total parenteral nutrition after liver transplantation. Lance 1994; 344:837-840.
57. Hasse JM, Blue LS. Liepa GU et al. Early enteral nutrition support in patients undergoing liver transplantation. JPEN 1995; 19:437-443.
58. Lowel JA. Liver transplant recipient and enteral feeding. (Letter). Surgery 1995; 119:357-358.
59. Pescovitz MD, Mehta PL, Leapman SB et al. Tube jejunostomy in liver transplant recipients. Surgery 1995; 117:642-647.
60. Shapiro JAM, Bain GV, Sigalet D, Kneteman NM et al. Rapid resolution of chylous ascites after liver transplantation using somatostating analog and total parenteral nutrition. Transplantation 1996; 61:140-1411.
61. Brougham TA, Murray NG. Nutrition in chronic liver disease and liver transplant. In: Latifi R, Dudrick SJ eds. Surgical Nutrition: Strategies in critically ill patients. RG Landes: Austin, Springer-Verlag: Heidelberg, 1995:155-170.

Chapter 13

Nutritional Support in Renal Transplantation

Susan T. Crowley, Richard Formica and Antonio Cayco

Because of the high prevalence and adverse impact of malnutrition in the pretransplant end stage renal disease (ESRD) population, the effects of renal transplantation on the nutritional state of the recipient is of interest. This Chapter reviews the available published literature on selected aspects of nutrition in the kidney transplant recipient (KTR). Because of their correlation with increased risk of morbidity and mortality and their high prevalence among KTRs, the treatments of protein malnutrition and dyslipidemia are examined. The impact of vitamin supplementation on modification of another potential cardiovascular risk factor in the KTR, hyperhomocysteinemia is also considered. Finally, because serious morbidity secondary to osteoporosis in the KTR has been recognized, bone metabolism in the KTR is discussed.

Protein Malnutrition and Nitrogen Balance

The attributable impact of the restoration of renal function on nitrogen balance in the KTR has been confounded by the requisite use of immunosuppressive therapy. The latter has been well known to independently influence protein turnover. In particular, glucocorticoids have been repeatedly shown to increase protein catabolic rate (PCR) and result in negative nitrogen balance. Over a decade ago, multiple investigators demonstrated that PCR doubled within 6-11 days post-renal transplantation, independent of protein intake and in parallel with changes in steroid dose. Studies reported that despite restoration of renal function and intake of at least 1.2 gm protein/kg IBW/day, glucocorticoid therapy had a major impact on PCR and was associated with dramatic nitrogen wasting.

Fortunately, it has been demonstrated that protein wasting is a not an invariable consequence of glucocorticoid therapy. In nondiabetic renal transplant patients, increased dietary protein intake (1.5 gm/kg/day) in the immediate postoperative period diet accompanied by 30-35 kcal/kg/day of caloric intake is able to effect neutral nitrogen balance, whereas more protein restricted patients develop negative nitrogen balance. Thus, dietary manipulation is effective in maintaining nitrogen balance, even in the immediate post-transplant period when high dose steroid therapy is typically utilized.

More aggressive dietary protein supplementation has been shown to be a means of further reducing negative nitrogen balance associated with high-dose steroids in the early post-transplant period. In one study, isocaloric diets of approximately 30 kcal/kg/day, adjusted for the differences in the protein content of the diets, were consumed by 12 nondiabetics, randomized to either the control (1 gm/kg/day) or experimental (3 gm/kg/day) diet. Over the study period of four weeks, both groups lost a mean of 3 kg, yet a net increase of 4.5 kg of lean body mass was

Nutritional Support in Cancer and Transplant Patients, edited by Rifat Latifi and Ronald C. Merrell.

noted in the experimental group. For both groups, nitrogen balance was directly proportional to the nitrogen intake. Regression analysis of daily protein intake and nitrogen balance suggested that neutral nitrogen balance was achieved at an intake of 1.3 gm protein/kg/day. Thus, in the setting of adequate caloric intake, it is possible to improve nitrogen balance by protein supplementation without suffering undue side effects such as clinically significant hyperkalemia or azotemia.

Dietary protein intake recommendations in the long term KTR requires a balance between the potentially beneficial and the potentially detrimental effects of a high protein diet. Diminished muscle mass, as evidenced by reduced forearm muscle circumference, has been demonstrated in a significant proportion (38%) of diabetic and nondiabetic patients even two years after successful renal transplantation, when consuming 1 gm/kg/day protein and 25-35 kcal/kg/day. This occurs despite improvement in other nutritional parameters (weight, serum albumin) and minimal steroid use. It has also been shown, by computerized tomographic assessment of mid-thigh muscle area, that marked muscle atrophy occurs in the otherwise stable kidney transplant patient. Given the apparent extent of muscle wasting and protein malnutrition in the stable KTR, dietary protein restriction for any reason would appear questionable.

Some investigators however, have postulated that chronic rejection is a forme-fruste of hyperfiltration induced vascular damage. Over the long term, therefore, a normal or high protein diet might potentially be detrimental to renal graft survival by exacerbating hyperfiltration. A short-term study examining the effect of a low protein (0.55 gm/kg/day) diet on a small cohort of KTRs revealed that urinary protein excretion significantly diminished during low protein dietary intake, and neutral nitrogen balance was achieved. However, significant declines in serum total protein, albumin, as well as transferrin occurred which would suggest that this was too restrictive a diet to maintain overall protein balance.

Whether dietary protein restriction in the long term renal transplant recipient is adequate to maintain nitrogen balance has been investigated by others as well. Among a small cohort of long term nondiabetic KTRs with moderate renal insufficiency consuming a diet containing 0.6 gm protein/kg/day with a caloric intake of 30 kcal/kg/day, no significant changes in inulin-measured glomerular filtration rate or serologic nutritional parameters were noted over a four week follow-up period. While nitrogen balance did not significantly change (baseline = -0.88 ± 1.14 gm/day; at three weeks = -1.59 ± 0.5 gm/day) nitrogen balance did correlate with protein intake ($r=0.42$, $p< 0.05$). Furthermore, caloric intake was positively correlated with nitrogen balance, with near-neutral nitrogen balance being achieved in subjects consuming greater than 25 kcal/kg/day, regardless of protein intake. Thus, there appears to be a threshold for caloric intake (28 kcal/kg/day) which must be met to maintain weight and neutral nitrogen balance.

Because it appears difficult to achieve a protein-restricted diet without compromising essential caloric intake, protein restriction should probably be avoided in the renal transplant recipient. A diet containing at least 1 gm/kg/day of protein and 25-35 kcal/kg/day appears to be the best compromise in the stable, long term renal allograft recipient (Table 13.1).

Dyslipidemia

Cardiovascular disease remains a leading cause of death in renal transplant recipients. In Europe, among long term KTRs, coronary artery disease is responsible for 32% of deaths, approximately equivalent to the number of deaths caused by infection. In the United States, 18% of all long-term KTRs die of coronary artery disease. Despite these findings, there are few studies elucidating the risk factors for cardiovascular disease in the kidney transplant patient.

Table 13.1. Suggested dietary protein intake in the KTR

	Time Period	
	Postoperative	**Long-Term**
Protein	1.5 gm/kg/d	1 gm/kg/d
Calories	30-35 kcal/kg/d	25-35 kcal/kg/d

A recent retrospective review of 54 long term KTRs demonstrated significantly greater LDL and TC levels in KTRs vs. normal controls. Furthermore, blood lipid levels in KTRs with coronary artery disease were significantly greater than in KTRs without coronary disease suggesting an association between blood lipids and atherogenesis in the transplant population, analogous to their relationship in the general population. In another study, a Cox proportional hazard analysis of vascular disease risk factors for a large cohort of both cyclosporine and non-cyclosporine treated KTRs, determined that low serum high density lipoprotein (HDL) level was an independent relative risk factor for ischemic heart disease. Thus, in addition to other traditional risk factors for coronary artery disease, such as diabetes mellitus and hypertension that KTRs may have, it seems reasonable to conclude that elevated serum lipids place these patients at increased risk of cardiovascular disease and that lipid lowering strategies may be beneficial (Table 13.2).

The incidence of hyperlipidemia post transplantation is high; approximating 40% in one large study of over 500 subjects. The cause of hyerlipidemia in the KTR has multiple factors (Table 13.3). Diabetes mellitus, insulin resistance and weight gain all play a role. Additionally, immunosuppressive medications have a role. Prednisone has been shown to increase serum lipids in patients with systemic lupus erythematosis in a dose-related fashion. For each 10 mg increase in dose, prednisone increased TC by 7.5 mg/dl. In KTRs, prednisone has also been shown to increase LDL and TC while reducing HDL.

Blood lipids are also affected by cyclosporine. In a trial evaluating the effectiveness of cyclosporine in treating psoriasis, TC and triglycerides were significantly increased during cyclosporine treatment and diminished after discontinuation of cyclosporine. Studies in KTRs have shown cyclosporine treatment to be associated with a significant increase in serum lipids. However, because of concomitant use of prednisone, the attributable effect of cyclosporine on blood lipids in these studies could not be independently assessed.

With regard to treatment of dyslipidemias, it has been well established that reduction of TC and LDL in hypercholesterolemic patients in the general population significantly reduces the number of coronary events and coronary deaths. This has been shown in both primary and secondary prevention trials. It has also been established that reduction of TC and LDL in patients in the general population who have coronary artery disease but, whose cholesterol is in the normal range by western standards, also reduces coronary artery disease endpoints. In all of these trials success was achieved through the combined use of dietary modification and HMG CoA reductase inhibitors.

The use of the HMG CoA reductase inhibitors, lovastatin and fluvastatin, has been shown to be safe and effective in reducing TC and LDL, and in raising HDL in KTRs receiving azathioprine and/or cyclosporine and prednisone. Of note, in patients being treated with cyclosporine, HMG CoA reductase inhibitors did not change cyclosporine blood levels. Side effects such as muscle aches, frank myositis or laboratory abnormalities in creatinine phosphokinase, AST or ALT were minimized by reducing the maximum dose of the statin in KTRs to

Table 13.2. Cardiovascular risk factors in the KTR

Hypertension
Dyslipidemia—low HDL
Diabetes Mellitus
Hyperhomocysteinemia?

Table 13.3. Causes of hyperlipidemia in the KTR

Medications:
Prednisone, Cyclosporine
Diabetes Mellitus
Insulin Resistance
Obesity

20 mg per day. It should be pointed out that although these medications have been shown to be safe and effective in lowering TC and LDL as well as in raising HDL in KTRs, long-term efficacy and a survival benefit have not yet been demonstrated in this population. It is still a reasonable assumption however, that serum lipid lowering therapy in the hyperlipidemic KTR is beneficial because there is no evidence to suggest that elevated serum lipids in the KTR behave differently than in the general population.

There is limited published research concerning the use of alternative lipid lowering agents in the renal transplant population (Table 13.4). A second line agent that has been shown to be very effective at reducing TC, LDL, and triglycerides, and can raise HDL in cyclosporine and noncyclosporine treated KTRs is nicotinic acid. While most patients in one study tolerated the dose of 1 gm twice a day without undue side effects, conclusions about safety must be limited due to the small number of subjects.

Similarly, experience with the use of the bile acid sequestrant cholestyramine, is limited. Bile acid sequestrants significantly reduce TC and LDL in hypercholesterolemic patients but, there has been concern about impairment of cyclosporine absorption by the sequestrants. In one small study in KTRs, cholestyramine taken as a single 4 gm dose at least 4 hours after the last dose of cyclosporine did not affect the absorption of cyclosporine from the gastrointestinal tract. However, no information on efficacy of lipid lowering was provided. The strict compliance required to properly use cholestyramine, the large amount needed, the somewhat unpleasant taste, and the constipation associated with its use diminish enthusiasm to evaluate this agent further.

Based on epidemiological evidence from the Greenland Eskimo population, who have a very low rate of CAD and a fish-rich diet, fish oil tablets have also been used in an effort to reduce blood lipids. In one study conducted in KTRs however, the administration of fish oil tablets had no effect on TC, LDL, or HDL. Fish oil did significantly reduce serum triglycerides and decrease platelet adhesion in response to ADP which may be beneficial in reducing atheromatous plaque thrombus formation. The only significant side effect from fish oil consumption was fishy eruction, which, overtime, the participants in the study reportedly grew accustomed to.

Table 13.4. Treatment of hyperlipidemia*

Agent	Note
HMG Co-A r.i.s (ex. lovastatin, fluvastatin)	Preferred agent Reduces TC, LDL No change in CYA level Myositis Max dose of statin 20 mg qd
Nicotinic Acid	Reduces TC, LDL, TG Increases HDL Safety?
Cholestyramine	Reduces TC, LDL Impaired CYA absorption? Compliance?
Fish Oil	Reduces TG Decreases platelet adhesion Fishy eruction

* in addition to dietary fat reduction

Vitamin Supplementation

In addition to lipid lowering agents, vitamin supplementation has been proposed as an intervention to lower the risk of coronary artery disease in patients at risk. Folic acid supplementation, in particular, has been the subject of considerable investigation. Folic acid is a requisite factor in one of the metabolic pathways of the sulfur-containing amino acid, homocysteine (HCY), which was first linked with accelerated atherosclerosis nearly three decades ago. In vitro evidence suggests that HCY can both directly and indirectly, via enhancement of lipid peroxidation, produce endothelial injury. Epidemiological investigations have demonstrated a strong correlation between elevated plasma HCY levels (hyperhomocysteinemia) and the frequency of vascular disease. In fact, hyperhomocysteinemia has been shown to be an independent risk factor for primary as well as secondary cardiovascular disease in the general population.

Increased levels of plasma HCY have been demonstrated in a variety of acquired disease states including folate deficiency and renal insufficiency. Present in 75% of ESRD patients in one study, hyperhomocysteinemia was 33 times more common in ESRD patients than in controls, and was far more common (2- to 15-fold) than traditional risk factors for cardiovascular disease. Longitudinal follow up to determine if HCY levels correlate with incident cardiovascular disease rates in ESRD is ongoing.

The effect of restoration of renal function through transplantation and of the administration of folic acid supplements on homocysteine metabolism in the KTR has been recently examined. In 27 long-term, non-cyclosporine treated KTRs with creatinines ranging from normal to 0.5 mmol/l, plasma HCY level in the KTR was inversely proportional to renal clearance. In addition, it was possible to reduce HCY levels via folate supplementation (1-5 mg/day) as similarly demonstrated in normal patients, dialysis patients, and in patients with renal insufficiency.

Immunosuppressive therapy may have an independent effect on HCY metabolism. Specifically, cyclosporine therapy can raise plasma HCY independently of renal function and may abrogate the HCY lowering effect of folic acid via interference with the folate dependent metabolism of homocysteine.

Despite the aforementioned studies, the clinical relevance of hyperhomocysteinemia to cardiovascular disease in the KTR population is unclear. In a previous multivariate analysis of cardiovascular risk factors in stable long-term renal transplant patients, HCY levels were not considered. Past cross-sectional comparison of HCY levels in KTRs with and without cardiovascular disease demonstrated significantly greater HCY levels in transplant patients with cardiovascular disease compared to patients without it. Since comparisons were uncontrolled for differences in time at which sampling of plasma HCY and the cardiovascular event occurred, and lacked matching for other cardiovascular risk factors, conclusions about causality could not be firmly drawn. Hence, the contributory role of hyperhomocysteinemia in the development of cardiovascular disease in the KTR requires further investigation.

Vitamin supplementation with alpha tocopherol, vitamin E, as a means of reducing cardiovascular disease risk has also been a subject of considerable investigation in the general population. In vitro, vitamin E supplementation increases the resistance of LDL to oxidation which may result in improved endothelial cell vasodilator function, decreased foam cell formation, and decreased chemotactic signals for monocytes—all proposed mechanisms which contribute to the atherogenic process. In addition, growing epidemiological evidence suggests that vitamin E supplementation reduces the risk of myocardial events and possibly reduces the risk of death from myocardial infarction in the general population. However, no data currently exists to support the efficacy or safety of supplemental vitamin E in the kidney transplant population.

Bone Metabolism

Kidney transplantation is the second most frequently used modality of renal replacement therapy in the United States, serving the needs of 27.1% of the end-stage renal disease population. As KTRs live longer, long term complications, such as post transplantation bone disease start to impact on patient morbidity.

Osteoporosis is a condition wherein bone density is reduced such that fracture risk is dramatically increased. Significant declines in vertebral bone mineral density (BMD) have been observed not only during the first years after kidney transplantation, but continuously for rates up to -1.7% per year for several years post transplantation. In a recently conducted cross-sectional study among long-term KTRs, the prevalence of osteoporosis at the hip or spine was 42%. Osteoporosis has been reported not only in KTRs but in heart and liver transplant recipients as well. These findings suggest that the loss of bone is quite prevalent and may be attributed not entirely to previous renal osteodystrophy but to transplantation itself.

Various mechanisms have been proposed to explain the pathogenesis of bone loss in KTRs (Table 13.5). Bone mass is governed by the two dynamic and opposing forces of bone resorption and bone formation. Previously, bone loss during the first year post transplantation, was attributed to a decrease in bone formation rate, or a low bone turnover state, as a result of corticosteroid therapy. Recently however, long-term KTRs more than a year after transplantation were found to have a high bone resorption and turnover state. Furthermore, their high bone resorption rate was correlated with reduced BMDs. The etiology of this accelerated resorption of bone tissue remains unknown. Despite the restoration of normal renal function and calcium homeostasis by a functioning allograft, a hyperplastic parathyroid gland could occasionally fail to involute and lead to persistent hyperparathyroidism. This residual hyperparathyroidism could cause the high bone resorption and turnover which would eventually

Table 13.5. Causes of bone loss in the KTR

Residual Hyperparathyroidism
Medications—
Corticosteroids
Cyclosporine
FK506

lead to bone loss. The immunosuppressive drugs, cyclosporine and FK506, have also been shown to cause a high turnover bone loss. Vitamin D deficiency does not appear to be a major cause of reduced bone mass in KTRs with decent allograft function. Several studies have consistently shown adequate stores of 25(OH)-Vitamin D and 1,25 (OH)2-Vitamin D in KTRs even in those with reduced BMD.

Several treatment regimens have been evaluated to reduce the high turnover bone loss associated with transplantation. The agents, calcitonin (40 IU IM daily) or disodium etidronate (400 mg/day p.o. for 15 days every three months), increased vertebral BMD in liver transplant patients compared to the untreated historical control group. In a randomized study, a regimen consisting of 40 µg of 25(OH) -Vitamin D3 and 3 grams of calcium taken daily significantly minimized the reduction in the lumbar, femoral and total body BMD in KTRs. However, these studies are limited by their relatively short duration of follow up. Further studies are needed to evaluate and compare the long-term efficacy and safety of these different regimens in the prevention and treatment of post transplant bone disease.

Summary

Malnutrition is highly prevalent in the pretransplant ESRD population and is associated with increased morbidity and mortality. Hence, attention to nutritional prescription in the perioperative period is particularly important to avoid compounding an already compromised nutritional state. The well known protein catabolic effect of glucocorticoid therapy can be abrogated by ensuring adequate caloric and dietary protein intake. Optimally, the protein and caloric intake in the perioperative period should be at least 1.5 gm protein/kg/day and 30-35 kcal/kg/day.

The optimal long-term dietary prescription of the KTR is controversial. Maintenance of nitrogen balance needs to be balanced by the theoretical risk of hyperfiltration induced by dietary protein loading. In summary, because it appears difficult to achieve a protein restricted diet without compromising essential caloric intake, protein restriction should probably be avoided in the long term renal transplant recipient. A diet containing 1 gm/kg/day and 25-35 kcal/day appears to be the best compromise in the stable renal allograft recipient.

A major cause of mortality in the KTR remains cardiovascular disease which has been linked to markers of malnutrition in other populations, specifically to alterations in lipid metabolism. In conjunction with a low fat diet, careful pharmacological therapy with HMG CoA reductase inhibitors appears to be safe and effective in the short-term treatment of dyslipidemias in the transplant patient. Since there is currently no evidence to suggest that lipids in the KTR behave differently than in the general population, it is reasonable to assume that KTRs will benefit from serum lipid reduction until a treatment outcomes study suggests otherwise.

An alternate nutritional parameter that may be associated with enhanced cardiovascular disease risk in the KTR, hyperhomocysteinemia, should also be considered. In non-cyclosporine treated KTRs, hyperhomocysteinemia can be modified by nutritional supplementation with

Table 13.6. Vitamin and mineral supplementation

Consider:
Folic acid 5 mg qd
Calcium 1-3 g/d
Vitamin D3
Vitamin E ?

folic acid. Since folate supplementation is relatively innocuous, modest repletion (5 mg/day) should be considered in this group. The utility of folate supplementation in the vitamin-replete cyclosporine treated KTR is less clear.

Bone disease remains a significant cause of morbidity in the long-term renal allograft recipient. Further studies are needed to evaluate and compare the long-term efficacy and safety of different regimens in the prevention and treatment of post transplant bone disease. Several therapeutic interventions hold promise for attenuating the debilitating bone-associated side effects of immunosuppressive medications.

Because of limited original research concerning nutrition in the KTR, the true impact of altered nutrition in the KTR is poorly understood. Additional investigations correlating measures of nutrition to specific renal transplant patient and allograft outcomes are needed.

References

1. Ikizler, TA, Hakim RM. Nutrition in end-stage renal disease. Kidney Intl 1996; 50:343-357.
2. Cogan MA, Sargent JA, Yarbrough SG et al. Prevention of prednisone-induced negative nitrogen balance:effect of dietary modification on urea generation rate in patients on hemodialysis receiving high-dose glucocorticoids. Ann Int Med 1981; 95:158-161.
3. Whittier FC, Evans DH, Dutton S et al. Nutrition in renal transplantation Am J of Kidney Dis 1985; 6(6):405-411.
4. Miller DG, Levine SE, D'Elia JA et al. Nutritional status of diabetic and nondiabetic patients after renal transplantation. Am J Clin Nutr 1986; 44:66-69.
5. Horber FF, Zurcher RM, Herren H et al. Altered body fat distribution in patients with glucocorticoid treatment and in patients on long-term dialysis. Am J Clin Nutrition 1986; 43:758-769.
6. Salahudeen AK, Hostetter TH, Raatz SK, Rosenberg ME. Effects of dietary protein in patients with chronic renal transplant rejection. Kidney Intl 1992; 41;183-190.
7. Windus DW, Lacson S, Delmez JA. The short-term effects of a low-protein diet in stable renal transplant recipients. Am J Kidney Dis 1991; 17(6): 693-699.
8. Kasiske BL, Guijarro C, Massy ZA et al. Cardiovascular disease after renal transplantation. J Am Soc Nephrol 1996; 7:158-165.
9. Hilbrands LB, Demacker PNM, Hoitsma AJ et al. The effects of cyclosporine and prednisone on serum lipid and (apo)lipoprotein levels in renal transplant recipients. J Am Soc Nephrol 1995;5:2073-2081.
10. Goldberg R, Roth D. Evaluation of fluvastatin in the treatmentof hypercholesterolemia in renal transplant recipients taking cyclosporine. Transplantation 1996;62:1559-1564.
11. Lal SM, Hewett JE, Petroski GF et al. Effects of nicotinic acid and lovastatin in renal transplant patients: A prospective, randomized, open-labeled crossover trial. Am J Kidney Dis 1995; 25:616-622.
12. Jensen RA, Lal SM, Diaz-Arias A et al. Does cholestyramine interfere with cyclosporine absorption? A prospective study in renal transplant patients. ASAIO J 1995; 41:M704-706.
13. Urakaze M, Hamazaki T, Yano S et al. Effect of fish oil concentrate on risk factors of cardiovascular complications in renal transplantation. Trans Proc 1989; 21:2134-2136.
14. Nygard O, Nordrehaug JE, Refsum H et al. Plasma homocysteine levels and mortality in patients with coronary artery disease. New Engl J Med 1997; 337:230-236.
15. Arnadottir M, Hultberg B, Vladov V et al. Hyperhomocysteinemia in cyclosporine-treated renal transplant recipients. Transplantation 1996; 61(3):509-512.

16. Kirkman RL, Strom TB, Weir MR et al. Late mortality and morbidity in recipients of long-term renal allografts. Transplantation 1982; 34:347-351.
17. Pichette V, Bonnardeaux A, Prudhomme L et al. Long-term bone loss in kidney transplant recipients: A cross-sectional and longitudinal study. Am J Kidney Dis 1996; 28:105-114.
18. Cayco AV, Wysolmerski J, Simpson C et al. Posttransplant bone disease: Evidence for a high bone resorption state. J Am Soc Nephrol 1997; 8:549A.
19. Valero MA, Lonaz C, Larrodera L et al. Calcium and biphosphonates in bone loss after liver transplantation. Calcif Tissue Int 1995; 57:15-19.
20. Talalaj M, Gradowska L, Marcinowska-Suchowierska E et al. Efficiency of preventive treatment of glucocorticoid-induced osteoporosis with 25-hydroxyvitamin D_3 and calcium in kidney transplant recipients. Transplant Proc 1996; 28:3485-3487.

Chapter 14

Total Parenteral Nutrition in Patients Undergoing Hematopoietic Cell Transplantation

Gretchen R. Kilmartin, Joel M. Rappeport and Wendy Holmes

Hematopoietic cell transplantation (HCT) has become a widely accepted treatment modality for many malignant and nonmalignant hematologic disorders. Its use, however, is associated with multiple toxicities resulting from high dose chemoradiation, prolonged immunosuppression with opportunistic infections, and the graft-versus-host disease (GVHD) reaction, all of which can negatively impact upon nutritional status. The nutritional management of HCT recipients can be very complex. HCT recipients often have alterations in metabolism and nutrient utilization, impaired gastrointestinal function and decreased ability to consume adequate oral nutrition. As a result, the use of total parenteral nutrition (TPN) has become a standard part of therapy for most patients.[1-3] Although TPN has been shown to help maintain nutritional status and prevent loss of lean body mass (LBM) in adult[1,4] and pediatric HCT recipients,[5,6] its efficacy in enhancing hematopoietic or immunologic recovery and improving outcome, have not been conclusively established. The provision of TPN has also been associated with its own set of metabolic complications and cost. In an effort to minimize these complications and improve nutritional response and outcome, recent attention has focused on the use of specialized nutrients like glutamine (GLN) which traditionally has not been included in standard TPN formulations. Additionally, enteral nutrition as an alternative to TPN therapy has also been investigated because of its role in both reducing complications and costs associated with TPN.

In this chapter, an overview of HCT and its associated complications which can affect nutritional status will be provided. Nutritional assessment, support and monitoring during both the immediate and long-term phase following HCT as well as novel therapies such as the use of GLN, branched chain amino acids (BCAA), arginine, taurine, and growth hormone (GH) will also be reviewed.

Hematopoietic Cell Transplantation

Indications for Use and Procedure Overview

HCT has become a viable therapeutic option for many patients with a variety of genetic and acquired illnesses due to the refinement of histocompatibility typing in humans and

Nutritional Support in Cancer and Transplant Patients, edited by Rifat Latifi and Ronald C. Merrell.

Table 14.1. Diseases treated with hematopoietic cell transplantation

Acquired Diseases	Genetic Diseases
Aplastic anemia	Thalassemia
Chronic myeloid leukemia	Severe combined immunodeficiency disease
Acute nonlymphocytic leukemia	Sickle cell anemia
Acute lymphoblastic leukemia	Wiskott-Aldrich syndrome
Myelodysplastic syndrome	Infantile malignant osteopetrosis
Multiple myeloma	Adrenal leukodystrophy
Chronic lymphocytic leukemia	Maroteaux-Lamy syndrome
Non-Hodgkins lymphoma	Metachromatic leukodystrophy
Hodgkin's disease	Globoid cell leukodystrophy
Breast cancer	Hurler's syndrome
Neuroblastoma	Hunter's syndrome
Germ-cell cancer	Diamond-Blackfan syndrome
	Gaucher disease
	Fanconi's anemia
	Glanzman's thrombesthemia

improved supportive medical therapy including isolation environments, antibiotics, transfusion support and TPN.[7-10] Diseases which have been successfully treated with HCT are presented inTable 14.1. Based on the primary disease and identification of a potential donor, one of three different types of HCT can be performed. Allogeneic HCT is the most common type and occurs between a patient and a human leukocyte antigen system (HLA) matched related or unrelated donor. Presumably the patient and donor are mismatched at minor non-HLA loci even if they are HLA identical. Another source of allogenic cells is from umbillical cord blood. Allogeneic HCT is most commonly used to treat genetic or acquired hematopoietic disorders, such as aplastic anemia and thalassemia, and hematological malignancies such as leukemia. A syngeneic HCT is performed between identical twins. This is the ideal transplant situation for non-genetic disorders, since the donor and recipient are matched at all genetic loci. Autologous HCT is the third type of HCT which involves harvesting the patient's own bone marrow for hematopoietic stem cells which are then cryopreserved for infusion at a later date. This type of transplant has been used to treat solid tumors, lymphoma and other diseases that do not originate in the bone marrow. Autologous HCT has also been used to treat hematologic malignancies when a suitable donor is not available.

Recently, autologous and allogenic peripheral blood stem cell transplantation (PBSCT) have become available as alternatives to bone marrow harvesting.[11] With this type of transplant, stem cells are harvested from the peripheral blood following cytokine stimualtion instead of the bone marrow. Advantages of PBSCT may include quicker hematopoietic recovery, decreased hospital length of stay, and reduced cost. While there are potential advantages to PBSCT, long term disease-free survival when compared with autologous HCT has not been established.

Many factors can effect the outcome of HCT including the stage and type of disease, genetic disparity between donor and recipient, pre-existing medical conditions, and the age of the patient. Additionally, the presence of pre-existing diabetes or glucose intolerance has been identified as an importnat co-factor for morbidity in these patients.[12]

HCT can be defined as "the intravenous infusion of hematopoietic pleuripotential cells to re-establish marrow function in patients with damaged or defective bone marrow".[7,10] Patients undergo an intense preparative or conditioning regimen prior to marrow infusion. Depending on the type of disease, either hematopoietically ablative dose chemotherapy or combined high dose chemotherapy and total body irradiation (TBI) are used to ablate both the immune system and bone marrow to create total immunosuppression and hematopoietic space for the new marrow to engraft. The donor hematopoietic stem cells are then infused into the recipient through a central catheter allowing the circulating cells to migrate to the hematopoietic microenvironment. The doses of chemotherapy and irradiation given in HCT are much higher than those given with conventional therapy. Therefore, without marrow rescue, mortality would result because of irreversible marrow aplasia and immunosuppression. Unlike solid organ transplants which are capable of functioning immediately after transplantation, it may take 3-5 weeks for full reconstitution of the hematopoietic system or engraftment to occur.[7] A longer period is required for reconstitution of the immune system. Until full marrow recovery has occurred, intensive supportive care is employed which may include reverse environmental isolation; red cell and platelet transfusions; gastrointestinal sterilization with the use of nonabsorbable antibiotics; and specialized nursing and nutritional care.

Complications of HCT

HCT is often accompanied by potential life threatening complications which include chemoradiation toxicity including hepatic veno-occlusive disease (VOD), nonengraftment or graft rejection, bacterial, fungal and viral infections, thrombocytopenia, GVHD, and recurrent disease when malignant disease is treated.[7,10,13] The frequency of these complications varies according to the type of HCT. Many of the toxicities have either additive effects or an interdependent relationship. Chemoradiation toxicity, GVHD, VOD and infection can have a profound impact on outcome as well as nutritional requirements and nutrient utilization, and will be breifly reviewed in the next few sections of this text. More extensive reviews on these subjects are available.[14-21]

Chemoradiation Toxicity

Chemotherapy and TBI induced toxicity occur during or immediately following conditioning and may persist for three to six weeks contributing significantly toward anorexia and poor oral nutritional intake.[2,13,22] Generally, patients who receive both chemotherapy and radiation experience more severe toxicity.[23] Chemotherapy and radiation therapy have a direct effect on both the central nervous system and gastrointestinal system, leading to nausea and vomiting which commonly occur during the first 2 weeks following HCT. Total body irradiation is now most commonly given in multiple fractions in order to minimize these effects. Recently, in some settings low dose irradiation precedes HCT in an attempt to achieve a mixed chimeric state. These mini-transplants may be associated with less toxicity. Conditioning with cytarabine and cyclophosphamide have been associated with a dose-dependent effect on the severity of nausea and vomiting.[2,24]

Oral mucositis resulting from chemoradiation therapy as well as herpes simplex and fungal infection is a very painful condition which usually occurs 4-10 days following conditioning therapy.[22,23,25] Methotrexate which is used as GVHD prophylaxis can further exacerbate mucositis.[26] Esophageal and, gastrointestinal mucositis may also be present, leading to dysphagia, pain, bleeding, nausea, vomiting and diarrhea. Opiates used for mucositis-related pain can also lead to gastric stasis, ileus, anorexia, nausea and vomiting. Xerostomia, dysgeusia and hypogeusia have also been reported.[26-28] Decreased taste thresholds for salty and sweet foods, and a metallic taste have been associated with high-dose cyclophosphamide use[27] and may persist long after HCT.[28-30]

Both high dose chemotherapy and TBI have a devastating effect on the gastrointestinal epithelium leading to villous atrophy, increased intestinal permeability, diarrhea, malabsorption of many nutrients including zinc, lactose and protein.[31-34] The intestinal damage is most significant in the first 2 weeks following conditioning and returns to normal within 3-4 weeks following HCT in the absence of acute GVHD or infection.[32] The severity of diarrhea has also been related to the dose of chemotherapy utilized and is further exacerbated when TBI is added to the conditioning regimen[2,23] even when the radiation is fractionated. Intestinal damage has also been shown to be greater and more prolonged in patients > 30 years old, independent of the type of conditioning regimen used.[32] Despite efforts to minimize these side effects with the use of antiemetics, antidiarrheals, gastrointestinal decontamination, and meticulous mouth care, most patients have minimal oral intake and are maintained on TPN during this catabolic period.

Graft-versus-Host Disease

GVHD is thought to be initiated by the contamination of T lymphocytes in the donor marrow infusion and recognition of foreign transplantation antigens, leading to graft rejection of the host by the graft.[35] Bacterial translocation from the GI tract related to increased intestinal permeability following conditioning therapy, has also been postulated in the initiation of GVHD.[32] Gastrointestinal decontamination with the use of nonabsorbable antibiotics has helped to reduce this bacterial load with an associated reduction in the incidence of GVHD.[36] The severity of GVHD has been associated with the degree of histocompatibility antigen disparity and therefore is seen most commonly following allogeneic HCT, particularly with unrelated transplants.[37] Mild cases of acute GVHD have also been reported in syngeneic and autologous transplants. Older patient age and sex mismatching between donor and recipient have also been associated with an increased risk of significant acute and chronic GVHD.[38] This is particularly significant when the donor is a multiparous female.

GVHD consists of two distinct syndromes. The appearance of acute GVHD usually coincides with engraftment, and may appear as early as 7-10 days following HCT. In 20-66% of allogeneic recipients acute GVHD occurs 20-70 days after HCT, despite GVHD prophylaxis.[38] The chronic phase of GVHD appears 70 days to 1 year following allogeneic HCT affecting 25-50% of long term survivors.[39] Chronic GVHD may appear as a progression, after a quiescent period following resolution of acute GVHD, or de novo.

Acute GVHD is manifested by symptoms that involve the skin, oral and esophageal mucosa, liver, gastrointestinal system, and lymphoid cells, and is associated with increased susceptibility to infection. Acute GVHD can develop in more than one organ system, affecting each to similar degrees. Skin involvement can vary from a mild macular erythematous rash to a severe total body toxic epidermal necrolysis syndrome. Nitrogen losses from the skin can be extensive in the severe form. Mild to moderate elevations of hepatic enzymes and bilirubin levels occur in hepatic GVHD, accompanied by cholestatic jaundice and hepatic dysfunction in more serious forms of the disease. Hypoalbuminemia is commonly due to intestinal protein loss rather than decreased synthetic production.[40,41] Fat malabsorption may occur in the presence of cholestasis which can complicate oral nutrition.

Intestinal acute GVHD has the most pronounced effect on nutritional status. Intestinal GVHD is most prominent in the ileum, cecum, and ascending colon, but can affect the upper gastrointestinal tract as well.[22,33,34,37] The symptoms of intestinal GVHD include: nausea, vomiting, abdominal cramping, secretory diarrhea, altered intestinal motility, and hypoproteinemia.[22,23,34,37,40-42] Secretory diarrhea, with its characteristic watery, green, guaic positive, mucoid consistency, is the most common manifestation. Diarrheal losses can vary between 500 ml/day to 2-10 l/d in the severe form of the disease. In a few cases somatostatin analogues like octreotide, have been shown to decrease stool volume.[22,43,44] Reductions of 30-50% in stool volume have been observed when octreotide is given by continous infusion.[22] Malabsorption of many nutrients is common including lactose, protein, fat, zinc, and other vitamins and minerals.[41,42] Steatorrhea occurs as a result of decreased ileal reabsorption of bile contributing to decreased bile salt availability for emulsification and absorption of long chain fats.[42] Steatorrhea may be further exacerbated in the presence of liver GVHD because of decreased bile salt synthesis and altered enterohepatic circulation. Nitrogen losses are severe due to the protein rich content of the diarrheal fluid,[41] and capillary leakage in other parts of the body.[40] Nitrogen-losses are further exacerbated in the treatment of acute GVHD, which includes the use of high dose steroids. Nitrogen balance may become extremely negative because of the catabolic effects of steroid therapy and the increased metabolic demands associated with acute GVHD.[45] Furthermore, prednisone, methotrexate and cyclosporine A (CSA), used as GVHD prophylaxis, may also impact on gastrointestinal toxicities. TPN is often a necessity when severe intestinal GVHD is present in order to impede loss of LBM. Generally oral intake can begin when stool output is less than 500ml/day with the use of a diet low in fat, fiber, and lactose.[42]

Chronic GVHD can affect the skin, liver, mouth, esophagus, eyes, lungs and vagina and less frequently affects the intestinal and musculoskeletal systems.[37] The presence of anorexia, dysphagia, mucositis, dysgeusia, xerostomia, esophageal strictures, cholestatic liver disease, and steatorrhea can adversely affect nutritional intake. Metabolic abnormalities may also occur such as increased work of breathing associated with chronic pulmonary GVHD. The prolonged use of steroids and other immunosuppressive agents in this population can also lead to altered body composition, characterized by skeletal muscle wasting and fluid retention, and diabetes. Immunosuppression may result in recurrent and prolonged infections. Calcium absorption can also be altered with long-term steroid use leading to osteoporosis and avascular necrosis which can limit exercise tolerance and restrict joint mobility. Patients with chronic GVHD may require specialized nutrition support if weight loss, hypoalbuminemia and/or insufficient intake and utilization of nutrients occur. Unlike acute GVHD, enteral feedings should be the preferred mode of nutrition support as intestinal malabsorption is less common and therapy may be required for prolonged periods.

Veno-Occlusive Disease

VOD of the liver occurs as a consequence of conditioning therapy leading to narrowing and fibrosis of the hepatic venules and secondary injury to the hepatocytes.[22,46] VOD of the liver is seen in 20-60% of HCT recipients and usually occurs during the first few weeks following transplantation.[14,22] Symptoms of VOD include increased bilirubin levels, hepatomegaly, weight gain and ascites, hepatic pain, and in severe cases, hepatic encephalopathy, and heptorenal syndrome. The severity of VOD varies from mild, reversible disease, to fatal disease associated with multiorgan failure.[46] The majority of patients develop mild VOD, while approximately 25% of patients will develop life-threatening disease.[47] In an effort to reduce both severity and incidence of VOD, prophylactic continous low-dose heparin infusion has been investigated with inconclusive results.[48-50] Despite this lack of unequivical efficacy many centers are using prophylactic heparin, and in our experience, while the overall incidence of VOD seems unchanged, there seems to be a decrease in the severity of VOD. Supportive treatment is aimed at maintaining fluid and electrolyte balance with use of diuretics and sodium and fluid restrictions. Protein restriction should be considered if hepatic encephalopathy is present. VOD and its treatment may have a significant impact on the ability to provide adequate nutritional support.

Infectious Complications

Infection occurs in most HCT recipients at some time during the post-transplant course as a result of marrow aplasia, the use of immunosuppressive drugs, and fetal recapitulation of the immune system.[51] Damage to the cutaneous and gastrointestinal "barriers" also contribute to infection. Bacterial, viral, and fungal infections are the most commonly observed infections following HCT.[13,52-54] Bacterial and fungal infections occur during the granulocytopenic phase early after transplantation. Viral and fungal opportunistic infections may occur 6-25 weeks following HCT due to prolonged immunoincompetence, frequently in the presence of GVHD. The presence of chronic GVHD results in opportunistic infections for prolonged periods of time, often years. These infections may include cytomegalovirus (CMV), herpes simplex, varicella zoster, fungal, adenovirus, *Pneumocystis carinii* infections, and interstitial pneunomias, and are associated with increased mortality. In the absence of chronic GVHD the immune system is reestablished by one-year post-transplant resulting in a decrease of infectious susceptibility.[51] The systemic effects of infection such as fever, may impact on nutritional status by increasing metabolic demands and nitrogen breakdown, and at the same time may suppress appetite. Nutrient utilization may also be altered during infection due to the systemic stress response precipitating hyperglycemia and glucosuria. The use of multiple antibiotics to treat infection can also cause anorexia, nausea, vomiting, diarrhea, renal dysfunction, and electrolyte imbalances, which may further complicate nutritional management.[2,30]

In an effort to prevent early infections, protected environments, application of topical antibiotics to skin and orifices, and gastrointestinal sterilization nonabsorbable antibiotics have been utilized. Later, prophylactic acyclovir, gangciclovir, trimethoprim-sulfamethoxazole and penicillin therapy may be utilized as well as replacement gamma globulin infusions. The use of sterile and low microbial diets to prevent food borne infection is controversial and may lead to further reductions in oral intake due to the unpalatable nature of these diets.[55-58] Because HCT recipients are at risk for the development of food borne illness with potentially serious consequences, many centers use a modified low microbial diet or neutropenic diet, omitting foods which may harbor pathogenic bacteria. Safe food handling practices are also taught to patients and families to further minimize risk of food borne illness post discharge.

Nutritional Support for HCT Recipients

The goal of nutrition support in these complex clinical settings is often maintenance rather than repletion of nutritional status, as the majority of patients are well nourished prior to HCT.[4,5,59,60] It is not uncommon for patients to purposely gain weight before admission in anticipation of poor oral intake and weight loss during their treatment. Throughout the course of HCT however, nutritional status is often compromised due to the cytotoxicity of conditioning therapy, acute and chronic complications that may arise, and the side effects of medications utilized in their treatment. Medication interactions are of particular importance. There are numerous medications which cause gastrointestinal discomfort, electrolyte and organ dysfunction, and may impact on nutritional status. Some examples include nonabsorbable antibiotics and oral magnesium supplementation which frequently contribute to diarrhea. As a result, HCT recipients often have decreased oral intake, increased nutrient requirements, and impaired utilization of nutrients, which necessitates individualized and often changing nutritional intervention. The nutritional support of these patients can be complex. An interdisciplinary approach involving dietitians, doctors, nurses and pharmacists is essential for provision of optimal nutritional support. In the next few sections of this text, nutritional assessment and nutritional support will be reviewed including TPN provision, composition, and associated complications, as well as the potential role of enteral nutrition.

Table 14.2. Suggested baseline nutrition assessment parameters

Height, growth history in children
Weight (usual and current), weight change in the last six months, % ideal body weight
Previous treatment-related side effects
Past medical history: renal/hepatic disease, family/personal history of diabetes, alcohol intake
Albumin, prealbumin, cholesterol and triglyceride levels
Diet history: Supplement use, food allergies and intolerances, modified diet, cultural dietary habits
Energy/activity level
Medications
Anthropometrics if serial measurements are planned

Nutrition Assessment

An initial nutrition assessment should be completed prior to conditioning therapy to detect pre-existing nutritional deficiencies and medical conditions that may affect medical nutrition therapy. The initial assessment should include laboratory and anthropometric data, as well as a complete diet history. Table 14.2 provides suggested guidelines for the initial nutrition assessment. Because nutrition assessment is an ongoing process, monitoring nutrition status and response to nutritional support is essential for optimal long-term nutritional management. Suggested monitoring parameters are provided in Table 14.3. A review of commonly used nutrition parameters and their limitations are provided below.

Weight changes often occurring over short periods are more reflective of changes in fluid balance and have been shown to correlate poorly with changes in body cell mass (BCM) or lean body mass (LBM).[2,3,30,45] Other anthropometric measurements (skin fold thickness, and limb circumference) are also affected by fluid balance and have been found to be insensitive and inaccurate in predicting body composition in HCT,[2] but can be useful if long term follow-up is planned.[62] Measurement of body composition using more sophisticated methods (bioelectrical impedance, isotope dilution, neutron activation analysis, and total body potassium) are more useful than obtaining weight,[4,61,62] but are impractical and expensive for routine use.

Biochemical indicators of nutritional status such as albumin, transferrin and prealbumin have also been studied in HCT. Albumin has been found to be an insensitive marker of nutritional status in HCT because of its long half life (20 days) and its sensitivity to infusion of blood products, volume changes and steroid therapy.[2,3,30,61] Transferrin which has a half life of 9 days, is more reflective of amino acid intake but is also affected by changes in hydration status. Prealbumin has been found to be a more sensitive marker of nutritional status due to its short half-life (2-3 days), and because it is not readily affected by changes in volume status or the infusion of blood products.[63] Prealbumin levels have also been shown to correlate well with changes in BCM in children and adult HCT recipients,[5,61,64] but can also be affected by non-nutritional factors such as fever and stress, and liver disease. Prealbumin levels must be interpreted with caution if steroid therapy is initiated or if renal disease develops, as both conditions may falsely elevate prealbumin levels.[65,66] Monitoring prealbumin levels for trends can still be useful during these conditions, especially if the steroid dose remains relatively stable. Nitrogen balance studies are useful in determining both degree of metabolic stress and protein requirements but require accurate collection of urine, stool and vomitus. Nitrogen balance has also been found to correlate with changes in BCM.[61] Limitations to the usefulness of this test include the cost and inaccuracies inherent in the collection of urine and stool.[67] Because of these limitations and the tendency for nitrogen balance to remain negative despite the provision of nutrition support, routine monitoring of nitrogen balance is not warranted. Clinical judgment also plays a key role in the interpretation of nutrition assessment parameters.

Table 14.3. Suggested guidelines for monitoring nutrition support

Daily	Biweekly	Weekly
Weight	Liver function tests	Prealbumin
Intake/Output	Calcium	
Blood/Urinary glucose		Triglycerides
Electrolytes		Cholesterol
Phosphorus, magnesium		
Calorie/Protein intake		

Caloric Requirements

Several investigators have studied caloric requirements in HCT using either indirect calorimetry or predictive equations.[45,68-73] Data from these studies are presented in Table 14.4. Indirect calorimetry has long been recognized as the most precise way to determine resting energy expenditure (REE) and estimate caloric requirements but may be impractical and expensive for routine repeated use. For adults, in the absence of indirect calorimetry, the Harris-Benedict equation has been used extensively to estimate basal energy expenditure (BEE), and has been found to correlate with REE in some studies.[68,69] Energy requirements have also been found to increase following HCT. During the first 1-3 weeks following marrow grafting, stress factors of 1.3-1.5 times the BEE have been used to estimate caloric needs.[30] Others recommend estimating caloric needs based on 30-35 kcal/kg ideal body weight (IBW) following HCT.[30,70] For children, energy requirements are routinely determined using the Harris-Benedict equation with stress factors of 1.6-1.8 times the BEE for the first 1-3 weeks following marrow grafting, and 1.4 times the BEE for maintenance.[74] The Seashore equation has also been frequently applied to children requiring treatment in intensive care settings.[75] Szeluga et al[45] found that 45-65 kcal/kg/day were required by children, and 30-50 kcal/kg/day were required by adolescents and adults to maintain zero nitrogen balance. Additionally, higher caloric requirements were observed in males and in patients with acute GVHD. These requirements may be related to body composition differences and the catabolic affect of high-dose corticosteroid therapy used in the treatment of GVHD, respectively.

While maintaining nitrogen balance is certainly optimal, negative nitrogen balance may be an unavoidable consequence of HCT and relative inactivity, regardless of the level of caloric or protein intake provided.[61,70-72] A trend towards lower caloric requirements is evident in recent studies that have used indirect calorimetry in adult recipients. One group of investigators[70] found respiratory quotients >1.0 and energy requirements significantly less than the 40/ kcal/kg, HCT recipients received based on a predictive equation,[45] suggesting that overfeeding was likely. Another group of investigators[71] observed similar nitrogen balances between groups and improved serum albumin levels in HCT recipients given 25 kcal/kg and .8g protein/kg versus 35 kcal/kg and 1.4 g protein/kg. Energy requirements may be elevated following HCT because of fever, infection and steroid use. However, the thermogenic effect of providing nutrition support should not he ignored as energy requirements have been shown to correlate with increased energy intake following HCT in a few studies.[45,68] Caution must be taken to avoid the negative effects of overfeeding with parenteral nutrition which include liver function test abnormalities and steatosis, hyperglycemia, fluid and metabolic imbalances, and compromised respiratory function.[71,76]

Table 14.4. Caloric requirements of hematopoietic cell transplant recipients receiving total parenteral nutrition

Study	Study Design	Length of Study	n	Results
Szeluga et al (45)	Predictive equation developed to estimate caloric intake needed to maintain zero nitrogen balance in allogeneic/autologous recipients	30 days	84	45-65 kcal/kg/day required by children 30-50 kcal/kg/day required by adults Higher needs found in males, acute GVHD, and patients who received the majority of their kcal intake from TPN
Geibeg et al (70)	Randomized, controlled, double-blind study comparing an increased nitrogen load in syngeneic/allogeneic recipients. Indirect calorimetry used to measure REE and compared to a predictive equation	19 days	28	Higher respiratory quotients found in all patients who received 40 kcal/kg/ 40 kcal/kg/IBW REE was significantly less than predicted energy intake
Taveroff et al (71)	Prospective, controlled trial, comparing metabolic effects of a reduced calorie and protein TPN in allogeneic/autologous recipients	24 days	15	Nitrogen balance was similar between groups Improved serum albumin, minimized sodium/potassium disturbances and greater weight loss were found in reduced kcal/protein TPN group
Hutchinson et al (69)	Prospective, controlled trial, comparing REE with BEE at 6 different time points following allogeneic HCT	25 days	13	Mean REE = 24 kcal/kg/day REE correlated with BEE REE increased by 5% three weeks after HCT (not significant)
Tomalis et al (73)	Prospective trial comparing REE with BEE at 6 time points following autologous HCT for breast cancer	22 days	9	REE did not correlate with BEE REE was not affected in the pre or post phase of HCT
Peters et al (68)	Prospective trial comparing REE with BEE at 6 time points following allogeneic HCT	21 days	6	REE correlated with BEE (23 kcal/kg/day) Peak rise in REE occurred 7 days following HCT

REE = resting energy expenditure
BEE = basal energy expenditure, estimated using the Harris-Benedict equation

Many of these studies used small numbers of patients and in some cases were not randomized or blinded, which limits the validity of these data. Further randomized, controlled, large-scale studies are needed to more accurately estimate caloric requirements for this population, especially for children. Caloric requirements may vary considerably between individuals based on their medical status, body cell mass, and caloric intake. A caloric intake of 30-35 kcal/kg/IBW is adequate for most adult recipients. Caloric requirements should be adjusted upwards to 35-40 kcal/kg/IBW in the presence of GVHD and infection, as clinically tolerated. However, the complications of hyperglycemia and hypertriglyceridemia associated with acute GVHD and concomitant steroid therapy often limit the provision of extra calories. Parenteral caloric requirements should be decreased following engraftment in the absence of infection and GVHD, and with the resumption of oral nutrition towards maintenance, or 25 kcal/kg/IBW.

Protein Requirements

A number of investigators have studied protein metabolism requirements in HCT which are presented in Table 14.5.[60,61,70,72,77,78] HCT recipients have been found to have increased protein requirements and negative nitrogen balance despite the provision of what appears to be adequate nutrition support.[61,70,72] Nitrogen losses occur following HCT due to the catabolic effects of cytoreductive therapy[70,72] and increased nitrogen losses from the intestinal tract and skin.[32,40] The magnitude of nitrogen loss and catabolism has been shown to increase dramatically following HCT. One group of investigators[70] found that urinary urea nitrogen (UUN) values more than doubled from baseline, reaching a mean value of 17g/day four days following HCT. Additionally, UUN values approached 20g/day by the second week following HCT, with the addition of IV methylprednisolone (.25mg/kg BID) as part of GVHD prophylaxis. This level of nitrogen loss has been associated with significant stress related catabolism.[79] Calculated nitrogen balance results have also been shown to consistently become more negative over time following HCT,[61,70,77] especially in male allogeneic HCT recipients, independent of the level of stress.[77] Sex differences in nitrogen balance are thought to be related to the greater proportion of muscle mass in men, suggesting that males may require greater amounts of protein to maintain nitrogen balance.

In two recent randomized, controlled studies, provision of a high protein TPN (2.0g/kg/IBW) compared with a standard protein TPN (1.5g/kg/IBW)[60,70] did not significantly improve nitrogen balance between groups at any time period. However, an overall group effect of significantly more positive nitrogen balance values was observed in one study,[70] suggesting that a higher nitrogen formula may impede loss of LBM. The nitrogen sparing effects of supplemental GLN, BCAA and arginine will be reviewed in a subsequent section of this text.

Although negative nitrogen balance appears to be an inevitable consequence of HCT, it is beneficial to provide increased protein to these catabolic patients to help repair damaged endothelial tissue. Provision of at least 1.5 g/kg/IBW is recommended for adult recipients.[2,30] In the presence of GVHD and steroid therapy, protein requirements may increase further, up to 2.0 g/kg/IBW, as intestinal and urinary nitrogen losses can be excessive. For children, protein needs are commonly estimated to be twice the normal recommended dietary allowance (RDA) for age. Protein provision may need modification if significant hepatic or renal dysfunction develops.

Table 14.5. Protein requirements of hematopoietic cell transplant recipients receiving total parenteral nutrition

Study	Study Design/Intent	n	Protein intake	Caloric Intake	Results
Keller et al (72)	Studied protein metabolism by measuring leucine kinetics following allogeneic HCT	6	1.8 g /kg /IBW	35 kcal/kg/IBW	Negative nitrogen balance occurs because of increased protein metabolism
Cheney et al (61)	Studied body composition changes using isotope dilution, also measured nitrogen balance analysis following allogeneic HCT	9	1.5 g /kg/IBW	BEE x 1.6	Negative nitrogen balance occurred throughout the study Nitrogen balance correlated with changes in BCM
Geibeg et al (70)	Randomized, controlled, double-blind study comparing an increased nitrogen dose on nitrogen balance in allogeneic and syngeneic HCT recipients	28	2.0 g /kg/IBW versus 1.6 g /kg/IBW	40 kcal/kg/IBW	No significant differences in nitrogen balance were observed between groups The higher nitrogen group maintained more positive nitrogen balance values, accounting for the overall group effects

continued on next page

Study	Study Design/Intent	n	Protein intake	Caloric Intake	Results
Constantino et al (60)	Randomized, controlled study comparing an increased nitrogen dose on nitrogen balance in HCT recipients	33	2.3 g/kg/IBW versus 1.4 g/kg/IBW	40 kcal/kg/IBW	Negative nitrogen balance occurred in both groups independent of the level of protein provided
Lenssen et al (78)	Randomized, controlled, double-blind study on the effect of BCAA (45% vs 23%) supplementation on nitrogen balance in allogeneic HCT recipients	19	1.5 g/kg/IBW	BEE X 1.5-1.65	No significant differences in nitrogen balance were observed between groups
Cheney et al (77)	Prospective study to investigate sex differences in nitrogen balance in allogeneic HCT recipients	40	1.5 g/kg/IBW	BEE X 1.5-1.65	Nitrogen balance was significantly lower in males, even when controlled for the level of stress

IBW = ideal body weight
BEE = basal energy expenditure, Harris-Benedict equation
BCM = body cell mass
BCAA = branched chain amino acids

Route of Nutritional Support

Enteral versus Parenteral

The benefits of enteral nutritional support (oral or tube feeding (TF)) over TPN, have been clearly established. These benefits include lower cost, fewer complications, improved nutrient utilization, and perhaps decreased bacterial translocation in critical illness and stress.[80-83] Enteral nutrition is often not feasible as the sole means of support in the acute period following HCT because of the well known side effects of cytoreductive therapy and /or gastrointestinal GVHD. Several investigators have demonstrated the inability to maintain nutritional status with only an oral diet because of gastrointestinal toxicity.[1,4,59,84] Often, placement of nasoenteric feeding tubes is not performed because of gastrointestinal toxicity, increased risk of infection associated with feeding tube placement, and formula contamination. The use of TPN is also more convenient because patients usually have permanent central access devices in place. However, some feel that TPN should not be routinely given to all HCT recipients.[4] Two recent prospective, randomized studies,[4,84] have investigated the use of either a specialized enteral feeding program or a combined enteral/parenteral program as an alternative to TPN in adult HCT recipients.

Szeluga et al[4] compared the effects of a specialized enteral feeding program versus TPN in 61 allogeneic and autologous HCT recipients. The specialized enteral feeding program consisted of supplementation with IV amino acids (.5 g/kg/day) until subjects were able to eat an equal amount of oral protein, individualized dietary counseling, supplements and/or tube feeding. Although the enteral group consumed significantly less and had a significant decrease in BCM, no significant differences in outcome between groups were found. These authors concluded that TPN should not be routinely prescribed for all HCT patients but should be reserved for those who are intolerant to enteral feeds. It should be noted that there were 7 patients (23%) who failed enteral feedings due to severe gastrointestinal intolerance and failed TF attempts, and were crossed over to receive TPN. Additionally, 22 patients (73%) in the enteral group required IV amino acid supplements for an average of 7 ± 5 days, due to failure to consume more than .5 g/kg/d of protein. These data suggest that while some pateints may be supported with enteral feeding alone, most patients require some form of supplemental parenteral nutrition.

Mudler et al[84] compared TPN to a combination of partial parenteral nutrition (PPN) and enteral nutrition (EN) (nasogastric tube feedings) in 22 autologous HCT patients. Both treatments were effective in maintaining body weight and nitrogen balance. There were no tube feeding related complications. The percentage of days with diarrhea was significantly lower in the PPN/EN group, suggesting that enteral nutrition may have had a beneficial effect on the recovery of intestinal mucosa function. A higher incidence of positive blood cultures were found in the PPN/EN group but did not reach statistical significance. Enhanced invasion of bacteria into the blood stream with feeding tube insertion was postulated as being responsible for the higher incidence of positive blood cultures. These authors concluded that a combined approach to feeding autologous HCT patients is an acceptable alternative to TPN.

Based on these limited data, the majority of adult patients, and presumably children who require nutrition support for growth as well as maintenance, clearly can not be supported on enteral nutrition alone in the acute period following HCT. Early enteral nutrition may be more realistic for use in autologous HCT recipients because these patients generally experience milder gastrointestinal toxicity. Allogeneic HCT recipients who are no longer neutropenic (generally > 30 days after HCT) with failure to thrive and/or chronic GVHD, and those who develop adult respiratory distress syndrome requiring intensive care unit treatment have been successfully treated with enteral nutrition support at some centers.[85,86] When the decision has been made to use the enteral route, feeding tube placement should be presented to patients and families in a positive manner to enhance acceptance. Emphasis should be placed on safety, ease of medication administration, and decreasing the pressure that patients feel to eat and drink adequately.

Total Parenteral Nutrition

Initiation and Composition

Initiation

Because of the profound effects of cytoreductive therapy, most patients require TPN during the first 30 days following HCT. Those who develop acute GVHD may require longer and more intensive supplementation with TPN. The timing of TPN initiation remains controversial. At most centers, TPN is initiated following conditioning therapy or the day following HCT (day 0) in patients who demonstrate an inability to consume > 50% of their caloric requirements orally. In patients with preexisting malnutrition and in pediatric patients, earlier initiation of TPN is suggested. Early provision of TPN in normally nourished HCT recipients however, may also be beneficial. Improved long-term survival was found in 137 HCT recipients given prophylactic TPN one week prior to HCT in a randomized, controlled study.[59] The interactions between nutrients in the TPN and cytoreductive therapy were speculated to have an increased antitumor effect and may have modulated lymphokine production leading to increased survival in these patients.

Composition

Specific guidelines for optimal macronutrient, electrolyte, vitamin and trace element provision in TPN are undefined. General recommendations have been extrapolated from studies done in other populations. Requirements vary considerably among patients depending on their clinical status and the medications used in their treatment.

Carbohydrate

Dextrose is the preferred calorie source and provides 3.4 kcal/gm. The amount of dextrose provided depends on caloric needs, but usually supplies 50-60% of total calories and can be given in concentrations of 15-30%.[87] Caution should be taken to avoid overfeeding with dextrose which can cause hyperglycemia, significant impairments in oxygenation, fatty infiltration of the liver, and cholestasis.[76] The maximum glucose utilization rate for adults (5 mg/kg/minute) should not be exceeded, to prevent overfeeding. In children, the maximum thresholds are as follows: infants 15 mg/kg/minute; toddlers, 10-15 mg/kg/minute; preschool/school age, 8-10mg/kg/minute and teens, 6-8 mg/kg/minute.[88]

Protein

Standard crystalline amino acids containing essential and nonessential amino acids are used in a variety of concentrations ranging from 3.5-15%. More concentrated amino acid solutions are helpful to minimize the volume of TPN, which may be necessary in VOD or renal failure. Recent studies have suggested that particular amino acids which are found in limited amounts or omitted completely from standard amino acid formulations such as glutamine (GLN), arginine (ARG), taurine, and branched chain amino acids (BCAA) have important immunologic roles and may alter the metabolic response in critical illness.[89,90] Use of these amino acids may improve the clinical and metabolic efficiency of TPN.

BCAA have been used in critical illness because of their proposed nitrogen sparing effects.[89,91] Their use when tested in a randomized, double blind study in 19 HCT recipients, did not result in improved nitrogen balance.[78] However, the number of patients used in this study was small and therefore the use of BCAA in HCT remains controversial. Additionally, these amino acid solutions are significantly more expensive than standard solutions. Therefore, routine administration of BCAA in HCT recipients is not warranted at the current time.

Glutamine is the most abundant amino acid in the plasma and free amino acid pool and is the preferred fuel for rapidly dividing cells such as enterocytes, colonocytes and lymphocytes.[92,93] Traditionally, GLN has been classified as a nonessential amino acid and therefore has not been included in standard enteral/TPN formulas. However, during periods of physiologic stress induced by critical illness, GLN levels fall markedly because of both increased utilization and inadequate GLN synthesis,[94,95] suggesting that supplementation with GLN may be beneficial. Based on this premise, many now consider GLN a "conditionally" essential amino acid in critical illness.[92] Recent studies in animals have shown that GLN supplemented parenteral and enteral nutrition improves intestinal mucosal integrity, reduces bacterial translocation, improves nitrogen balance and improves survival after chemo- and radiation therapy when compared with GLN free nutrition support.[96-99] GLN supplemented TPN has also resulted in improved nitrogen balance in surgical patients.[100-101]

Because HCT induces a state of physiologic stress and has a profound negative effect on gastrointestinal function, the use of GLN supplemented TPN in HCT recipients has recently been investigated in three, randomized, controlled, double blind trials.[102-104] Significant improvements in nitrogen balance, a decreased incidence of clinical infection, and a shortened length of hospital stay were found in 45 allogeneic HCT recipients randomized to receive TPN supplemented with GLN.[102] Similarly, a decrease in the length of hospital stay was found in 29 autologous and allogeneic HCT recipients randomized to receive GLN supplemented TPN.[103] These latter investigators did not find significant differences in nitrogen balance and incidence of clinical infection which may have been related to the generally sicker population studied. GLN supplemented TPN has also been shown to reduce the fluid retention and edema commonly associated with HCT.[103-104] Schletinga et al[104] observed reductions in both fluid retention and edema, and decreased incidence of infection in 20 allogeneic HCT recipients who received GLN supplemented TPN. These authors speculated that restoration of normal body fluid compartments may have attenuated microbial invasion and infection.

The mechanism for GLN's proposed actions in HCT are unknown. Zeigler et al[102] postulate that GLN may promote repair of the mucosal barrier resulting in decreased bacterial translocation and infection; alter the function of immune cells; help to maintain tissue antioxidant levels; and reduce the catabolic state associated with HCT. These effects may indirectly or directly decrease the risk of infection, length of stay, and improve nitrogen balance.

GLN-supplemented TPN is not readily available for use in most transplant centers because of its cost and its instability in solution. The use of oral GLN is being investigated at some centers. In theory, because GLN is the preferred fuel of the enterocyte and other rapidly dividing cells, the provision of oral GLN should be more beneficial than parenteral GLN by helping to improve gastrointestinal integrity and may help oral mucosal recovery as well. However, a recent pilot study investigated the use of supplemental oral GLN (16 g/day) versus a control solution in 24 autologous recipients and found no difference in severity of mucositis or the number of days with diarrhea.[105] It should be noted that the actual dose of oral glutamine received during this study was markedly less than the dose of IV glutamine given in other studies.[102,103] Because most patients were unable to tolerate GLN in liquid form, these authors felt that providing GLN in some other form of food may be a possible option. Oral GLN diluted with water or juice has been fairly well tolerated at our center. Intolerance is usually related to nausea and vomiting. These problems may be avoided if proper timing of oral GLN supplementation is determined.

GLN has been supplemented in TPN at levels of .29-.57 g/kg/day and has been shown to be safe and well tolerated.[106] Supplementation with GLN in doses of .5 g/kg or one-third of the amino acid load in patients with normal renal and liver function has been recommended. Continued research efforts are needed to determine the optimal route and dose of GLN in HCT recipients.

Arginine's role as an immunomodulator has also been recognized. Improvements in nitrogen balance, wound healing, reductions in infection rate and stimulation of T-lymphocyte response have been observed in animal studies.[107] Improvements in immune function, and a moderate improvement in nitrogen balance were also observed in a randomized, prospective study in surgical oncology patients receiving arginine-supplemented enteral nutrition.[108] However, these effects have not been investigated in HCT recipients.

Taurine is the most abundant intercellular amino acid in white blood cells and platelets.[109] Like GLN, taurine is not a standard part of TPN amino acid profiles. Desai et al[109] observed depletion of taurine in 41 HCT recipients after conditioning therapy. The magnitude of the depletion was greater in patients requiring TPN, but also occurred in patients who did not receive TPN. The clinical significance of taurine depletion in HCT remains unclear, but these authors speculate that it may have a negative effect on recovery from chemoradiotherapy-induced myelosuppression. Dietary supplementation with taurine in animal studies has been shown to improve survival and white cell recovery after TBI.[110] In adult patients with hepatobiliary disorders taurine supplementation has been shown to promote bile acid secretion[111] and may have a role in the prevention of TPN induced cholestasis.[112]

Fat

Intravenous (IV) fat is a calorically dense nutrient. IV lipid emulsions generally supply 10-30% of total calories, and should not exceed 30% of total calories, or 1 g/kg/day in adults, and 3 g/kg/day in children.[87,113] IV lipid emulsions contain a significant amount of linoleic acid (an omega-6 fatty acid), which is needed to prevent essential fatty acid deficiency. However, recent experimental evidence suggests that large amounts of omega-6 fatty acids are immunosuppressive.[114] The use of IV lipid emulsions was recently investigated in a randomized, controlled study in 492 HCT recipients.[113] Significant differences in infection rate were not found between the group of patients receiving the low lipid dose (6-8% of total calories) versus the standard lipid dose (25-30% of total calories). Future studies are needed to validate these findings. However, the provision of moderate amounts of IV lipid (< 30% of total calories from fat) appears to be appropriate. Omega-3 fatty acids and structured lipids which have been shown to reduce the inflammatory response in critical illness without suppressing immunologic function,[115,116] are not currently available for use in TPN in the United States.

Electrolytes

Electrolyte replacement is a standard part of the TPN formulation. Requirements may vary considerably among patients. Hypomagnesemia, hypophosphatemia, and hypokalemia are commonly observed following HCT because of medications used (diuretics, amphotericin B, cyclophosphamide, corticosteroids), high volumes of dextrose containing IV solutions, decreased intake, and excessive gastrointestinal losses related to TBI or GVHD induced diarrhea. Hypomagnesemia has also been associated with CSA induced renal wasting of magnesium.[117] Magnesium deficiency has also been associated with CSA induced hypertension, tremors, seizures, depression and ataxia.[118] Supplementation with 15 mEq/1 of magnesium in adults and

0.3 mEq/kg in children has been recommended,[119] but further supplementation may be required. Magnesium supplementation (oral and IV) may continue long after TPN is discontinued in allogeneic HCT recipients, because of CSA administration.

The use of sodium-containing antibiotics can contribute considerably to the total sodium load. Therefore, additional supplementation may not be necessary. Sodium intake may need to be limited in the presence of renal disease or VOD. Patients receiving potassium-sparing diuretics, like aldactone, may need to limit parenteral potassium supplementation. CSA can also be associated with hyperkalemia.[120]

Vitamins and Trace Elements

Standard multivitamin (MVI) preparations for use in TPN are in accordance with the established recommendations from the American Medical Association (AMA).[121] These recommendations, while higher than the Recommended Daily Allowances (RDAs) because of their reduced bioavailability and stability in parenteral solutions, are in question because they were based on studies in nonstressed, normometabolic patients. Vitamin K and trace elements such as selenium, copper, chromium, zinc, manganese, and molybdenum must be added in addition to the standard MVI preparation. Iron is not routinely added to standard TPN trace element preparations because it is not compatible with IV lipids in solution. Additionally, iron supplementation is avoided in HCT because of the potential for iron overload syndrome precipitated by multiple blood transfusions.

Many factors can influence vitamin requirements and utilization in HCT. The use of nonabsorbable antibiotics for gastrointestinal sterilization in HCT, and the use of broad spectrum antibiotics in neutropenic cancer patients have been associated with vitamin K deficiency.[122,123] Supplemental doses of vitamin K provided in TPN (10 mg/week adults and 5 mg/week children) should be adequate to prevent vitamin K deficiency.[119] Decreased plasma alpha-tocopherol and beta-carotene levels have also been observed in 19 HCT recipients following conditioning therapy.[124] Both vitamin E and beta-carotene are potent antioxidants, and deficiencies may play a role in early post-transplant organ toxicity.[124,125] Beta-carotene is not part of standard parenteral vitamin preparations and while the vitamin E provision in TPN meets the recommendations for parenteral supplementation, it may be inadequate to compensate for the oxidative damage which results from conditioning therapy. Damage by free radicals may be further exacerbated by the provision of IV lipids which are high in polyunsaturated fatty acids. A recent case report of a 44 year old HCT recipient who developed severe VOD and was successfully treated with 400 mg of vitamin E and 20 gm of GLN suggests potential benefits of supplemental antioxidant therapy.[125] Some centers provide additional vitamin C to promote tissue recovery via collagen biosynthesis following cytoreductive therapy, although this has not been substantiated in clinical trials. Future randomized, controlled studies are needed to delineate the optimal doses, and benefit of these therapies in HCT.

Folic acid requirements are mediated by the need for hematopoietic reconstitution, losses secondary to shortened red cell survival, and skin exfoliation in GVHD. Folinic acid may be required to bypass the effects of methotrexate and trimethoprim-sulfamethoxazole on folic acid metabolism in the setting of a tenuously engrafting marrow. One to two mg/day have been required by cancer patients and critically ill surgical patients to maintain folate levels.[126,127] Folic acid (1 mg/day) and a standard multivitamin supplement should continue long after HCT, because of dietary deficiencies related to the omission of raw fruit and vegetables, which are rich sources of folate as well as continued "stress" hematopoiesis.

Trace elements such as zinc, copper, and selenium have important metabolic and immunologic roles. Zinc and copper are coenzymes in metabolic pathways and nucleic acid synthesis which aid in wound healing,[128,129] immunological function,[130,131] and hematopoiesis.[132,133]

Zinc deficiency has also been associated with decreased taste acuity,[134] an important feature in reestablishing oral feedings. Selenium is an antioxidant that works in conjunction with vitamin E to stabilize cell membranes. Trace element deficiencies occur in HCT recipients receiving TPN primarily because of increased intestinal losses, redistribution between circulation and tissues, hypoproteinemia, and altered intake.[135] With the exception of zinc, many of these deficiencies can be avoided with use of standard trace element preparations in TPN. In one study,[135] lower serum zinc levels were found in 13 allogeneic HCT recipients who received TPN and tended to be lower in those patients who died during the 4 week follow up. Zinc may need to be aggressively replaced in HCT recipients with excessive gastrointestinal losses (vomiting and diarrhea). The AMA guidelines for parenteral zinc supplementation recommend 17 mg/kg for 24 hour stool output.[136] Zinc as well as vitamin C and arginine supplementation may also be of benefit in severe cases of skin GVHD to promote wound healing,[137] although optimal amounts have not been determined in HCT recipients.

Trace element toxicity should also be monitored. Manganese is essential for growth and is a part of various enzyme systems which regulate macronutrient metabolism.[138] Both manganese and copper are excreted in bile, therefore in cholestatic patients elimination of these trace elements is poor, which can lead to toxicity. Manganese toxicity has been observed in a case report of a allogeneic HCT recipient with cholestasis.[139] Elevated manganese levels have also been found in other allogeneic HCT recipients with cholestasis.[140] Removal of manganese and copper from TPN has been suggested when persistent cholestasis is present. Further studies in HCT recipients are needed to determine both optimal vitamin and trace element requirements, and toxicity levels.

TPN-Related Complications

There are multiple metabolic and mechanical complications associated with TPN which have been well documented.[141,142] However, advances in knowledge and practice have resulted in more appropriate administration and monitoring of TPN to minimize complications. An understanding of the potential complications which are more common to HCT recipients and how to manage these complications will be reviewed in more detail in the next section of this text.

Hyperglycemia

In allogeneic HCT, the use of steroids for the treatment of GVHD, can lead to hyperglycemia. Steroid use in combination with CSA may further impair glucose metabolism in HCT recipients.[143] Impaired glucose tolerance has also been associated with chemotherapy/TBI induced damage to the insulin secreting pancreatic beta cells.[143] Stress and infection can also alter metabolism of dextrose resulting in hyperglycemia. Hyperglycemia should be avoided, as it is associated with impaired lymphocytic function and immunity, leading to increased susceptibility to infection.[76] An increased incidence of *Candida albicans* infections have also been correlated with hyperglycemia.[144] Additionally, hyperglycemia can precipitate electrolyte abnormalities because of the diuretic effect associated with glucosuria. In order to avoid the deleterious effects of hyperglycemia, daily monitoring of blood glucose (BG) values is essential. Insulin should be added directly to the TPN or given as a separate insulin drip to maintain BG < 200 mg/dl. Dextrose concentrations may also need to be decreased and replaced with an increased proportion of IV lipid as tolerated.

Hypertriglyceridemia

Hypertriglyceridemia is commonly seen in allogeneic HCT recipients and is related to both steroid therapy and CSA. Alterations in lipid metabolism have been observed in both renal transplant and HCT recipients receiving CSA.[145-147] Cyclosporine is lipophilic and more than 80% of the drug in plasma is bound to high and low density lipoprotein fractions.[148] When high triglyceride levels prevail, corresponding increases in CSA levels occur which can lead to a toxic accumulation of unbound CSA in the brain. IV lipid emulsions can safely be given to HCT recipients with close monitoring of lipid levels. The lipid infusion should be decreased or given on an intermittent schedule if serum triglyceride levels exceed 600 mg/dl. Heparin may be effective in helping to clear triglyceride levels. Insulin may also help clear triglyceride levels in those patients with concomitant hyperglycemia.

Liver Function Test Abnormalities

Altered liver function tests are a common manifestation of prolonged TPN use. Steatosis, or fatty liver, is characterized by elevations in transaminase values during the first 1-2 weeks following the initiation of TPN.[149] High infusions of parenteral dextrose and overfeeding of total calories and fat have been linked to the development of steatosis. Transaminase levels often return to normal spontaneously even with continued TPN therapy. Elevations in bilirubin levels and alkaline phosphatase are commonly observed 2-3 weeks after TPN initiation and are associated with the development of cholestasis. Absence of oral intake, sepsis and malignant disease are a few of the most common etiologies for the development of cholestasis. The development of VOD, drug toxicity, and hepatic GVHD can also cause alterations in liver function tests, which can complicate interpretation. Some commonly prescribed methods to combat TPN associated liver injury include; reducing nonprotein calorie intake, cycling TPN over 12-18 hours,[6,149-151] resumption of some oral/enteral intake, and removal of copper and manganese from TPN solutions in patients with persistent cholestasis. GLN and taurine supplementation have also been postulated to help decrease TPN associated liver function abnormalities.

Fluid Imbalance

Fluid overload is common following HCT because of the large amount of medications administered and the cytoreductive induced fluid shifts including the development of ascites. It is advantageous to concentrate the TPN solution as much as possible from the onset. Use of concentrated amino acid solutions such as Novamine (15%) may allow for further reductions in TPN volume, especially when VOD is present. Daily maintenance fluid requirements for adults and children > 40 kg = 1500 ml/m^2; and for children < 10 kg = 100 ml/kg; for children between 11-20 kg = 1000 ml + 50 ml/kg; and for children between 21-40 kg = 1500 ml + 20 ml/kg. Fluid requirements are often higher than maintenance because of insensible losses from the skin and gastrotintestinal tract, and also in the presence of fever. Careful monitoring of intake and output and weight is essential for monitoring fluid status.

Transition To Enteral Nutrition

Most patients resume some oral intake with the resolution of mucositis and engraftment, generally 3-4 weeks after HCT. During this period, oral intake is often still poor due to continued decreased taste acuity and anorexia. TPN should continue until greater than 50% of nutritional requirements are met by oral intake for greater than two days, in the absence of significant diarrhea and/or GVHD. In patients who require prolonged parenteral support, TPN requirements should be decreased to compensate for increasing oral intake. TPN may also be cycled for 12-18 hours during the night. Cyclic TPN allows for greater mobility, greater psychological benefits and may help to stimulate appetite,[150] especially in young children. Early satiety has been associated with the administration of TPN in healthy volunteers.[152] In a recent randomized, controlled study HCT recipients who received hydration only (5% dextrose), resumed oral intake sooner than those receiving TPN.[153] Cyclic TPN has been shown to be safe and well tolerated in HCT recipients.[6] Additional benefits may include improved nitrogen balance, and improvements in TPN associated liver function test abnormalities.[150]

Long-Term Nutritional Support

The nutritional management of HCT recipients may continue long after discharge from the hospital, often up to 6 months. In the immediate period following hospital discharge, anorexia associated with taste changes and nausea are common problems. Poor adherence to mouth care regimens leading to the development of oral thrush may exacerbate these problems. In other patients, bacterial, viral or parasitic infections may occur in the gastrointestinal tract. TPN use after discharge may continue in those patients who are unable to consume > 50% needs orally. However, in patients who have minimal gastrointestinal symptoms, enteral tube feedings should be preferentially considered.

Allogeneic HCT recipients are more prone to long-term nutrition related problems than recipients of autologous grafts because of the complications of chronic GVHD. In a recent retrospective study,[154] weight loss (28%), oral sensitivity (23%), xerostomia (18%), anorexia (8%), reflux symptoms and diarrhea (7%), steatorrhea (5%), and dysgeusia (3%) were observed in 192 allogeneic recipients one year following HCT.

Continued outpatient nutrition monitoring in this population is important for the prevention and treatment of late malnutrition. Nutrition support should be initiated in patients with weight loss >10% body weight or <90% IBW, documented poor oral intake (< 50% of estimated needs) and gastrointestinal malabsorption. The mode of nutrition support used should be based on medical status. With chronic GVHD, every effort should be made to use the enteral route as intestinal malabsorption is less common. Percutaneous endoscopically placed gastrostomy (PEG) tubes are a viable option for long-term feeds in these situations.

Of particular potential concern are children in whom delayed growth and growth hormone (GH) deficiency are observed long-term complications following allogeneic HCT.[155] In addition to poor nutritional intake, growth retardation has been attributed to the use of TBI (total body irradiation),[156-157] central nervous system irradiation given prior to HCT, and the development of chronic GVHD. The effect of TBI on growth velocity is related to both the total dose and fractionation of TBI administered, as well as patient age at the time of treatment. Conditioning with fractionated TBI has been shown to have a relative sparing effect on growth.[156] While adequate nutritional intake is essential to meet the continually changing nutritional needs of children to promote normal growth and development, there may also be a role for GH replacement therapy. Short-term GH replacement therapy in GH deficient children has been shown to improve growth velocity in some studies,[156,157] although adequacy of dietary intake was not assessed in these studies. Further validation in patients treated for longer time periods with GH as well as the impact of dietary intake on growth velocity are needed before routine use is warranted.

Efficacy of TPN

Survival advantages observed in HCT recipients over the last 20 years can largely be attributed to improvements in medical therapy as well as supportive care which includes the use of TPN. However, establishing a direct link between prophylactic conventional TPN use and improved hematologic recovery and outcome may never be possible because of ethical reasons. Supportive TPN use has been effective in preventing malnutrition in adult and pediatric HCT recipients.[1,4-6] Therefore, by preventing malnutrition we may indirectly reduce complications and improve outcome, as decreased immune function and poor outcome have long been associated with malnutrition.[158-161] Only one study to date,[59] has documented the efficacy

Table 14.6. Suggested guidelines for nutritional care of adult hematopoietic cell transplant recipients

	Caloric Intake (kCals/kg/IBW)	Protein Intake (g/kg IBW)	Electrolyte and Vitamin Considerations	Enteral/Oral Diet Considerations
Immediate post-transplant period	30-35 kCals	At least 1.5	16-64 mEq magnesium/day; consider zinc supplement if diarrhea > 500 ml/day	Low lactose, low microbial/neutropenic diet, individualize based on tolerance, supplemental isotonic, high protein tube feeds
Engraftment	25-30 kCals in the absence of GVHD	1.0-1.2	As needed; when off TPN, oral magnesium replacement with CSA use, multivitamin and 1 mg folate/day	Low lactose, low microbial/neutropenic high calorie, high protein diet
Acute GVHD	35-40 kCals, as clinically tolerated	1.5-2.0	Remove copper and manganese from TPN when total billirubin > 10-15 mg/dl; consider zinc supplementation for diarrhea > 500 ml/day; consider vitamin C, L-arginine and zinc supple-metation for severe skin GVHD	If stool output is < 500 ml; low lactose, low fat (if steatorrhea is present), low fiber, high protein, high calorie diet
Chronic GVHD	30-35 kCals	1.2-1.5	Calcium supplementation with chronic steriod use	High calorie, high protein, carbohydrate controlled diet; low sodium, isotonic high protein tube feeds
VOD	25-30 kCals	Restrict if hepatic encephalopathy develops	Restrict sodium and fluid, concentrate TPN, decrease potassium in TPN with potassium sparing diuretics, remove copper and manganese from TOPN when total bilirubin > 15-20 mg/dl	Low sodium, fluid restricted diet

* First three weeks following HCT
IBW = ideal body weight TPN = total parenteral nutrition CSA = cyclosporin A GVHD = graft -versus-host disease VOD = veno-occlusive disease

of prophylactic conventional TPN support. In this prospective, randomized, controlled study, improved long-term outcome was observed in 71 allogeneic recipients given TPN beginning a week prior and for 4 weeks following HCT, versus a control group (n = 66) which received IV hydration and an oral diet. These investigators also found that TPN was ultimately required by 40 of the 66 control patients because of nutritional depletion, resulting in a significant increase in length of stay in this subgroup. In a more recent trial,[102] GLN supplemented TPN was found to decrease hospital length of stay, decrease incidence of clinical infection and improve nitrogen balance in allogeneic HCT recipients, although 100 day survival rates were similar in both groups.

In contrast, Szeluga et al[4] found no significant differences in survival between enterally and parenterally fed allogeneic HCT recipients. In this study however, 73% of the control group required additional intravenous amino acids given through a peripheral catheter to meet nutritional requirements. These authors have suggested that TPN should only be used for patients failing enteral feeding trials because of the increased complications and cost-associated with TPN use. Clearly establishing outcome data on the benefits of TPN use in this population, as well as many other populations, is a difficult task because of ethical reasons. Additionally, outcome may be influenced by a number of other factors.

Summary

The nutritional management of HCT recipients has become an important, if not an essential part of supportive care. A summary of suggested recommendations for nutritional care are provided in Table 14.6. Although nutrition support with conventional TPN has not been shown to conclusively improve hematologic recovery or long-term outcome, TPN remains a standard part of supportive therapy because of the profound effects of treatment on the alimentary system which precludes adequate oral intake in most patients. As our knowledge expands on the unique needs of this population, perhaps nutritional management will focus more on the use of specialized nutrients which may help to both minimize TPN-associated complications and improve nutritional response and outcome. The use of TPN-supplemented with GLN, BCAA, taurine, and ARG warrants further study. Preliminary studies in HCT and other populations however, suggest that these nutrients may help to improve our ability to nourish HCT recipients safely and effectively. Other novel therapies such as the use of GH replacement therapy, may be advantageous for children with delayed growth velocity following TBI. GH therapy has also been shown to decrease the catabolism associated with steroid therapy in normal individuals,[162,163] and therefore may also have a potential role in HCT recipients. Furthermore, the use of enteral nutrition alone or in combination with TPN in the early post-transplant period should be encouraged in selected HCT recipients, even if only small amounts are provided because of the overwhelming benefits of enteral nutrition.

Acknowledgements

Special thanks to Deborah Ford Flanel, Associate Director of Clinical Nutrition, Yale-New Haven Hospital, for her support and guidance in facilitating the completion of this Chapter.

References

1. Schmidt GM, Blume KG, Bross KJ et al. Parenteral nutrition in bone marrow transplant reicipients. Exp Hemat 1980; 8:506-11.
2. Keenan AMM. Nutritional support of the bone marrow transplant patient. Nurs Clin North Am 1989; 24:383-93.

3. Aker SN, Lenssen P, Darbinian J et al. Nutritional assessment in the marrow transplant patient. Nutr Support Services 1983; 3:22-37.
4. Szeluga DJ, Stuart RK, Brookmeyer R et al. Nutritional support of bone marrow transplant recipients: A prospective, randomized clinical trial comparing total parenteral nutrition to an enteral feeding program. Cancer Research 1987; 47:3309-16.
5. Uderzo C, Rovelli A, Bonomi M et al. Total parenteral nutrition and nutritional assessment in leukemic children undergoing bone marrow transplantation. Eur J Cancer 1991; 27:758-62.
6. Reed MD, Lasarus HM, Herzig RH et al. Cyclic parenteral nutrition during bone marrow transplantation in children. Cancer 1983; 51:1563-70.
7. Rappeport JM. Indications and applications of bone marrow transplantation. In: Hoffman R, Benz EJ, Shattil SJ, Furie B, Cohen HJ eds. Hematology basic principles and practice. New York: Churchill Livingstone, 1991, 230-42.
8. Weinstein HJ, Rappeport JM, Ferrara JLM. Bone marrow transplantation. In: Nathan DG, Oski FA eds. Hematology of infancy and childhood, 4th edition. Philadelphia: W.B. Saunders Company, 1993:317-44.
9. Parkman RP. The application of bone marrow transplantation to the treatment of genetic diseases. Science 1986; 232:1373.
10. Armitage JO. Bone marrow transplantation. N Eng J Med 1994; 12:827-38.
11. Henon PR. Peripheral blood stem cell transplantations: Past, present, and future. Stem Cells 1993; 11:154-72.
12. Schouten HC, Maragos K, Vose J et al. Diabetes mellitus or an impaired glucose tolerance as a potential complicating factor in patients treated with high-dose therapy and autologous bone marrow transplantation. Bone Marrow Transplantat 1990; 6:333-5.
13. Sullivan KM. Complications of bone marrow transplantation. In: Hoffman R, Benz EJ, Shattil SJ, Furie B, Cohen HJ eds. Hematology basic principles and practice. New York: Churchill Livingstone, 1991:242-49.
14. Bearman SI. The syndrome of hepatic veno-occlusive disease after marrow transplantation. Blood 1995; 85:3005-20.
15. Vogelsang GB, Hess AD. Graft-versus-host disease: new directions for a persistent problem. Blood 1994; 84:2061-67.
16. Copelan EA, Deeg HJ. Conditioning for allogeneic marrow transplantation in patients with lymphohematopoietic malignancies without the use of totalbody irradiaiton. Blood 1992; 80:1648-58.
17. Deeg H, Storb R, Thomas E. Bone marrow transplantation. A review of delayed complications. Br J Haematol 1984; 57:185-208.
18. Kim T, Khan F, Galvin J et al. Comparision of total body irradiaiton techniques for bone marrow transplant. Int J Rad Oncol Biol/Phys 1980; 6:779-84.
19. Vogelsang GB, hess AD, Santos GW. Acute graft-versus-host disease: clinical characteristics in the cyclosporin era. Medicine 1988; 67:163-74.
20. Meyers JD. Infection in recipients of marrow transplants. In: Remington JS, Swartz MN eds. Current topics in infectious dieseases. New York: McGraw Hill, 1985: 261-92.
21. Zaia JA. Viral infections associated with bone marrow transplantation. Hematol Oncol Clin North Am 1990; 4:603-23.
22. Shuhart MC, McDonald GB. Gastrointestinal and hepatic complications. In: Forman SJ, Blume KG, Thomas ED eds. Boston: Blackwell scientific publications, 1994:454-81.
23. Wolford JL, McDonald GB. A problem-oriented approach to intestinal and liver disease after marrow tranplantation. J Clin Gastroenterol 1988; 10:419-33.
24. Riddell S, Appelbaum FR, Buckner CD et al. High-dose cytarabine and total body irradiaiton with or without cyclophosphamide as a preparative regimen for marrow transplantation for acute leukemia. J Clin Onc 1988; 6:576-82.
25. Rappeport JM. Principles of bone marrow transplantation. In: Peterson DE, Elias EG, Sonis ST eds. Head and neck management of the cancer patient. The Hague, Netherlands: Martin Wighoff Publishers, 1986:129-39.
26. Carl W, Higby DJ. Oral manifestations of bone marrow transplantation. Am J Clin Oncol 1985; 8:81-87.
27. Boock CA, Reddick JE. Taste alterations in bone marrow transplant patients. J Am Diet Assoc 1991; 91:1121-22.

28. Mattsson T, Arvidson K, Heimdahl A et al. Alterations in taste acuity associated with allogeneic bone marrow transplantation. J Oral Pathol Med 1992; 21:33-7.
29. Lenssen P, Sherry ME, Cheney CL et al. Prevalence of nutrition-related problems among long-term survivors of allogeneic marrow transplantation. J Am Diet Assoc 1990; 90:835-42.
30. Herrmann VM, Petruska PJ. Nutrition support in bone marrow transplant recipients. NCP 1993; 8:19-27.
31. Donaldson SS, Lenon RA. Alterations of nutritional status: impact of chemotherapy and radiation therapy. Cancer 1979; 43:2036-52.
32. Fegan C, Poynton CH, Whittaker JA. The gut mucosal barrier in bone marrow transplantation. Bone Marrow Transplant 1989; 5:373-77.
33. McDonald GB, Shulman HM, Sullivan KM et al. Intestinal and hepatic complications of human bone marrow transplantation. Part I. Gastroenterology 1986; 90:460-77.
34. McDonald GB, Shulman HM, Sullivan KM et al. Intestinal and hepatic complications of human bone marrow transplantaion. Part II. Gastroenterology 1986; 90:770-84.
35. Billingham RE. The biology of graft-versus-host disease reactions. Harvey Lectures 1967; 21:62.
36. Schmeiser T, Kurrle E, Arnold R et al. Application of antimicrobial prophylactic treatment in the prevention of infection and graft-versus-host desease in allogeneic bone marrow transplantation. Exp Hematol 1984; 12:105-06s.abstract.
37. Sullivan KM. Graft-versus-host disease. In: Forman SJ, Blume KG, Thomas ED ed. Bone marrow transplantation. Boston: Blackwell Scientific Publications, 1994:339-62.
38. Weisdorf D, Hakke R, Blazar B et al. Risk factors for acute graft-versus-host disease in histocompatible donor bone marrow transplantation. Transplantation 1991; 51:1197-1203.
39. Atkinson K, Horowitz M, Gale RP et al. Consensus among bone marrow transplanters for diagnosis, grading and treatment of chronic graft-versus-host disease. Bone Marrow Transplantation 1989; 4:2247-54.
40. Guiot HFL, Biemond J, Klasen E et al. Protein loss during acute graft-versus-host disease: diagnostic and clinical significance. Eur J Haematol 1987; 38:187-96.
41. Weisdorf SA, Salati LM, Longsdorf JA et al. Graft-versus-host disease of the intestine: A protein losing enteropathy characteried by fecal alpha1-antitrypsin. Gastroenterology 1983; 85:1076-81.
42. Gauvreau JM, Lenssen P, Cheney CL et al. Nutritional management of patients with intestinal graft-versus-host disease. Research 1981; 79:673-77.
43. Ely P, Dunitz J, Rogosheske J et al. Use of a somatostatin analogue, octreotide acetate, in the management of acute gastrointestinal graft-versus-host ddisease. Am J Med 1991; 90:707-10.
44. Bianco JA, Higano C, Singer J et al. The somatostatin analog octreotide in the management of the secretory diarrhea of acute intestnal graft-versus-host disease in patients after bone marrow transplantation. Transplantation 1990; 49:1194-95.
45. Szeluga DJ, Stuart RK, Brookmeyer R et al. Energy requirements of parenterally fed bone marrow transplant recipients. JPEN 1985; 9:139-43.
46. McDonald GB, Hinds M, Fisher L et al. Veno-occlusive disease of the liver and multiorgan failure after bone marrow transplantation: a cohort study of 355 patients. Ann Intern Med 1993; 188:255-67.
47. Bearman SI, Anderson GL, Mori M et al. Veno-occlusive disease of the liver: development of a model for predicting fatal outcome after marrow transplantation. J Clin Onc 1993; 11:1729.
48. Marsa-Vila L, Gorin NC, Laporte JF et al. Prophylactic heparin does not prevent liver veno-occlusive disease following autologous bone marrow transplantation. Eur J Haematol 1991; 47:346.
49. Attal M, Huguet F, Rubie H et al. Prevention of hepatic veno-occlusive disease after bone marrow transplantation by continuous infusion of low-dose heparin: a prospective, randomized trial. Blood 1992; 79:2834.
50. Cahn JY, Flesch M, Brion A et al. Prevention of veno-occlusive disease of the liver after bone marrow transplantation: heparin or no heparin? Blood 1992; 80:2149.
51. Parkman R. Immunological reconstitution following bone marrow transplantation. In: Foreman SJ, Blume KG, Thomas ED eds. Boston: Blackwell Scientific Publications, 1994:504-12.
52. Wingard JR. Prevention and treatment of bacterial and fungal infections. In: Foreman SJ, Blume KG, Thomas ED eds. Boston: Blackwell Scientific Publications, 1994:363-75.
53. Zaia JA. Cytomegalovirus infection. In: Foreman SJ, Blume KG, Thomas ED eds. Boston: Blackwell Scientific Publications, 1994:376-403.

54. Bowden RA. Other viruses after marrow transplant. In: Foreman SJ, Blume KG, Thomas ED eds. Boston: Blackwell Scientific Publications, 1994:443-53.
55. Weisdorf SA, Schwarzenberg SJ. Nutritional support of bone marrow transplantation recipients. In: Forman SJ, Blume KG, Thomas ED ed. Bone marrow transplantation. Boston: Blackwell Scientific Publicaitons, 1994:327-36.
56. Aker SN, Cheney CL. The use of sterile and low microbial diets in ultraisolation environments. JPEN 1983; 7:390-97.
57. Aker SN. Nutrition dialogue. In: Charuhas PM ed. Marrow transplant nutrition network. Seattle WA: Caremark, 1995; 3:3.
58. Driedger L, Burstall CD. Bone marrow transplantation: dietitian's experience and perspective. J Am Diet Assoc 1987; 87:1387-88.
59. Weisdorf SA, Lysne J, Wind D et al. Positive effect of prophylactic total parenteral nutrition on long-term outcome of bone marrow transplantation. Transplantation 1987; 43:833-8.
60. Constantino AM, DeFrancesco A, Massarenti M et al. Aspetti nutrizionali nel trapianto di midollo osseo autologo. Minerva gastroenterol Dietol 1993; 39:159-65.
61. Cheney CL, Gittere Abson K, Aker SN et al. Body composition changes in marrow transplant recipients receiving total parenteral nutrition. Cancer 1987; 59:1515-19.
62. Streat SJ, Beddae AH, Hill GL. Measurements of body fat and hydration of the fat free body in health and disease. Metabolism 1985; 34:509-518.
63. Bernstein LH, Spickerman AM, Pleban W et al. The assessment and monitoring of nutritional support. Monograph PA-3. Fullerton, Calif: Beckman Instruments, Inc. 1988;1-9.
64. Cheney CL, Lenssen P, Aker SN. Prealbumin as a nutritional marker in marrow transplantation. J Am Coll Nutr 1987; 6:450.(abstract)
65. Winkler MF, Gerrior SA, Pomp A. Use of retinolbinding protein and prealbumin as indicators of the response to nutrition therapy. J Am Diet Assoc 1989; 89:684-7.
66. Brose L. Prealbumin as a marker of nutritional status. Nutrition Forum 1990; 11:372-375.
67. Kopple JD. Uses and limitations of the balance technique. JPEN 1987; 11:79S-85S.
68. Peters E, Beck J, LeMaistre C. Changes in resting energy expenditure (REE) during allogeneic bone marrow transplantation. Am J Clin Nutr 1990; 5:521.
69. Hutchinson ML, Clemans GW, Springmeyer SC et al. Energy expenditure estimation in recipients of marrow transplants. Cancer 1984; 54:734-38.
70. Geibig CB, Owens JP, Mirtallo JM et al. Parenteral Nutrition for Marrow Transplant Recipients: evaluation of an increased nitrogen dose. JPEN 1991; 15:184-88.
71. Taveroff A, McArdle H, Rybka WB. Reducing parenteral energy and protein intake improves metabolic homeostasis after bone marrow transplantation. Am J Clin Nutr 1991; 54:1087-92.
72. Keller U, Kraenslin ME, Gratwohl A et al. Protein Metabolism assessed by I-13C leucine infusions in patients undergoing bone marrow transplantation. JPEN 1990; 14:480-83.
73. Tomalis L, Kennedy MJ, Caballero B. Resting energy expenditure in pateints undergoing autologous bone marrow transplant. JPEN 1994; 18:27S.
74. Spruce W. Supportive care in bone marrow transplantation. In: Johnson L ed. Bone marrow transplantation in children. New York: Raven Press, 1990:69-85.
75. Seashore J. Nutritional support of children in the intensive care unit. Yale J Biol Med 1984; 57:111-34.
76. Wolfe RR, O'Donnell TF Jr, Stone MD et al. Investigation of factors determining the optimal glucose infusion rate in total parenteral nutrition. Metab 1980; 29:892-900.
77. Cheney CL, Lenssen P, Aker SN et al. Sex differences in nitrogen balance following marrow grafting for leukemia. J Am Coll Nutr 1987; 6:223-30.
78. Lenssen P, Cheney CL, Aker SN et al. Intravenous branched chain amino acid trial in marrow transplant recipients. JPEN 1987; 11:112-18.
79. Cerra F. Assessment of nutritional and metabolic status. In: Berger K ed. A pocket manual of surgical nutrition. St. Louis: C.V. Mosby, 1984:43.
80. Wilmore DW, Smith RJ, O'Dwyer ST et al. The gut: a central organ after surgical stress. Surgery 1988; 104:917-23.
81. Alverdy JC, Aoys I, Moss GS. Total parenteral nutrition promotes bacterial translocation from the gut. Surgery 1988; 104:185-90.
82. Katz DP, Kwetan V, Askanazi J. Enteral nutrition: potential role in regulating immune function. Curr Opin Gastroenterol 1990; 6:199-203.

83. Moore GA, Moore EE, Jones TN et al. TEN versus TPN following major abdonimal trauma, reduced septic morbidity. J Trauma 1989; 29:916-23.
84. Mudler POM, Bouman JG, Gietema JA et al. Hyperalimentation in autologous bone marrow transplantation for solid tumors: comparison of total parenteral versus partial parenteral plus enteral nutrition. Cancer 1989; 64:2045-52.
85. Ringwald-Smith K, Krance R, Stricklin L. Enteral nutrition support in a child after bone marrow transplantation. NCP 1995; 10:140-43.
86. Miller J. Enteral nutrition in marrow transplant patients. In: Charahus PM, Lenssen P, Roake J eds. Marrow transplant network. Seattle WA, Caremark Inc. 1994; 2:1.
87. Cunningham BA, Lenssen P, Aker SN et al. Nutritional considerations during marrow transplantation. Nurs Clin North Am 1983; 18:585-96.
88. Foster-Fair J. Nutrition needs of pediatic patients with speciic diseases: pediatric bone marrow transplant (presented at A.S.P.E.N. Clinical Congress, 1994 by Nancy L. McCorkle and Sandra L. Davies). In: Charuhas PM, Colovos T, Lenssen P, Roake J eds. Marrow transplant network. Seattle, Caremark; 1994; 2:2.
89. Cerra FB, Upson D, Angelico R et al. Branched chain amino acids support postoperative protein synthesis. Surgery 1982; 92:192-99.
90. Grant JP. Nutritional support in critically ill patients. Ann Surg 1994; 220:610-16.
91. Echenique MM, Bistrian BR, Moldower LL et al. Improvements in amino acid use in the critically ill patient with parenteral formulas enriched with branched chain amino acids. Surg Gynecol Obstet 1984; 159:233-41.
92. Lacy JM, Wilmore DW. Is glutamine a conditionally essential amino acid? Nutr Rev 1990; 48:297-309.
93. Hardy PE, Fedorak RN, Thomson ABR et al. Glutamine and its effects on the intestine. Can J Gastroenterology 1991; 5:94-102.
94. Newsholme EA, Crabtree B, Andrawi MSM. Glutamine metabolism in lymphocytes: Its biochemical, physiological and clinical importance. QJ Exp Physio 1985; 70:473-89.
95. Windmueller HG. Glutamine utilization by the small intestine. Adv Enzymol 1982; 53:201-32.
96. O'Dwyer ST, Scott T, Smith RJ et al. 5-Fluorouracil toxicity on small intestinal mucosa but not white blood cells is decreased by glutamine. Clin Res 1987; 35:369A.
97. Klimberg VS, Souba WW, Dolson DJ et al. Prophylactic glutamine protects the intestinal mucosa from radiation injury. Cancer 1990; 66:62-8.
98. Fox AD, Krupke SA, DePaula J et al. Effect of glutamine supplemented enteral diet on methotrexate-induced enterocolitis. JPEN 1988; 2:325-31.
99. Fox AD, Krupke SA, DePaula JA et al. Glutamine supplemented diets prolong survival and decrease mortality in experimental enterocolitis. JPEN 1987; 12:8S.
100. Stehle D, Zander J, Mertes N et al. Effect of parenteral glutamine peptide supplements on muscle glutamine loss and nitrogen balance after major sugery. Lancet 1989:231-3.
101. Hammerqvist F, Wernerman J, Ali R et al. Addition of glutamine to total parenteral nutrition after elective abdominal surgery spares free glutamine in muscle, counteracts the fall in muscle protein synthesis and improves nitrogen balance. Ann Surg 1989: 209:455-61.
102. Ziegler TR, Young LS, Benfell K et al. Clinical and metabolic efficacy of glutamine supplemented parenteral nutrition after bone marrow transplantation. Ann Int Med 1992; 116:82-8.
103. Schloerb PR, Amare M. Total parenteral nutrition with glutamine in bone marrow transplantation and other clinical applications. JPEN 1993; 17:407-43.
104. Scheltinga MR, Young LS, Benfell K et al. Glutamine-enriched intravenous feedings attenuate extracellular fluid expansion after a standard stress. Ann Surg 1991; 24:385-92.
105. Jebb SA, Marcus R, Elia M. A pilot study of oral glutamine supplementation in pateints receiving bone marrow transplants. Clin Nutr 1995; 14:162-65.
106. Zeigler TR, Benfell K, Smith RF et al. Safety and metabolic effects of L-glutamine administration in humans. JPEN 1990;14:137S-46S.
107. Barbul A. Arginine: biochemistry, physiology, and therapeutic implications. JPEN 1986; 10:227-38.
108. Daly JM, Reynolds J, Thom A et al. Immune and metabolic effects of arginine in the surgical patient. Ann Surg 1988; 208:512-22.
109. Desai TK, Maliakkal J, Kinzie JL et al. Taurine deficiency after intensive chemotherapy and/or radiation. Am J Clin Nutr 1992; 55:708-11.

110. Abe M, Takahashi M, Takeuchi K et al. Studies on the significance of taurine in radiation injury. Radiat Res 1968; 33:563-73.
111. Wang W-Y, Liaw K-Y. Effect of a taurine-supplemented diet on conjugated bile acids in biliary sugical patients. JPEN 1991; 15:294-97.
112. Guertin F, Roy CC, Lepage G et al. Effect of taurine on total parenteral nutrition associated cholestasis. JPEN 1991; 15:247-51.
113. Lenssen P, Bruemmer B, Aker S et al. Relationship between IV lipid dose and incidence of bacteremias and fungemias in 492 marrow transplant (MT) patients. JPEN 1994; 18:22S.
114. Jensen GL, Mascioli EA, Seidner DL et al. Parenteral infusion of long-and medium-chain triglycerides and reticuloenothelial system function in man. JPEN 1990; 14:467-74.
115. Endres S, Ghorbani R, Kelley VE et al. The effect of dietary supplementation with n-3 polyunsaturated fatty acids on the synthesis of interleukin-1 and tumor necrosis factor by mononuclear cell. N Eng J Med 1989; 320:265.
116. Goodwin JS, Webb DR. Regulation of the immune response by prostaglandins. Clin Immunol Immunopathol 1980; 15:106-22.
117. June CH, Thompson CB, Kennedy MS et al. Profound hypomagnesemia and renal magnesium wasting associated with the use of cyclosporine for marrow transplantation. Transplantation 1985; 39:620-624.
118. June CH, Thompson CB, Kennedy MS et al. Correlation of hypomagnesemia with the onset of cyclosporine associated hypertension in marrow transplantation. Transplantation 1986; 41:47-51.
119. Cunningham BA. Parenteral management. In: Lenssen P, Aker PA eds. Nutritional assessment and management during marrow transplantation: a resource manual. Seattle: Fred Hutchenson Cancer Research Center, 1985,45-61.
120. Karb VB. Electrolyte abnormalities and drugs which commonly cause them. J Neurosci Nurs 1989; 21:125-29.
121. American Medical Association/Nutrition Advisory Group. Multivitamin preparation for parenteral use. A statement by the Nutrition Advisory Group. JPEN 1979: 3:258-62.
122. Carlin A, Walker WA. Rapid development of vitamin K deficiency in an adolescent boy receiving total parenteral nutrition following bone marrow transplantation. Nutr Rev 1991; 49:179-83.
123. Conly J, Suttie J, Reid E et al. Dietary deficiency of phylloquinone and reduced serum levels in febrile neutropenic cancer patients. Am J Clin Nutr 1989; 50:109-13.
124. Clemens MR, Ladner C, Ehninger G et al. Plasma vitamin E and B-carotene concentrations during radiochemotherapy preceding bone marrow transplantation. Am J Clin Nutr 1990; 51:216-19.
125. Nattakom TV, Charlton A, Wilmore DW. Use of vitamin E and glutamine in the successful treatment of severe veno-occlusive disease following bone marrow transplantation. Nutr Clin Prac 1995; 10:16-18.
126. Kirkemo AK, Burt ME, Brennan MF. Serum vitamin level maintenance in cancer patients on total parenteral nutrition. Am J Clin Nutr 1982; 35:1003-009.
127. Bradley JA, King RF, Schorah CJ et al. Vitamins in intravenous feeding: a study of water-soluable vitamins and folate in critically ill patients receiving intravenous nutrition. Br J Surg 1978; 65:492-94.
128. Hallbook T, Lanner E. Serum zinc and healing of venous leg ulcers. Lancet 1972; 2:780-2.
129. Greaves MW, Skillen W. Effects of long-continued ingestion of zinc sulphate in patients with venous leg ulcers. Lancet 1970; 2:889-91.
130. Bogden JD, Oleske JM, Munves E et al. Zinc and immunocompetence in the elderly: baseline data on zinc nutriture and immunity in unsupplemented subjects. Am J Clin Nutr 1987; 46:101-9.
131. Castillo-Duran C, Heresi G, Fisberg M et al. Controlled trial of zinc supplementation during recovery from malnutrition effects of growth and immune function. Am J Clin Nutr 1987; 45:602-8.
132. Mason KE. A conspectus of research on copper metabolism and requirements of man. J Nutr 1979; 109:1979-2066.
133. Aggett PJA. Trace elements in human health. Practitioner 1984; 228:935-8.
134. Ruz M, Cavan KR, Bettger WJ et al. Erythrocytes, erythrocyte membranes, neutrophils and platelets as biopsy material for the assessment of zinc status in humans. Br J of Nutr 1992; 68:515-27.
135. Antila HM, Salo MS, Kirvela O et al. Serum trace element concentrations and iron metabolism in allogeneic marrow transplant recipients. Ann Med 1993; 24:55-9.
136. American Medical Association, Committee on Foods and Nutrition. Guidelines for essential trace element preparations for parenteral use. JAMA 1979; 241:2051-54.

137. Gottschlich MM, Warden GD. Vitamin supplementation in the patient with burns. J Burn Care Rehabil 1990; 11:275-79.
138. Solomons NW. Manganese. In: Baumgartner TG ed. Clinical guide to parenteral micronutrition, 2nd ed. Gainesville: Fujisawa, 1991:268-75.
139. Fredstrom S, Rogosheske J, Gupta P et al. Extrapyramidal symptoms in a HCT recipient with hyperintense basal ganglia and elevated manganese. Bone Marrow Transplant 1995; 15:989-992.
140. Fredstrom S. Trace element toxicities in parenteral nutrition during HCT. Marrow Transplant Nutrition Network 1996; 3:1.
141. Ang SA, Kaly JM. Potential complications and monitoring of patients receiving TPN. In: Rombeau JL, Caldwell MD eds. Parenteral Nutrition, vol 2. Philadelphia: W.B. Sanders Company, 1984: 331-43.
142. Lenssen P. Monitoring and ocmplications of parenteral nutrition. In: Skipper A ed. Dietitian's Handbook of enteral and parenteral nutrition. Rockville: A.S.P.E.N. Publishers Inc., 1989:293-310.
143. Smedmyr B, Wibell L, Simonsson B et al. Impaired glucose tolerance after autologous bone marrow transplantation. Bone Marrow Transplant 1990; 6:89-92.
144.Curry CR, Quie PG. Fungal septicemia in patients receiving parenteral hyperalimentaion. N Engl J Med 1971; 285:1221-25.
145. Harris KP, Russell GI, Parvin SD et al. Alterations in lipid metabolism attributable to cycloporin A in renal transplant recipients. Br Med J Clin Res Ed 1986; 292:16.
146. Nemunaitis J, Deeg HJ, Yel GC. High cyclosporine concentrations after bone marrow tranplantation associated with hypertriglyceridemia. Lancet 1986; 2:744-5.
147. Carreras E, Villamor N, Reverter JC et al. Hypertriglyceridemia in bone marrow transplant recipients: Another side effect of cyclosporin A. Bone Marrow Transplantation 1989; 4:385-88.
148. DeKlipper A, Sennesael J, Lamote J et al. Cyclosporin leukoencephalopathy induced by intravenous lipid solution. Lancet 1992; 339:1114.
149. Quigley EMM, Marsh MN, Shaffer JL et al. Hepatobiliary complications of total parenteral nutrition. Gastroenterology 1993; 104:286-301.
150. Hyltander A. Intravenous feeding kinetics and its effects on energy and nitrogen metabolism. Nutr Clin Prac 1994; 9:51-7.
151. Fisher RL. Hepatobiliary abnormalities associated with total parenteral nutrition. Gastroenterol Clin North Am 1989; 18:645-66.
152. Gil KM, Skeie B, Kvetan V et al. Parenteral nutrition and oral intake: effect of glucose and fat infusions. JPEN 1991; 15:426-32.
153. Charuhas PM, Fosberg KL, Bruemmer B et al. Effect of parenteral nutrition (PN) on resumtion of oral intake in 258 marrow transplant (MT) patients following hospital discharge. JPEN 1994; 18:26s.
154. Lenssen P, Sherry ME, Cheney CL et al. Prevalance of nutrition related problems among long-term survivors of allogeneic marrow transplantation. J Am Diet Assoc 1990; 90:835-42.
155. Shalet SM, Didi M, Ogilvy-Stuart et al. Growth and endocrine function after bone marrow transplantation. Clin Endocrinology 1995; 42:333-39.
156. Huma Z, Boulad F, Black P et al. Growth in children after bone marrow transplantation for acute leukemia. Blood 1995; 86:819-24.
157. Giovanna G, Bozzola M, Locatelli F et al. Role of busulfan and total body irradiation on growth of prepubertal children receiving bone marrow transplantation and results of treatment with recombinant human growth hormone. Blood 1995; 86:825-31.
158. Bistrian BR, Blackburn GL, Scrimshaw NS et al. Cellular immunity in semistarved states in hospitalized adults. Am J Clin Nutr 1975; 28:1148-55.
159. Scrimshaw NS, Taylor CE, Gordon JE. Interactions of nutrition and infection. Geneva: World Health Organization, 1968.
160. Windsor JA, Hill GL. Weight loss with physiologic impairment: a basic indicator of surgical risk. Ann Surg 1988; 207:290-96.
161. Windsor JA, Hill GL. Risk factors for postoperative pneumonia: the importance of protein depletion. Ann Surg 1988; 208:209-214.
162. Horber FF, Haymond MW. Human growth hormone prevents the protein catabolic side effects of prednisone in humans. J Clin Invest 1990; 86:265-72.

Index

M

N

O

P

R

S

T

V